Applied Mathematical Sciences

Volume 113

Springer
New York
Berlin
Heidelberg
Barcelona
Budapest
Hong Kong
London
Milan
Paris
Santa Clara
Singapore
Tokyo

Applied Mathematical Sciences

1. *John:* Partial Differential Equations, 4th ed.
2. *Sirovich:* Techniques of Asymptotic Analysis.
3. *Hale:* Theory of Functional Differential Equations, 2nd ed.
4. *Percus:* Combinatorial Methods.
5. *von Mises/Friedrichs:* Fluid Dynamics.
6. *Freiberger/Grenander:* A Short Course in Computational Probability and Statistics.
7. *Pipkin:* Lectures on Viscoelasticity Theory.
8. *Giacoglia:* Perturbation Methods in Non-linear Systems.
9. *Friedrichs:* Spectral Theory of Operators in Hilbert Space.
10. *Stroud:* Numerical Quadrature and Solution of Ordinary Differential Equations.
11. *Wolovich:* Linear Multivariable Systems.
12. *Berkovitz:* Optimal Control Theory.
13. *Bluman/Cole:* Similarity Methods for Differential Equations.
14. *Yoshizawa:* Stability Theory and the Existence of Periodic Solution and Almost Periodic Solutions.
15. *Braun:* Differential Equations and Their Applications, 3rd ed.
16. *Lefschetz:* Applications of Algebraic Topology.
17. *Collatz/Wetterling:* Optimization Problems.
18. *Grenander:* Pattern Synthesis: Lectures in Pattern Theory, Vol. I.
19. *Marsden/McCracken:* Hopf Bifurcation and Its Applications.
20. *Driver:* Ordinary and Delay Differential Equations.
21. *Courant/Friedrichs:* Supersonic Flow and Shock Waves.
22. *Rouche/Habets/Laloy:* Stability Theory by Liapunov's Direct Method.
23. *Lamperti:* Stochastic Processes: A Survey of the Mathematical Theory.
24. *Grenander:* Pattern Analysis: Lectures in Pattern Theory, Vol. II.
25. *Davies:* Integral Transforms and Their Applications, 2nd ed.
26. *Kushner/Clark:* Stochastic Approximation Methods for Constrained and Unconstrained Systems.
27. *de Boor:* A Practical Guide to Splines.
28. *Keilson:* Markov Chain Models—Rarity and Exponentiality.
29. *de Veubeke:* A Course in Elasticity.
30. *Shiatycki:* Geometric Quantization and Quantum Mechanics.
31. *Reid:* Sturmian Theory for Ordinary Differential Equations.
32. *Meis/Markowitz:* Numerical Solution of Partial Differential Equations.
33. *Grenander:* Regular Structures: Lectures in Pattern Theory, Vol. III.
34. *Kevorkian/Cole:* Perturbation Methods in Applied Mathematics.
35. *Carr:* Applications of Centre Manifold Theory.
36. *Bengtsson/Ghil/Källén:* Dynamic Meteorology: Data Assimilation Methods.
37. *Saperstone:* Semidynamical Systems in Infinite Dimensional Spaces.
38. *Lichtenberg/Lieberman:* Regular and Chaotic Dynamics, 2nd ed.
39. *Piccini/Stampacchia/Vidossich:* Ordinary Differential Equations in $\mathbf{R}^n$.
40. *Naylor/Sell:* Linear Operator Theory in Engineering and Science.
41. *Sparrow:* The Lorenz Equations: Bifurcations, Chaos, and Strange Attractors.
42. *Guckenheimer/Holmes:* Nonlinear Oscillations, Dynamical Systems and Bifurcations of Vector Fields.
43. *Ockendon/Taylor:* Inviscid Fluid Flows.
44. *Pazy:* Semigroups of Linear Operators and Applications to Partial Differential Equations.
45. *Glashoff/Gustafson:* Linear Operations and Approximation: An Introduction to the Theoretical Analysis and Numerical Treatment of Semi-Infinite Programs.
46. *Wilcox:* Scattering Theory for Diffraction Gratings.
47. *Hale et al:* An Introduction to Infinite Dimensional Dynamical Systems—Geometric Theory.
48. *Murray:* Asymptotic Analysis.
49. *Ladyzhenskaya:* The Boundary-Value Problems of Mathematical Physics.
50. *Wilcox:* Sound Propagation in Stratified Fluids.
51. *Golubitsky/Schaeffer:* Bifurcation and Groups in Bifurcation Theory, Vol. I.
52. *Chipot:* Variational Inequalities and Flow in Porous Media.
53. *Majda:* Compressible Fluid Flow and System of Conservation Laws in Several Space Variables.
54. *Wasow:* Linear Turning Point Theory.
55. *Yosida:* Operational Calculus: A Theory of Hyperfunctions.
56. *Chang/Howes:* Nonlinear Singular Perturbation Phenomena: Theory and Applications.
57. *Reinhardt:* Analysis of Approximation Methods for Differential and Integral Equations.
58. *Dwoyer/Hussaini/Voigt (eds):* Theoretical Approaches to Turbulence.
59. *Sanders/Verhulst:* Averaging Methods in Nonlinear Dynamical Systems.
60. *Ghil/Childress:* Topics in Geophysical Dynamics: Atmospheric Dynamics, Dynamo Theory and Climate Dynamics.

(continued following index)

P.D. Hislop I.M. Sigal

Introduction to Spectral Theory

With Applications to Schrödinger Operators

Springer

P.D. Hislop
Department of Mathematics
University of Kentucky
Lexington, KY 40506-0027
USA

I.M. Sigal
Department of Mathematics
University of Toronto
Toronto, Ontario M5S 1A1
Canada

Editors

J.E. Marsden
Control and Dynamical Systems 104–44
California Institute of Technology
Pasadena, CA 91125
USA

L. Sirovich
Division of Applied Mathematics
Brown University
Providence, RI 02912
USA

Mathematics Subject Classification (1991): 81Q05, 35J10, 35Q55

Library of Congress Cataloging-in-Publication Data

Hislop, P.D., 1955–
Introduction to spectral theory : with applications to Schrödinger operators / P.D. Hislop, I.M. Sigal.
p. cm.—(Applied mathematical sciences ; v. 113)
Includes bibliographical references (p. –) and index.
ISBN 0-387-94501-6 (hardcover : alk. paper)
1. Schrödinger operators. 2. Spectral theory (Mathematics)
I. Sigal, Israel Michael, 1945– . II. Title. III. Series:
Applied mathematical sciences (Springer-Verlag New York Inc.) ; v. 113.
QA1.A647 vol. 113
[QC174.17.S3]
510 s—dc20
[515′7223] 95-12926

Printed on acid-free paper.

Production managed by Frank Ganz; manufacturing supervised by Jacqui Ashri.
Photocomposed pages prepared from the author's LaTeX file using Springer-Verlag's "svsing.sty" macro.
Printed and bound by R.R. Donnelley & Sons, Harrisonburg, VA.
Printed in the United States of America.

9 8 7 6 5 4 3 2 1

ISBN 0-387-94501-6 Springer-Verlag New York Berlin Heidelberg

Contents

Introduction and Overview

This book presents some basic geometric methods in the spectral analysis of linear operators. The techniques, applicable to many questions in the theory of partial differential operators, are developed so that we can address several central problems in the mathematical analysis of Schrödinger operators. Many of the ingredients of this analysis, such as the notion of *spectral stability*, *localization* and *geometric resolvent equations*, *spectral deformation*, and *resonances* have applications in other areas of mathematical physics and partial differential equations. Important examples of these applications include spectral geometry [His], the theory of the wave equation [DeBHS], random differential operators [CH1, FrSp], quantum field theory [BFS, JP, OY], and nonlinear equations [Si4]. One of the subjects occupying an important place in this book is the semiclassical analysis of eigenvalues and resonances. This area has seen rapid development since the early 1970's, and here we present many of the key results in textbook form.

Our approach to spectral analysis differs from the standard one. We emphasize the use of *geometric methods* that take advantage of the local action of a partial differential operator. This leads naturally to the geometric characterization of the *discrete* and *essential* parts of the spectrum. The spectrum of a linear operator is studied through properties of the resolvent, the Riesz projections, and Weyl and Zhislin sequences. These objects form the basis of what is sometimes called *geometric spectral analysis*. Broadly stated, the idea is that spectral properties of the operator can be obtained from an analysis of the behavior of the operator acting on families of functions whose supports satisfy certain geometric conditions. The methods developed from this approach apply to many differential operators appearing in quantum mechanics and geometric analysis. Although we do not study directly the continuous spectrum of self-adjoint operators and scattering theory in this book, many of the techniques developed here are applicable to these questions as well.

Our goal is to develop the necessary mathematics in a self-contained way with minimal reference to the general theory. The background mathematics have been chosen so that we can discuss the applications as soon as possible. This approach reflects our general view that the heart of modern analysis lies in its applications to other disciplines.

In order to illustrate the geometric techniques presented in the text, we have chosen our examples from the theory of *Schrödinger operators*. This theory, originating with quantum mechanics, is now becoming an area of mathematical analysis in its own right, with applications to problems in nonlinear partial differential equations and geometric analysis. A certain amount of physical intuition is useful in the study of Schrödinger operators. The applications chosen in this text provide a gradual introduction to the physical concepts in a mathematical context.

A basic tenet of quantum theory is that quantum systems, such as atoms, molecules, solids, and to some extent nuclei and even stars, are described by linear differential operators called *Schrödinger operators*. We refer the reader to the standard text by Landau and Lifshitz [LL] for an account of this theory. The Schrödinger operator for a quantum system is the linear partial differential operator

$$H \equiv -\frac{\hbar^2}{2m}\Delta + V, \tag{0.1}$$

acting on the Hilbert space $L^2(\mathbb{R}^n)$. The Laplace operator Δ is the second-order differential operator that, in Cartesian coordinates on $\mathbb{R}^n$, is given by

$$\Delta \equiv \sum_{i=1}^{n} \frac{\partial^2}{\partial {x_i}^2}. \tag{0.2}$$

The constant m is the reduced mass of the system, and the constant $h = 2\pi\hbar$ is called Planck's constant. The real-valued function V is called the *potential*. This family of linear operators, for various potentials V, describes the different quantum systems mentioned above.

Elements of the Hilbert space $L^2(\mathbb{R}^n)$, called *wave functions*, represent various states of the system. The time-evolution of a wave function for a quantum system with Schrödinger operator H is controlled by *Schrödinger's equation*,

$$i\hbar\frac{\partial}{\partial t}\psi_t = H\psi_t. \tag{0.3}$$

One of the principal goals of quantum mechanics is to describe the space-time evolution of wave functions. For any reasonable initial state of the system ψ_0, the solution of (0.3) is given by the formula

$$\psi_t = U_H(t)\psi_0. \tag{0.4}$$

The mapping $U_H(t)$: $\psi_0 \to \psi_t$ is called the *evolution operator* for the Schrödinger equation. For a solution to exist, this operator must remain bounded for all time. In quantum theory, the probability density of a wave function is given by $|\psi|^2$. The corresponding probability is conserved under the time evolution. This requires that

$$\|\psi_0\| = \|U_H(t)\psi_0\| = \|\psi_t\|. \tag{0.5}$$

Furthermore, in order to have a unique solution to (0.3), the evolution operator must satisfy the condition that for all $s, t \in \mathbb{R}$,

$$U_H(s)U_H(t) = U_H(s+t). \tag{0.6}$$

Notice that if we set $s = -t$ in (0.6), then we obtain the relation

$$U_H(-t) = U_H(t)^{-1}, \tag{0.7}$$

which reflects the principle of time reversal invariance.

These conditions on the evolution operator $U_H(t)$ are, in fact, equivalent to the fundamental property, called *self-adjointness*, of the Schrödinger operator H in (0.3). The condition of self-adjointness of H is a necessary and sufficient condition for the existence and uniqueness of a solution to the Cauchy problem for the Schrödinger equation (0.3) satisfying (0.5) and (0.6). The self-adjointness of the Schrödinger operator H guarantees that the evolution operator $U_H(t)$ forms a one-parameter group of unitary operators.

The theory of the Schrödinger equation began with Schrödinger's groundbreaking papers [Sch], published in 1926. In these papers, he formulated the basic equation of quantum mechanics, computed the bound states of the hydrogen atom, and developed a simple form of perturbation theory, which he then applied to study the Stark and Zeeman effects. The mathematical theory of quantum mechanics and the Schrödinger operator can be traced to the book of J. von Neumann [vN], first published in 1932. In this book, von Neumann presented the Hilbert space framework of quantum mechanics and proved the equivalence of the matrix approach of Heisenberg and the partial differential equation approach of Schrödinger. One of the central mathematical contributions of the book is the basic spectral theory of unbounded self-adjoint operators. von Neumann emphasized the importance of self-adjointness for solving the eigenvalue problem for the Schrödinger operator.

The framework presented in von Neumann's book was developed and applied to specific quantum systems in the 1950's. One of the main, early achievements was the the pioneering work of T. Kato, who initiated the rigorous study of individual Schrödinger operators. Kato proved the self-adjointness of atomic Hamiltonians, thus establishing the existence of the evolution operator for atoms and molecules [K]. Because of the connection between the spectral properties of a Schrödinger operator H and the dynamics of the quantum system, Kato's results generated much research into the *spectral theory* of the linear operators describing quantum mechanical systems. The next fundamental step, understanding the spectrum of N-electron atomic systems, was made in the works of Hunziker, Van Winter, and Zhislin in the 1960's. This opened the way to the study of N-body Schrödinger operators.

Several themes dominated the mathematical physics of quantum mechanics throughout the next decades. These include scattering theory for quantum mechanical particles, the stability of matter, and the theory of bound states. Scattering theory concerns the description of the space-time asymptotic behavior of scattered particles and the recovery of characteristics of the scatterer from this information. We refer to the books by Amrein, Jauch, and Sinha [AJS], Reed and Simon [RS3], and Sigal [Sig] for a description of quantum mechanical scattering theory and other references. The 1960's and the 1970's saw the resolution of problems in the quantum description of bulk matter. We refer to the review articles of Lieb [Li] and Fefferman [Fef]. Two of the main questions, the existence of the thermodynamic limit and the stability of matter, were resolved on the basis of two basic principles of quantum mechanics, the Uncertainty Principle and the Pauli Principle. Hence, it was shown that quantum theory alone accounts for the stability of matter around us.

The 1970's and the 1980's was a period of intense research on the Schrödinger equation. This period witnessed a sharpening of our understanding of the relation between classes of potentials and self-adjointness, the discovery of estimates on the number of eigenvalues in terms of the potential, and the proof of exponential bounds on eigenfunctions. The properties of Schrödinger operators for other quantum systems, such as quantum particles in external electric and magnetic fields or under the effect of random potentials, were also explored. We refer the reader to review articles by Herbst [He1], Hunziker [Hu1], and Simon [Sim13]. Scattering theory for two- and many-body systems was extensively studied and many of the major open problems were resolved.

This period also saw the beginning of rigorous understanding of two basic concepts of modern physics, *quantum tunneling* and *resonances*, which are dealt with in this book. One of the principal tools for the investigation of these characteristic properties of Schrödinger operators is the *semiclassical approximation*. The semiclassical approximation has been used by theoretical physicists from the beginning of quantum theory (see [LL]). It has also been extensively developed by mathematicians, especially in the context of high-frequency wave phenomena. The investigations into resonances and tunneling have brought about many advances and refinements in this method. The semiclassical approximation addresses the fundamental issue of the relation between quantum and classical mechanics. It is based on the observation that Planck's constant h, appearing as a coefficient of the differential operator in (0.1), controls the quantum effects. If h could be taken arbitrarily small, the quantum effects would reveal themselves as small fluctuations about well-known classical behavior and hence would be easier to compute. The *semiclassical approximation* consists of replacing $\hbar$ by a parameter and studying the asymptotic behavior of the spectrum of the Schrödinger operator and the solutions of the Schrödinger equation as this parameter tends to zero. The quantum system in this approximation is considered as a perturbation of a classical system.

One of our goals for this book is to emphasize the importance of *quantum resonances*. Although resonances in quantum mechanical systems play an important role in determining the space-time behavior of the system, the general, mathematical theory has been developed only in recent years. In addition to their importance in quantum phenomena, it has also become clear that, in the theory of linear and nonlinear partial differential equations, eigenvalues *and* resonances must be considered together to form a stable set with respect to perturbations. It is interesting to note the similarities between quantum and classical resonances. Originally introduced by Poincaré, the theory of classical resonances was developed through the works of Kolmogorov, Arnold, and Moser. Resonances in classical systems are known to play an important role in the dynamical behavior of the systems. The concept of resonance contributes greatly to the understanding of nonintegrable systems that are close to integrable ones. An interesting characteristic of classical resonance theory is the occurrence of exponentially small separatrix splittings, first discussed by Poincaré (see, for example, [ST]). As we will see, exponentially small splittings are also a facet of quantum resonance theory. These similarities are quite intriguing and have not yet been fully investigated.

This book emphasizes the geometric approach to the spectral analysis of linear differential operators on Hilbert space. The applications to Schrödinger operators are centered on the theory of bound states and quantum resonances. Although we restrict our examples to two-body potentials, many of the ideas described here apply to N-body quantum systems. Some of the specific topics presented in the text include

- the exponential decay of eigenfunctions (Chapter 3);
- the semiclassical behavior of eigenvalues (Chapter 11);
- the relation between quantum tunneling and the spectrum of Schrödinger operators (Chapter 12);
- the Aguilar–Balslev–Combes–Simon theory of quantum resonances (Chapter 16);
- the general theory of spectral stability (Chapter 19);
- the existence of quantum resonances in the semiclassical regime (Chapters 20–21);
- exponential bounds on resonance widths (Chapter 22).

In addition, many recent developments in the theory of quantum resonances are discussed.

The book is roughly divided into three parts. The first part concerns the decomposition of the spectrum of a linear operator into the *essential spectrum* and the *discrete spectrum*. Basic properties of the spectrum, the resolvent, and Hilbert spaces are discussed in Chapters 1 and 2. Our first main application to Schrödinger operators occurs in Chapter 3. There, we study Schrödinger operators with potentials that are larger than a constant M_0 at infinity and that have an eigenvalue E_0 less than M_0. We present the method of Agmon [Ag1] to prove that the eigenfunctions corresponding to the eigenvalue E_0 decay at least exponentially fast as $\|x\| \to \infty$. Physically, this decay results from the fact that in the region where

$E_0 < M_0$, the conservation of energy is violated. On the classical level, the orbits of a particle with energy E_0 do not enter this region. Consequently, it is called the *classically forbidden region*. The quantum mechanical wave function, however, is not zero there. This effect is called *quantum tunneling*. The main result of Chapter 3 is that any wave function corresponding to the energy E_0 tunnels into the *classically forbidden region* but its amplitude decays exponentially there. Chapter 3 also serves to introduce many of the techniques of estimation, which are used repeatedly in the applications.

After this application, we continue to study the properties and characterizations of the discrete and essential spectrum for general closed operators and for the special, but important, case of self-adjoint operators on a Hilbert space. Self-adjoint operators are discussed in general in Chapter 5, where characterization of their spectrum is given. Our main tool for the analysis of the discrete spectrum, the Riesz projections, is detailed in Chapter 6. We emphasize the characterization of the essential spectrum in terms of the behavior of the given operator on certain sequences of functions. This leads to the Weyl criterion for the essential spectrum, which is presented in Chapter 7. For partial differential operators, one can take advantage of the local action of the operator, and the geometric ideas of *local compactness* and *Zhislin sequences* play a central role in the identification of the essential spectrum. These notions are discussed in Chapter 10. The underlying idea is that the essential spectrum is determined by the behavior of the operator acting on functions supported near infinity. A nice application of these geometric notions is Persson's theorem, which provides a formula for the bottom of the essential spectrum. This is proved in Chapter 14 after some sufficient conditions for the self-adjointness of Schrödinger operators are developed.

The second part of the text is devoted to understanding various notions of *stability* for the discrete and essential spectrum. The question of stability for discrete eigenvalues under various types of perturbations will occupy us in several chapters. Both the usual perturbation theory framework, analytic and asymptotic, and a more geometric framework, are presented. *Geometric perturbation theory* is developed in the context of semiclassical analysis. This theory of perturbations is based on localization; operators are small perturbations because they are localized in regions of configuration space or phase space that are in some sense forbidden. The most common reason that certain regions of space are forbidden is due to energy conservation. These methods are first applied in Chapter 11 to study the semiclassical behavior of the low-lying eigenvalues of Schrödinger operators with potentials that grow at infinity. We then present, in Chapter 12, a more detailed computation of the difference of two eigenvalues in a situation dominated by quantum tunneling. These two chapters follow the work of Simon [Sim5, Sim6].

The characteristic property of the essential spectrum, from which its name is derived, is its robustness under various perturbations. We first discuss relatively bounded perturbations, which, according to the fundamental Kato-Rellich theorem, preserve the self-adjointness of the unperturbed operator. We then refine this to the notion of *relatively compact perturbations*. Weyl's theorem, which we prove in Chapter 14, states the invariance of the essential spectrum under relatively compact perturbations. We apply this to Schrödinger operators with potentials vanishing at infinity in order to compute the essential spectrum. *Spectral stability* for the discrete spectrum is introduced in Chapter 15. There we discuss the standard theory of stability of an isolated eigenvalue with respect to *analytic* perturbations. The notion of a type A analytic family of operators is also presented in this chapter.

The main themes of *spectral deformation*, *spectral stability*, and *nontrapping estimates* are developed in the third part of the book. These are illustrated by the theory of quantum resonances. The basic theory of quantum resonances, as developed by Aguilar, Balslev, Combes, and Simon, is presented in detail. This theory, dating from the early 1970's, provides the basis of most recent studies of resonances in quantum mechanical systems. This theory is applied in Chapter 20 to the *shape resonance model* in the semiclassical regime. This provides a good example of geometric perturbation theory and other methods discussed earlier. Quantum resonances also provide examples of *asymptotic*, nonanalytic perturbation theory. In order to deal with these situations, we must broaden the notion of *spectral stability* introduced in Chapter 15. We do this in Chapter 19, in which the general notion of stability is presented, along with some well-known examples from quantum mechanics such as the anharmonic oscillator and the Zeeman effect. The main result of this chapter is a *stability criterion* for discrete eigenvalues due to Vock and Hunziker [VH]. Some geometric techniques, including localization formulas and geometric resolvent equations, that are useful in the verification of the stability criteria are presented in the last section of Chapter 19. The Aguilar–Balslev–Combes–Simon theory of resonances provides the initial motivation for the theory of *spectral deformation*. The general theory is described in Chapter 17 in the form developed by Hunziker [Hu2]. The details of the application to Schrödinger operators is presented in Chapter 18. This method is often useful for dealing with problems involving embedded eigenvalues.

The *shape resonance model* in the theory of quantum resonances is perhaps the oldest in quantum mechanics, dating back to the late 1920's. It was introduced to model the emission of an alpha particle from an atomic nucleus. This model is described in detail in Chapter 20. In order to establish the criteria for spectral stability presented in Chapter 19, we need the technique of *geometric perturbation theory*. In this technique, one uses *localization formulas*, involving the resolvent operators, to localize the perturbations in regions for which *a priori* estimates can be computed. These estimates follow either from quantum tunneling results, as described in Chapters 3 and 12, or from *quantum nontrapping* results. Nontrapping estimates imply the absence of bound states or resonances for quantum systems, and the decay of energy for waves. In Chapter 21, we show how certain nontrapping assumptions on the potential imply the resolvent estimates necessary for the stability criteria. Many of the objects of study in geometric perturbation theory involve exponentially small quantities (such as the splitting of eigenvalues discussed in Chapter 12). A technique for proving the exponential smallness of resonance widths is given in Chapter 22. Finally, we conclude with a chapter giving other examples of geometric perturbation theory and other topics in the quantum theory of resonances.

The guiding idea of this text is to develop the essential mathematics as directly as possible and then to illustrate the mathematics with examples drawn from research articles. To keep the book self-contained, we give background material on basic functional analysis in four appendices. The reader should be familiar with the basics of real analysis and point set topology. Measure theory and Lebesgue integration theory (with the exception of the Lebesgue dominated convergence theorem) are not explicitly used in the text. However, we assume that the reader is familiar with the idea of the completion of a normed linear space and will accept the fact that $L^2(\mathbb{R}^n)$ is the completion of $C_0^\infty(\mathbb{R}^n)$. Various properties of the Banach spaces $L^p(\mathbb{R}^n)$, such as the density of certain sets of "nice" functions, basic inequalities, and the Lebesgue dominated convergence theorem, are reviewed in Appendix 2. Certain facts about Sobolev spaces over $\mathbb{R}^n$ and the Fourier transform are given in Appendix 4.

Exercises are intertwined with the text in each chapter. These problems range from routine verification of statements and immediate extensions of the ideas presented in the text, to more challenging and involved problems, especially in the later chapters. We certainly encourage the reader to work the problems while studying the text.

References to original research papers related to the problems discussed here are given. We have limited ourselves to works published after 1980, except in certain cases of direct relevance. We refer the reader to the book by Cycon, Froese, Kirsch, and Simon [CFKS] and the books by Reed and Simon [RS1, RS2, RS3, RS4] for many of the earlier references.

For more information concerning functional analysis and perturbation theory, we refer the reader to the books of Akhiezer and Glazman [AG], the book of Kato [K], the four-volume series by Reed and Simon [RS1, RS2, RS3, RS4], and Volumes 3 and 4 of the work by Thirring [Th3, Th4]. Other aspects of the semiclassical analysis of the Schrödinger equation can be found in the books of Helffer [Hel], Maslov [Mas], and Robert [R1].

Acknowledgements

This book developed out of a course on mathematical physics given at the University of California, Irvine, during the academic year 1985–1986. Versions of it were used at the University of Kentucky; at the Centre de Physique Théorique in Marseille, France, for a DEA course; and for a mathematical physics course at the Université de Paris VII. We would like to thank J. M. Combes, S. DeBièvre, W. Hunziker, J. Marsden, and B. Simon for their comments. This book is based on research of PDH, partially supported by NSF grants DMS 91-06479 and 93-07438, and of IMS, partially supported by NSERC Grant NA 7901. The authors gratefully acknowledge this support.

1

The Spectrum of Linear Operators and Hilbert Spaces

We assume that the reader has the basic understanding of Banach spaces and linear operator theory contained in Appendices 1 and 3. We suggest that the reader review these appendices before beginning this chapter. We begin with the main topic of our interest: the *spectrum of a linear operator*. We introduce the *resolvent* of a linear operator on a Banach space. This operator is crucial to the definition of the spectrum. We define the spectrum and give some of its properties. We then specialize to Hilbert spaces and develop their basic characteristics in this and the following chapter.

1.1 The Spectrum

Let A be a linear operator on a Banach space X with domain $D(A) \subset X$. Let us recall from Appendix 3 that A is *invertible* if there is a bounded operator, which we call A^{-1}, such that $A^{-1} : X \to D(A)$, $AA^{-1} = 1_X$, and $A^{-1}A = 1_{D(A)}$ (where 1_X is the identity on X). We distinguish the invertibility of the operator A from the existence of *inverses* for A, which may be unbounded. By considering the invertibility of the operator $A - \lambda$, $\lambda \in \mathbb{C}$ (here, λ means $\lambda \cdot 1_X$), we obtain a disjoint decomposition of $\mathbb{C}$ into two sets that characterize many properties of A.

Definition 1.1. *Let A be a linear operator on X with domain $D(A)$.*

(1) *The spectrum of A, $\sigma(A)$, is the set of all points $\lambda \in \mathbb{C}$ for which $A - \lambda$ is not invertible.*

(2) *The resolvent set of* A, $\rho(A)$, *is the set of all points* $\lambda \in \mathbb{C}$ *for which* $A - \lambda$ *is invertible.*

(3) *If* $\lambda \in \rho(A)$, *then the inverse of* $A - \lambda$ *is called the resolvent of* A *at* λ *and is written as* $R_A(\lambda) \equiv (A - \lambda)^{-1}$.

Let us note that by definition,

$$\sigma(A) \cup \rho(A) = \mathbb{C} \tag{1.1}$$

and

$$\sigma(A) \cap \rho(A) = \phi. \tag{1.2}$$

Theorem 1.2. *The resolvent set* $\rho(A)$ *is an open subset of* $\mathbb{C}$ *(and hence* $\sigma(A)$ *is closed), and* $R_A(\lambda)$ *is an analytic operator-valued function of* λ *on* $\rho(A)$.

Remark 1.3. A map $\lambda \in \rho \subset \mathbb{C} \to A(\lambda) \in \mathcal{L}(X)$ is said to be (norm) *analytic* at $\lambda_0 \in \rho$ if $A(\lambda)$ has a power series expansion in $(\lambda - \lambda_0)$ (with coefficients in $\mathcal{L}(X)$) that converges in the norm with nonzero radius of convergence; that is, there exist bounded operators A_n such that

$$A(\lambda) = \sum_{n=0}^{\infty} (\lambda - \lambda_0)^n A_n.$$

Proof of Theorem 1.2.

(1) To show that $\rho(A)$ is open, we show that there is an ϵ-ball about each point $\lambda_0 \in \rho(A)$ which is contained in $\rho(A)$. Let $\lambda \in \mathbb{C}$ be such that $|\lambda - \lambda_0| < \|(A - \lambda_0)^{-1}\|^{-1}$. Then, by decomposing $A - \lambda$ as

$$A - \lambda = (A - \lambda_0)[1 - (A - \lambda_0)^{-1}(\lambda - \lambda_0)], \tag{1.3}$$

we see that it is invertible by Theorems A3.29 and A3.30. Thus, it suffices to take $\epsilon < \|(A - \lambda_0)^{-1}\|^{-1}$.

(2) To show analyticity at $\lambda_0 \in \rho(A)$, it follows from (1.3) that $(A - \lambda)^{-1} = (A - \lambda_0)^{-1}(1 - (\lambda - \lambda_0)(A - \lambda_0)^{-1})^{-1}$. Now $|\lambda - \lambda_0| \|(A - \lambda_0)^{-1}\|^{-1} < 1$, so using the Neumann series for $(1 - T)^{-1}$ (see Theorem A3.30), we get

$$\begin{aligned} (A - \lambda)^{-1} &= (A - \lambda_0)^{-1} \sum_{k=0}^{\infty} (\lambda - \lambda_0)^k (A - \lambda_0)^{-k} \\ &= \sum_{k=0}^{\infty} (\lambda - \lambda_0)^k (A - \lambda_0)^{-k-1}. \end{aligned} \tag{1.4}$$

One can easily check that the expansion is norm-convergent, and so we get analyticity. □

Let us examine $\sigma(A)$, which are those $\lambda \in \mathbb{C}$ for which $A - \lambda$ is not invertible. There are basically three reasons why $A - \lambda$ fails to be invertible:

(1) $\ker(A - \lambda) \neq \{0\}$;

(2) $\ker(A - \lambda) = \{0\}$, and $\text{Ran}(A - \lambda)$ is dense so that $(A - \lambda)$ has a densely defined inverse but is unbounded;

(3) $\ker(A - \lambda) = \{0\}$, but $\text{Ran}(A - \lambda)$ is not dense; in this case $(A - \lambda)^{-1}$ exists and may be bounded on $\text{Ran}(A - \lambda)$ but is not densely defined; therefore, it cannot be *uniquely* extended to a bounded operator on X.

According to these three situations, we classify $\sigma(A)$.

Definition 1.4.

(1) *If* $\lambda \in \sigma(A)$ *is such that* $\ker(A - \lambda) \neq \{0\}$, *then* λ *is an eigenvalue of* A *and any* $u \in \ker(A - \lambda)$, $u \neq 0$, *is an eigenvector of* A *for* λ *and satisfies* $Au = \lambda u$. *Moreover,* $\dim(\ker(A - \lambda))$ *is called the (geometric) multiplicity of* λ *and* $\ker(A - \lambda)$ *is the (geometric) eigenspace of* A *at* λ. *(Note that* $\ker(A - \lambda)$ *is a linear subspace of* X.*)*

(2) *The discrete spectrum of* A, $\sigma_d(A)$, *is the set of all eigenvalues of* A *with finite (algebraic) multiplicity and which are isolated points of* $\sigma(A)$.

(3) *The essential spectrum of* A *is defined as the complement of* $\sigma_d(A)$ *in* $\sigma(A)$: $\sigma_{\text{ess}}(A) \equiv \sigma(A) \setminus \sigma_d(A)$.

Remark 1.5.

(1) If $\dim X < \infty$, then the only reason why $(A-\lambda)$ is not invertible is if $(A-\lambda)$ is not one-to-one, that is, $\sigma(A) = \sigma_d(A)$. Of course, all multiplicities are finite, but the geometric multiplicity need not equal the algebraic multiplicity for any eigenvalue. Recall that A is diagonalizable if and only if the algebraic multiplicity equals the geometric multiplicity for each eigenvalue (see, for example, [HiSm]).

Problem 1.1. Verify Remark 1.5; show that if $\dim X < \infty$, the above cases (2) and (3) for lack of invertibility of $A - \lambda$ do not occur.

(2) $\sigma_{\text{ess}}(A)$ and $\sigma_d(A)$ provide a disjoint and complete decomposition of $\sigma(A)$. Note that $\sigma_{\text{ess}}(A)$ may contain eigenvalues of A. For example, an eigenvalue of infinite multiplicity or one that is a limit of a sequence of eigenvalues (so it is not isolated) belongs to $\sigma_{\text{ess}}(A)$.

(3) The set of all points $\lambda \in \sigma(A)$ such that λ is not an eigenvalue but $\text{Ran}(A-\lambda)$ is not dense is called the *residual spectrum* of A. We will see later that for a wide class of operators (including many that occur in mathematical physics) this set is empty.

1.2 Properties of the Resolvent

We want to collect in this section some basic properties of the resolvent of A, $R_A(\lambda) = (A - \lambda)^{-1}$, $\lambda \in \rho(A)$.

Proposition 1.6. *Let A be a linear operator on X. Then for $\mu, \lambda \in \rho(A)$,*

(1) $R_A(\lambda)R_A(\mu) = R_A(\mu)R_A(\lambda)$;

(2) $R_A(\lambda) - R_A(\mu) = (\lambda - \mu)R_A(\lambda)R_A(\mu)$, which is the *first resolvent identity.*

Proof. We first prove (2) by writing the identity:

$$\begin{aligned} R_A(\lambda) - R_A(\mu) &= R_A(\lambda)(A - \mu)R_A(\mu) - R_A(\lambda)(A - \lambda)R_A(\mu) \\ &= R_A(\lambda)R_A(\mu)(\lambda - \mu), \end{aligned}$$

which holds on all X. If we interchange μ and λ and compare the result, we get the commutativity of $R_A(\lambda)$ and $R_A(\mu)$. □

Although we have defined $\sigma(A)$, it remains to show that $\sigma(A)$ is nonempty. This can be done using the resolvent and some elementary complex analysis if A is bounded. There is some subtlety here, for there are examples of unbounded operators with empty spectrum (as there are examples of unbounded operators with empty resolvent sets)!

Theorem 1.7. *Let $A \in \mathcal{L}(X)$. Then $\sigma(A)$ is nonempty and is a closed subset of $\{z \mid |z| \leq \|A\|\}$.*

Proof. We consider a formal expansion of $R_A(\lambda)$, $\lambda \in \rho(A)$:

$$R_A(\lambda) = (A - \lambda)^{-1} = -\lambda^{-1}\left[1 + \sum_{k=0}^{\infty} \lambda^{-k} A^k\right]. \tag{1.5}$$

If $|\lambda^{-1}|\|A\| < 1$ (i.e., if $|\lambda|$ is sufficiently large), the power series converges and the right side of (1.5) defines a bounded operator. As in the proof of Theorem A3.30, one shows that this is the inverse of $(A - \lambda)$. Hence, $\{z \mid |z| > \|A\|\} \subset \rho(A)$. Now suppose $\sigma(A)$ is empty. This would mean that $R_A(\lambda)$ is an entire analytic function by Theorem 1.2. Furthermore, from (1.5), it follows that $\lim_{|\lambda|\to\infty} \|R_A(\lambda)\| = 0$, and hence $R_A(\lambda)$ would be a bounded entire function. Therefore, by Liouville's theorem (see, for example, [Ma]), $R_A(\lambda) = 0$ for all λ. This implies that $(A - \lambda)R_A(\lambda) = 1 = 0$, a contradiction. □

Proposition 1.8. *Let $\Omega \subset \mathbb{C}$ contain only isolated eigenvalues of A. Then $R_A(z)$ is a meromorphic function on Ω with poles exactly at the eigenvalues of A in Ω.*

Proof. $R_A(z)$ is analytic on $\Omega \setminus \{\text{eigenvalues of } A \text{ in } \Omega\}$. If ψ is an eigenfunction with eigenvalue $\lambda \in \Omega$, then

$$R_A(z)\psi = (z - \lambda)^{-1}\psi,$$

and so $R_A(z)$ has a pole (or an essential singularity) at $z = \lambda \in \Omega$. □

It is also appropriate to mention here the *second resolvent identity*. Whereas the first identity, Proposition 1.6, compares the resolvent of a fixed operator at two points in its resolvent set, the second compares the resolvent of two operators at a common point in their resolvent sets.

Proposition 1.9. (Second resolvent identity). *Let A and B be two closed operators with $z \in \rho(A) \cap \rho(B)$. Then*

$$\begin{aligned} R_A(z) - R_B(z) &= R_A(z)(A - B)R_B(z) \\ &= R_B(z)(B - A)R_A(z). \end{aligned} \tag{1.6}$$

Proof. One applies the identity

$$a^{-1} - b^{-1} = a^{-1}(b - a)b^{-1},$$

taking note that the right side of (1.6) defines a bounded operator. □

1.3 Hilbert Space

All of our discussion so far has concerned linear operators on a Banach space X. To go further, we must introduce more structure on X. We use the abbreviations LVS for linear vector space and $NLVS$ for normed linear vector space, as introduced in Appendix 1.

Definition 1.10. *Let V be an LVS. An inner product $\langle \cdot, \cdot \rangle$ on V is a map from $V \times V \to \mathbb{C}$ satisfying, for any $x, y, z \in V$, $a \in \mathbb{C}$:*

(1) $\langle ax + y, z\rangle = a\langle x, z\rangle + \langle y, z\rangle$ *(linear in the first entry);*

(2) $\langle x, y\rangle = \overline{\langle y, x\rangle}$ *(complex symmetric);*

(3) $\langle x, x\rangle \geq 0$ *and* $\langle x, x\rangle = 0$ *if and only if* $x = 0$ *(positive definiteness).*

Any LVS with an inner product is called an inner product space (IPS).

Examples 1.11.

(1) $X = \mathbb{R}^n$ and define, for $x, y \in \mathbb{R}^n$: $\langle x, y\rangle = \sum_{i=1}^n x_i y_i$. Then this is an inner product (note that it is real, so condition (2) is replaced by $\langle x, y\rangle = \langle y, x\rangle$). Moreover, this inner product depends on the choice of the basis for $\mathbb{R}^n$. Here we have taken the components of x and y relative to the same fixed basis.

(2) $X = C([0, 1])$ and define for $f, g \in X$: $\langle f, g\rangle = \int_0^1 f(t)\overline{g(t)}dt$; then this defines an inner product on X.

(3) $X = l^2$ and define, for any $x, y \in l^2$: $\langle x, y\rangle \equiv \sum_{k=0}^{\infty} x_i \overline{y_i}$. This sum always converges, and $\langle \cdot, \cdot\rangle$ is an inner product of l^2.

Problem 1.2. Prove the assertions in Examples 1.11.

Let V be an IPS with inner product $\langle \cdot, \cdot\rangle$. Then for each $v \in V$, define

$$\|v\| \equiv \langle v, v\rangle^{\frac{1}{2}}. \tag{1.7}$$

Then $\|\cdot\|$ is a *norm* on V. We say that it is the norm induced by the IP.

Definition 1.12. *An inner product space that is complete in the norm induced by the inner product (as in 1.7) is called a Hilbert space.*

Consequently, each Hilbert space is a Banach space with additional structure coming from the inner product.

Examples 1.13.

(1) $X = l^2$ is a Hilbert space as follows from Proposition A1.18.

(2) $X = \mathbb{R}^n$ or $\mathbb{C}^n$ are trivially Hilbert spaces.

(3) It follows from Remark A1.14 that $C([0, 1])$ equipped with the norm given in part (2) of Examples 1.11 is not a Hilbert space: there are Cauchy sequences in the norm induced by the IP with no limit in $C([0, 1])$. We will, however, later show how to "complete" this space so that it becomes a Hilbert space.

(4) $X = l^p$, $p \neq 2$, is not a Hilbert space.

(5) $X = C([0, 1])$ with the sup norm $\|f\|_\infty = \sup_{0\leq x\leq 1} |f(x)|$, provides an example of a Banach space that is not a Hilbert space (i.e., there is no IP that induces this norm).

(6) $X = L^2(\mathbb{R}^n)$, where L^2 is defined in Appendix 2, is a Hilbert space with the inner product $\langle f, g\rangle \equiv \int f(x)\overline{g(x)}dx$. Note that by the Schwarz inequality (Theorem 1.14) this is finite for all $f, g \in L^2$. The norm induced by this $\|f\| \equiv \langle f, f\rangle^{1/2}$ is just the L^2-norm of Appendix 2.

(7) $X = H^s(\mathbb{R}^n)$, the Sobolov space of order s defined in Section 2 of Appendix 4, is a Hilbert space with an inner product:

$$\langle f, g\rangle_s \equiv \sum_{|k|\leq s} \sum_{i=1}^{n} \langle \partial^{k_i} f/\partial x_i^{k_i}, \partial^{k_i} g/\partial x_i^{k_i}\rangle,$$

where the symbol on the right is the usual L^2-inner product. The Sobolov norm is defined by $\|f\|_s = \langle f, f\rangle_s^{1/2}$.

There are many connections between the IP and its induced norm. For example, a norm induced by an IP always satisfies the *parallelogram law:*

$$\|x+y\|^2 + \|x-y\|^2 = 2\|x\|^2 + 2\|y\|^2. \tag{1.8}$$

Problem 1.3. Prove the parallelogram law (1.8) for a norm induced by an IP.

There is a converse to this statement. Suppose X is an NLVS over $\mathbb{R}$ and that the norm satisfies the parallelogram law (1.8). If we define

$$\langle x, y\rangle \equiv \frac{1}{2}(\|x\|^2 + \|y\|^2 - \|x-y\|^2), \tag{1.9}$$

then $\langle \cdot, \cdot \rangle$ is an IP for X.

Problem 1.4. Prove that (1.9) defines an IP on X (we are working over the reals here).

Finally, we prove an inequality between the norm and IP which is used quite often.

Theorem 1.14 (Schwarz inequality). *For any $x, y \in X$, an IPS, we have*

$$|\langle x, y\rangle| \leq \|x\| \, \|y\|. \tag{1.10}$$

Problem 1.5. Prove the Schwarz inequality. *Hint*: Use the fact that $\|ax+by\| \geq 0$, with $b = -\langle y, x\rangle |\langle x, y\rangle|^{-1}$, and the discriminant rule for quadratic equations, or minimize with respect to a.

As a final topic in this introduction, we discuss two notions of convergence in Hilbert spaces. The first comes from the topology induced from the norm (as discussed earlier here) and the second from the convergence with respect to linear functionals (as we will see more clearly in Chapter 2).

Definition 1.15. *Let X be a Hilbert Space. A sequence $\{u_n\}$, $u_n \in X$, converges strongly to $u \in X$ if $\lim_{n\to\infty} \|u_n - u\| = 0$. A sequence $\{y_n\}$, $y_n \in X$, converges weakly to $y \in X$ if for each $f \in X$, $\lim_{n\to\infty} |\langle y_n, f\rangle - \langle y, f\rangle| = 0$.*

Example 1.16. Let u_n be a smooth function of fixed, compact support localized about $x_n = n$, so that supp $(u_n) \cap K = \phi$ for any compact set K and for all n large enough. For example, let $u \in C_0^\infty(\mathbb{R}^n)$ be such that supp $u \subset B_1(0)$. We localize this function about $x_n = n$ by setting $u_n(x) \equiv u(x-n)$; see Figure 1.1. Then for any $v \in X$, $\lim_{n\to\infty}\langle v, x_n\rangle = 0$, and so the sequence $\{u_n\}$ converges weakly to zero. Note, however, that $||u_n|| = ||u||$, and so u_n cannot converge strongly to zero.

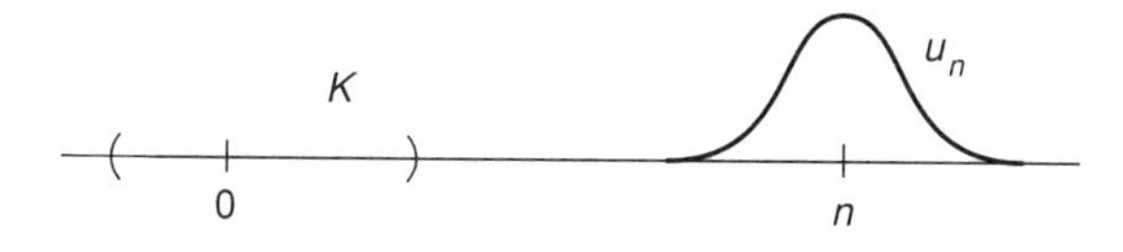

FIGURE 1.1. Wave packet u_n localized about n.

2

The Geometry of a Hilbert Space and Its Subspaces

Among the various subsets of a Hilbert space $\mathcal{H}$, the subsets with a linear structure play a distinguished role. In this chapter, we explore some simple geometry associated with these subsets.

2.1 Subspaces

Definition 2.1. *A subset W in a Hilbert space $\mathcal{H}$ is called a subspace if it is closed under addition and scalar multiplication, that is, if $x, y \in W$ and $a \in \mathbb{R}$ or $\mathbb{C}$, then*

$$ax + y \in W.$$

Examples 2.2.

(1) Let $\mathcal{H} = \mathbb{R}^2$. Any straight line through the origin forms a (linear) subspace, whereas a straight line not passing through the origin does not.

(2) Let $\mathcal{H} = L^2(\mathbb{R}^n)$. Choose any $g \in \mathcal{H}$, $g \neq 0$. Then $\{f \in \mathcal{H} | \langle f, g \rangle = 0\}$ forms a subspace.

Proposition 2.3. *A closed subspace of a Hilbert space $\mathcal{H}$ is again a Hilbert space.*

Problem 2.1. Prove Proposition 2.3.

Problem 2.2. If $A \in \mathcal{L}(\mathcal{H})$ (a bounded operator), then $\ker A$ is a closed subspace of $\mathcal{H}$.

We now want to study some properties and relations among closed subspaces. Henceforth, M will denote a *closed* subspace of a Hilbert space $\mathcal{H}$.

Lemma 2.4. *For any $x \in \mathcal{H}$ there exists a unique vector $y \in M$ such that if $d \equiv \inf_{z \in M} \|x - z\|$, the distance from x to M, then $d = \|x - y\|$.*

Proof. If $x \in M$, then choose $y = x$. If $x \notin M$, then there exists a sequence $\{y_n\}$, $y_n \in M$, such that

$$\lim_{n \to \infty} \|x - y_n\| = d,$$

by definition of d. We will show that this implies $\{y_n\}$ is a Cauchy sequence. Consequently, there exists $y \in M$ such that $y = \lim_{n \to \infty} y_n$. Then, by continuity, $d = \lim_{n \to \infty} \|x - y_n\| = \|x - y\|$.

To show that $\{y_n\}$ is a Cauchy sequence, we begin with the parallelogram law (1.8):

$$\begin{aligned} \|y_n - y_m\|^2 &= \|(y_n - x) - (y_m - x)\|^2 \\ &= 2\|y_n - x\|^2 + 2\|y_m - x\|^2 - \|y_n + y_m - 2x\|^2. \end{aligned} \tag{2.1}$$

Now as $d = \inf_{y \in M} \|x - y\|$, it follows that

$$\|\frac{1}{2}(y_n + y_m) - x\| \geq d,$$

because $y_n + y_m \in M$. Returning to (2.1), we get

$$\|y_n - y_m\| \leq 2\|y_n - x\|^2 + 2\|y_m - x\|^2 - 4d^2, \tag{2.2}$$

and so as $m, n \to \infty$, the right side of (2.2) approaches $4d^2 - 4d^2 = 0$. The uniqueness of y follows from Theorem 2.8 ahead. □

For any $x \in \mathcal{H}$, the vector $y \in M$ for which $d = \|x - y\|$ has a clear geometric interpretation. The vector y is the projection of x onto M; any $x \in \mathcal{H}$ can be resolved into a vector $y \in M$ and a vector w perpendicular to M (see Theorem 2.8).

Definition 2.5. *Let M be a closed subspace of a Hilbert space $\mathcal{H}$. The orthogonal complement of M in $\mathcal{H}$, denoted $M^\perp$, is the set of vectors in $\mathcal{H}$ which are orthogonal to M:*

$$M^\perp = \{x \in \mathcal{H} | \langle x, m \rangle = 0 \ \ \forall m \in M\}. \tag{2.3}$$

Examples 2.6.

(1) Let $\mathcal{H} = \mathbb{R}^2$ as in Examples 2.2. If M is a straight line through the origin, then $M^\perp$ is also a straight line through the origin and is the line that is at a right angle to M, see Figure 2.1.

(2) Let $\mathcal{H} = L^2(\mathbb{R}^n)$, and let $M = \{\lambda g \mid \lambda \in \mathbb{C},\ g \in \mathcal{H} \text{ and } g \neq 0\}$. Then M is a closed subspace, and the set described in part (2) of Examples 2.2 is $M^\perp$.

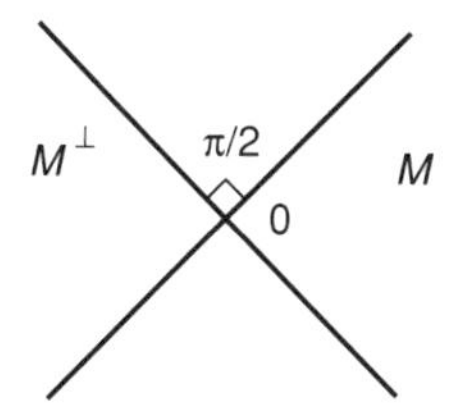

FIGURE 2.1. The subsets M and $M^\perp$ in $\mathbb{R}^2$.

Proposition 2.7. *The subset $M^\perp$ is a closed subspace of $\mathcal{H}$ and is therefore a Hilbert space; $M \cap M^\perp = \{0\}$.*

Problem 2.3. Prove Proposition 2.7.

We know from our experience with finite–dimensional spaces, such as $\mathbb{R}^n$, that given a closed subspace M (a hyperplane passing through the origin) and its complement $M^\perp$, we can resolve any vector w into a sum of two unique vectors, one lying in M (this is the closest vector of M to w) and one lying in $M^\perp$ (see Figure 2.2). This property is common to all Hilbert spaces and is one reason why they are relatively easy to handle.

Theorem 2.8 (The projection theorem). *Let M be a closed subspace of $\mathcal{H}$, and let $M^\perp$ be its orthogonal complement. Then any $x \in \mathcal{H}$ can be written uniquely as $x = y + z$ with $y \in M$ and $z \in M^\perp$.*

Proof. Let y be the projection of x on M, and set $z \equiv x - y$. We show that $z \in M^\perp$. Let $t \in \mathbb{R}$ and $m \in M$. If $d = \|x - y\| = \inf_{n \in M} \|x - n\|$, then

$$d^2 \leq \|x - (y + tm)\|^2 = \|x - y\|^2 + t^2\|m\|^2 - 2\,t\,\mathrm{Re}\langle x - y, m\rangle,$$

and so for all t,

$$t^2\|m\|^2 - 2\,t\,\mathrm{Re}\langle x - y, m\rangle \geq 0.$$

This implies that $\langle z, m\rangle = \langle x - y, m\rangle = 0$, for, if not, take $t = \|m\|^{-2}\mathrm{Re}\langle z, m\rangle$ and obtain a contradiction. For uniqueness, see Problem 2.4. □

Problem 2.4. Prove that the decomposition in Theorem 2.8 is unique. Finish the proof of Lemma 2.4. (*Hint*: Suppose $x = y_i + z_i$, $i = 1, 2$, with $y_i \in M$, $z_i \in M^\perp$, and study $0 = (y_1 - y_2) + (z_1 - z_2)$ to show $y_1 = y_2$ and $z_1 = z_2$.)

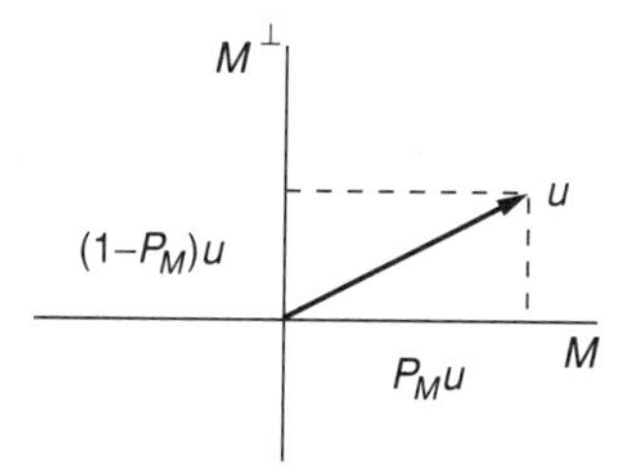

FIGURE 2.2. The orthogonal decomposition of a vector in $\mathbb{R}^2$ relative to a subspace M.

It follows from Theorem 2.8 that given a closed subspace M in $\mathcal{H}$, we have a unique decomposition of each vector in $\mathcal{H}$. We express this fact by saying that $\mathcal{H}$ is the *direct sum* of M and $M^\perp$ (which are in themselves Hilbert spaces), and we write $\mathcal{H} = M \oplus M^\perp$. In general, a Hilbert space $\mathcal{H}$ is a *direct sum* of Hilbert spaces $\mathcal{H}_1$ and $\mathcal{H}_2$ if each vector $x \in \mathcal{H}$ can be written uniquely as $x = x_1 + x_2$ with $x_i \in \mathcal{H}_i$ and x_1 orthogonal to x_2.

We noted that if $M \subset \mathcal{H}$ is a closed subspace, then $M^\perp$ is also closed. One can check that $(M^\perp)^\perp = M$. Moreover, $M^\perp$ can be defined as given in (2.3) even if M is not closed. It is still true that $M^\perp$ is closed and then $(M^\perp)^\perp = \bar{M}$, the closure of M.

Proposition 2.9. *A subset W is dense in $\mathcal{H}$ if $W^\perp = \{0\}$.*

Proof. Use the facts that (i) $W^\perp = (\bar{W})^\perp$ and (ii) the projection theorem. □

2.2 Linear Functionals and the Riesz Theorem

Let X be a Banach space over $\mathbb{C}$ (or $\mathbb{R}$).

Definition 2.10. *A bounded linear functional (LF) f on X is a bounded linear operator from X to $\mathbb{C}$ (or $\mathbb{R}$).*

Examples 2.11.

(1) Let $X = C([0, 1])$. Then the map $f \to \int_0^1 f(x)dx$ is a bounded linear functional on X.

(2) Let $X = L^2(\mathbb{R}^n)$. Let $g \in L^2(\mathbb{R}^n)$, $g \neq 0$, fixed. Then it follows from the Hölder inequality, Theorem A2.4, that the map

$$f \in L^2(\mathbb{R}^n) \to \|fg\|_1 = \langle f, g\rangle$$

defines a bounded linear functional on $L^2(\mathbb{R}^n)$. One of the major results of Hilbert space theory is that *every* bounded linear functional on $L^2(\mathbb{R}^n)$ has this form, which we prove ahead.

Lemma 2.12. *For any bounded LF f on a Hilbert space $\mathcal{H}$, the null space of f,*

$$Null(f) = \{x \in \mathcal{H} | f(x) = 0\},$$

is a closed subspace of codimension one (by the codimension of a subspace M in $\mathcal{H}$, we mean dim $M^\perp$ in $\mathcal{H}$) provided $f \neq 0$.

Proof. We leave it as a problem to prove that Null(f) is a closed subspace. Assume $f \neq 0$, then Null$(f) \neq \mathcal{H}$. By the projection theorem, there exists $x_0 \in \mathcal{H}$ such that $x_0 \in \text{Null}(f)^\perp$. By the linearity of f,

$$f(x - f(x)f(x_0)^{-1}x_0) = f(x) - f(x) = 0.$$

If y is any vector in $\text{Null}(f)^{\perp}$, then

$$y - f(y)f(x_0)^{-1}x_0 = 0,$$

and so $\text{Null}(f)^{\perp}$ is one-dimensional. □

Problem 2.5. Prove that $\text{Null}(f)$ is a closed subspace of $\mathcal{H}$.

Theorem 2.13 (The Riesz lemma or representation theorem). *For any bounded LF f on a Hilbert space $\mathcal{H}$, there exists a unique $y_f \in \mathcal{H}$ such that*

$$f(x) = \langle x, y_f \rangle \;\; \forall x \in \mathcal{H}.$$

Furthermore, $\|f\| = \|y_f\|$ (where $\|f\|$ is the operator norm of f and $\|y_f\|$ is the Hilbert space norm).

Proof.

(1) Assume $f \neq 0$, otherwise take $y_f = 0$. By Lemma 2.12, there exists $x_0 \in \text{Null}(f)^{\perp}$, $x_0 \neq 0$. Set $y_f = \overline{f(x_0)}\|x_0\|^{-2}x_0$. Now given any $x \in \mathcal{H}$, it is easy to see that $[x - f(x)f(x_0)^{-1}x_0] \in \text{Null}(f)$. Hence, by a computation:

$$\begin{aligned}
\langle x, y_f \rangle &= f(x_0)\|x_0\|^{-2} \langle x, x_0 \rangle \\
&= f(x_0)\|x_0\|^{-2}[\langle x - f(x)f(x_0)^{-1}x_0, x_0 \rangle \\
&\quad + f(x)f(x_0)^{-1}\|x_0\|^2] \\
&= f(x).
\end{aligned}$$

(2) Uniqueness is left as Problem 2.6. To prove the statement about norms, note that

$$\|f\| = \sup_{x \in \mathcal{H}} \|x\|^{-1}|f(x)| = \sup_{x \in \mathcal{H}} \|x\|^{-1}|\langle x, y_f \rangle| \leq \|y_f\|,$$

by the Schwarz inequality. On the other hand,

$$\|y_f\|^2 = f(y_f) \leq \|f\| \, \|y_f\|,$$

so $\|y_f\| \leq \|f\|$. □

Problem 2.6. Prove that the vector representative for a bounded LF is unique.

2.3 Orthonormal Bases

We can push the notion of orthogonality in a Hilbert space much further. Given a nonzero vector $u_1 \in \mathcal{H}$, we construct M_1, the one-dimensional subspace containing u_1, and $M_1^\perp$. We may then take $u_2 \in M_1^\perp$, $u_2 \neq 0$, and, restricting ourselves to $M_1^\perp$, construct $M_2^\perp \equiv \{w \in M_1^\perp \mid \langle u_2, w\rangle = 0\}$. If $\mathcal{H} = \mathbb{R}^N$ and we were to continue this way, we would soon exhaust $\mathcal{H}$ and obtain a set of N mutually orthogonal vectors that span the space. If $\mathcal{H}$ is not finite–dimensional, we can, with appropriate care, still find a set of orthogonal vectors that span the space.

Definition 2.14. *A collection $\{x_\alpha\}$ of vectors in a Hilbert space $\mathcal{H}$ is called an orthonormal set if $\langle x_\alpha, x_\beta\rangle = \delta_{\alpha\beta}$.*

Theorem 2.15 (Pythagorean theorem). *Let $\{x_i\}_{i=1}^N$ be an orthonormal set; then for all $x \in \mathcal{H}$,*

$$\|x\|^2 = \sum_{i=1}^N |\langle x, x_i\rangle|^2 + \|x - \sum_{i=1}^N \langle x, x_i\rangle x_i\|^2. \tag{2.4}$$

Proof. Write $x = y + z$, where $y = \sum_{i=1}^N \langle x, x_i\rangle x_i$ and $z = x - y$. We check that $\langle y, z\rangle = 0$:

$$\begin{aligned}\langle y, z\rangle &= \sum_{i=1}^N |\langle x, x_i\rangle|^2 - \sum_{i,j=1}^N \overline{\langle x, x_i\rangle}\langle x_i, x_j\rangle\langle x, x_j\rangle \\ &= \sum_{i=1}^N |\langle x, x_i\rangle|^2 - \sum_{i,j=1}^N \overline{\langle x, x_i\rangle}\langle x, x_j\rangle\delta_{ij} = 0.\end{aligned}$$

Consequently, $\|x\|^2 = \langle y + x, y + z\rangle = \|y\|^2 + \|z\|^2$. □

An immediate consequence of the Pythagorean theorem is the following inequality, known as *Bessel's inequality.* For any orthonormal set $\{x_i\}_{i=1}^N$ and any vector $x \in \mathcal{H}$, we have

$$\sum_{i=1}^N |\langle x, x_i\rangle|^2 \leq \|x\|^2.$$

Geometrically, the first term in the decomposition (2.4) is the length of the projection of x along the subspace spanned by $\{x_i\}_{i=1}^N$, and the second term is the length of the projection of x onto the orthogonal subspace. We now ask whether we can find a sufficiently large set of orthonormal vectors in $\mathcal{H}$ such that the second term in (2.4) is zero.

Definition 2.16. *An orthonormal set B in a Hilbert space $\mathcal{H}$ is called an orthonormal basis (ONB) if there exists no other orthonormal set B' in $\mathcal{H}$ such that B is a proper subset of B'.*

An ONB B in $\mathcal{H}$ is a maximal orthonormal set. Clearly, if B is an ONB, then $B^{\perp} = \{0\}$. A main theorem in Hilbert space theory is the following.

Theorem 2.17. *Every Hilbert space has an orthonormal basis.*

Idea of the Proof. The proof is complicated by the fact that the ONB need not be a countable set. Consequently, the proof uses ideas from set theory and, in particular, transfinite induction (Zorn's lemma). One constructs a partial order relation on the set $\mathcal{O}$ of all orthonormal sets in $\mathcal{H}$. The relation is simply set inclusion. One next considers chains that are subsets of $\mathcal{O}$ which are totally ordered. One shows that each chain has a maximal element. This is simply the union of all elements in the chain. It now follows from Zorn's lemma that there exists a maximal element in $\mathcal{O}$. □

We now obtain a general expansion theorem.

Theorem 2.18. *Let $B = \{x_\alpha\}$ be any ONB in $\mathcal{H}$. Then for any $x \in \mathcal{H}$,*

$$x = \sum_{\alpha} \langle x, x_\alpha \rangle x_\alpha,$$

where the sum converges absolutely, and

$$\|x\|^2 = \sum_{\alpha} |\langle x, x_\alpha \rangle|^2.$$

Conversely, if $\{c_\alpha\}$ is a collection of complex numbers satisfying

$$\sum_{\alpha} |c_\alpha|^2 < \infty,$$

then $\sum_\alpha c_\alpha x_\alpha$ converges absolutely to a vector in $\mathcal{H}$.

Here we have indexed the ONB by an arbitrary set that is not necessarily countable. In this case, the sums appearing in Theorem 2.18 have to be defined. We say that a set of vectors $\{x_\alpha\}$, with index set $\mathcal{J}$, is *summable* if for any $\epsilon > 0$, there exists a finite set of indices $\mathcal{J}_\epsilon$ such that $\| \sum_{\alpha \in \mathcal{J}_1} x_\alpha \| < \epsilon$, for any finite set of indices $\mathcal{J}_1$ disjoint from $\mathcal{J}_\epsilon$. We refer the reader to the book by Halmos [Hal] for further details on summability. In all of our applications, the Hilbert spaces have a countable ONB, and we deal with countable index sets.

Definition 2.19. *A Hilbert space is separable if it has a countable orthonormal basis.*

Problem 2.7. Prove that Definition 2.19 is equivalent to Definition A1.21.

Proof of Theorem 2.18 for separable Hilbert spaces.

Let $x \in \mathcal{H}$, $x \neq 0$. Consider the sequence in $\mathbb{C}$, $\{\langle x, x_i \rangle\}$. By the Pythagorean theorem, Theorem 2.15,

$$\sum_{i=1}^{n} |\langle x, x_i \rangle|^2 \leq \|x\|^2, \tag{2.5}$$

for any n. Since $\sum_{i=1}^{n} |\langle x, x_i\rangle|^2$ is a monotonically increasing sequence, it converges (by (2.5)) as $n \to \infty$ and

$$\sum_{i=1}^{\infty} |\langle x, x_i\rangle|^2 \leq \|x\|^2.$$

Now define a sequence of vectors in the Hilbert space $\mathcal{H}$ by

$$x^{(n)} = \sum_{i=1}^{n} \langle x, x_i\rangle x_i.$$

This is a Cauchy sequence, for if $n > m$, we have

$$\|x^{(n)} - x^{(m)}\|^2 = \| \sum_{i=m+1}^{n} \langle x, x_i\rangle x_i \|^2 = \sum_{i=m+1}^{n} |\langle x, x_i\rangle|^2,$$

and, by the above convergence, this converges to zero as $n, m \to \infty$. Consequently, $\exists\, x' \in \mathcal{H}$ such that $x' = \lim_{n\to\infty} x^{(n)}$. By construction, $\langle x', x_i\rangle = \langle x, x_i\rangle$, and so for all i:

$$\langle x' - x, x_i\rangle = 0,$$

which says that $x' - x$ is orthogonal to the ONB $\{x_i\}$. But by maximality of the ONB, we must have $x' - x = 0$, for otherwise $\{x' - x, x_1, x_2, \cdots\}$ would be an ONB properly containing $\{x_i\}$. Consequently, $x = x'$, the following sum converges:

$$x = \sum_{i=1}^{\infty} \langle x, x_i\rangle x_i,$$

and

$$\|x\|^2 = \lim_n \|x^{(n)}\|^2 = \sum_{i=1}^{\infty} |\langle x, x_i\rangle|^2.$$

Here, we used the fact that $\|x^{(n)}\| \to \|x\|$. □

Problem 2.8. Prove the remaining statements in Theorem 2.18, including the fact that $\|x^{(n)}\| \to \|x\|$.

Separable Hilbert spaces allow a complete classification up to isometric isomorphism. By an isometric isomorphism, we mean a map $\Phi : \mathcal{H}_1 \to \mathcal{H}_2$, between two Hilbert spaces $\mathcal{H}_i$, $i = 1, 2$, over $\mathbb{C}$ or $\mathbb{R}$, which is (1) linear,

$$\Phi(\lambda x + y) = \lambda\Phi(x) + \Phi(y),$$

for all $x, y \in \mathcal{H}_1$, $\lambda \in \mathbb{C}$ or $\mathbb{R}$; (2) onto and one-to-one (hence invertible); and (3) an isometry, so for any $x \in \mathcal{H}_1$,

$$\|\Phi(x_1)\| = \|x_1\|,$$

where the norm on the left side is in $\mathcal{H}_2$ and on the right is in $\mathcal{H}_1$.

Theorem 2.20. *Let $\mathcal{H}$ be a separable Hilbert space over the field F ($F = \mathbb{C}$ or $\mathbb{R}$). If $\mathcal{H}$ has an ONB consisting of $N < \infty$ elements, then $\mathcal{H}$ is isometrically isomorphic to $\mathbb{C}^N$ if $F = \mathbb{C}$, or to $\mathbb{R}^N$ if $F = \mathbb{R}$. If $\mathcal{H}$ has a countable ONB (which is not finite), then $\mathcal{H}$ is isometrically isomorphic to $l^2(\mathbb{C})$ or $l^2(\mathbb{R})$, if $F = \mathbb{C}$ or $\mathbb{R}$, respectively.*

Proof. Consider the case for which $\mathcal{H}$ has a countable ONB that is not finite, say $\{x_i\}$. For any $x \in \mathcal{H}$, the sequence $\{\langle x, x_i\rangle\} \in l^2$, by Theorem 2.18. Define a map $U : \mathcal{H} \to l^2$ by

$$U : x \in \mathcal{H} \to \{\langle x, x_i\rangle\} \in l^2.$$

Clearly, U is linear since the inner product is linear in the first entry. Moreover, the map U is well defined on all $\mathcal{H}$. By the second part of Theorem 2.18, it is onto. By the first part of that theorem, it is an isometry:

$$\|Ux\|^2 = \sum_{i=1}^{\infty} |\langle x, x_i\rangle|^2 = \|x\|^2.$$

The case when $\mathcal{H}$ has a finite basis proceeds similarly. □

Problem 2.9. Complete the proof of Theorem 2.20.

3

Exponential Decay of Eigenfunctions

We take a pause from our development of the theory of linear operators to present a first application to Schrödinger operators. Let us recall from the Introduction that a Schrödinger operator is a linear operator on the Hilbert space $L^2(\mathbb{R}^n)$ of the form $H = -\Delta + V$, where $-\Delta \equiv -\sum_{i=1}^n \partial^2/\partial x_i^2$ and the potential V is a real–valued function. The general problem we study here is as follows. Suppose that L is a linear operator on $L^2(\mathbb{R}^n)$ with eigenvalue λ and corresponding eigenfunction ψ, that is, a function $\psi \in L^2(\mathbb{R}^n)$ such that

$$L\psi = \lambda\psi.$$

Since $\psi \in L^2(\mathbb{R}^n)$, it has some average decay as $\|x\| \to \infty$. How is this decay determined by the operator L? In the case that L is a Schrödinger operator, we would like to know how the behavior of the potential V, as $\|x\| \to \infty$, determines the decay of an eigenfunction. This can be answered very nicely provided we content ourselves with upper bounds on the rate of decay. We will also use this discussion to introduce various geometric ideas concerning Schrödinger operators. These ideas will play important roles in the later chapters on semiclassical analysis.

3.1 Introduction

We consider the behavior of eigenfunctions $\psi(x)$ of the Schrödinger operator

$$H = -\Delta + V \text{ on } L^2(\mathbb{R}^n),$$

in the limit as $\|x\| \to \infty$. We will present a simplified version of the general theory developed by Agmon [Ag1]. As for H, we assume that $V(x)$ is real and well

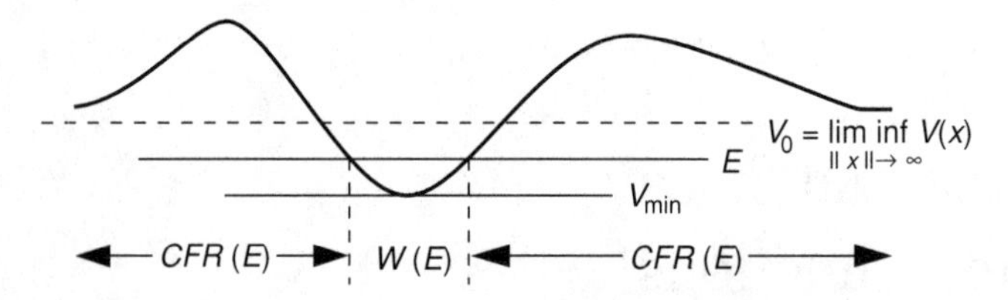

FIGURE 3.1. A potential well $W(E)$ and the classically forbidden region CFR(E).

enough behaved so that $\sigma(H) \subset \mathbb{R}$. We will give sufficient conditions for this later in Chapters 8 and 13. We assume that H has a discrete eigenvalue $E < \inf \sigma_{\text{ess}}(H)$, with an eigenfunction $\psi \in L^2(\mathbb{R}^n)$, so that $H\psi = E\psi$. We will reformulate the condition $E < \inf\ \sigma_{\text{ess}}(H)$ as a geometric condition on the potential in the main theorem.

The behavior of $\psi(x)$ as $\|x\| \to \infty$ can be inferred from the following physical situation. Suppose that $\liminf_{\|x\|\to\infty} V(x) > V_{\min}$ and that V is sufficiently deep to support a bound state (i.e., an eigenfunction) at energy E:

$$V_{\min} < E < \liminf_{\|x\| \to \infty} V(x) \equiv V_0. \tag{3.1}$$

Furthermore, we suppose that outside some compact set $W(E)$, we have $E - V < 0$. We refer to this region $\mathbb{R}^n \setminus W(E)$ as the *classically forbidden region for energy E*, denoted by CFR(E); see Figure 3.1. A classical particle with energy E is forbidden to enter $\mathbb{R}^n \setminus W(E)$ by the conservation of energy. The quantum mechanical wave function, however, penetrates this region. This effect is called *quantum tunneling*. We want to describe the rate of decay of ψ in this region.

We expect the decay to be very rapid as can be seen from simple one-dimensional models. In Figure 3.2, we show a square potential well:

$$V(x) = \begin{cases} V_0 > 0, & |x| > a, \\ 0, & |x| < a. \end{cases}$$

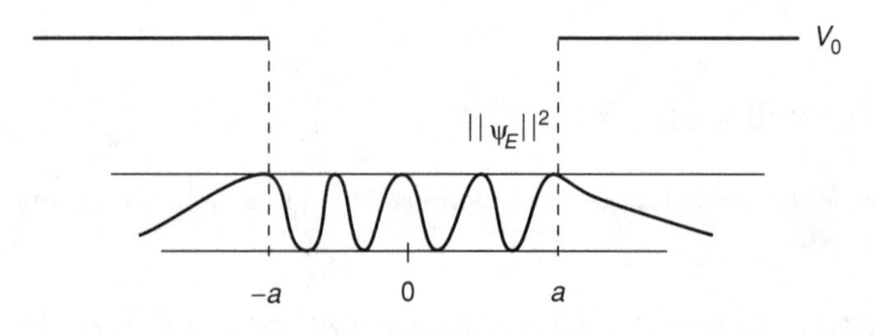

FIGURE 3.2. A square well potential in one dimension.

The Schrödinger equations in each region are

$$|x| < a, \qquad \frac{-\hbar^2}{2m}\psi'' = E\psi,$$

$$|x| > a, \qquad \frac{-\hbar^2}{2m}\psi'' + V_0\psi = E\psi,$$

which can be solved explicitly. (Recall that $h = 2\pi\hbar$.)

We are interested in energies E satisfying (3.1): $0 < E < V_0$. The eigenfunctions are

$$|x| < a, \qquad \psi_1(x) = c_1 \cos kx + c_2 \sin kx, \qquad \text{where } k = \left(\frac{2mE}{\hbar^2}\right)^{\frac{1}{2}}, \tag{3.2}$$

and

$$|x| > a, \qquad \psi_2(x) = c_0 e^{-\kappa|x|}, \qquad \text{where } \kappa \equiv \left[\frac{2m}{\hbar^2}(V_0 - E)\right]^{\frac{1}{2}}. \tag{3.3}$$

Using solutions (3.2) and (3.3) and the condition that the wave function and its first derivative must be continuous at $x = \pm a$, we can solve for the allowable energy levels. An eigenvalue E must satisfy

$$\sin^{-1}\left(\frac{E}{V_0}\right)^{\frac{1}{2}} = \frac{1}{2}(n\pi - 2ka) \tag{3.4}$$

for $n \in \mathbb{N}$. For any solution of (3.4), the eigenfunction in the classically forbidden region is

$$\psi_E(x) = c_0 e^{-\left[\frac{2m}{\hbar^2}(V_0 - E)\right]^{\frac{1}{2}}|x|}, \tag{3.5}$$

which shows pointwise exponential decay. A more complicated and realistic model is shown in Figure 3.1. In this case, we cannot solve the Schrödinger equation exactly, but we can use semiclassical WKB methods to approximate the wave function. The result for $|x| > a$ is

$$\psi_E(x) \underset{h \downarrow 0}{\sim} c_0 e^{-\frac{\sqrt{2m}}{\hbar}\int_a^{|x|}\sqrt{V(s)-E}\,ds}, \tag{3.6}$$

which generalizes (3.5). Our task is to show that under the conditions described here, eigenfunctions are bounded above by exponential factors of this type. To do so, we must first find an appropriate generalization of the WKB factor occurring in (3.6).

A function $\psi \in L^2(\mathbb{R}^n)$ satisfies an L^2 *upper bound* if $e^F\psi \in L^2(\mathbb{R}^n)$, for some function $F : \mathbb{R}^n \to \mathbb{R}^+$. This function measures the decay rate of ψ as $\|x\| \to \infty$. If $F = F(\|x\|)$, then F gives an isotropic decay rate for ψ. The Agmon method provides an *anisotropic* decay rate. Note that the L^2 upper bound is *a priori* weaker than a pointwise bound on ψ. It suggests that as $\|x\| \to \infty$, $\psi \to 0$ faster than $e^{-F(x)}$. In many cases, this pointwise bound can be proved; we will give one such theorem ahead.

3.2 Agmon Metric

We introduce the necessary geometrical object, very much like a metric, which we will use to measure the exponential decay of eigenfunctions. Let us recall that for a real, n-dimensional manifold $\mathcal{M}$, the *tangent space* at a point $x \in \mathcal{M}$, denoted by $T_x(\mathcal{M})$, can be considered to be the real linear vector space $\mathbb{R}^n$. A *metric* is an assignment of an inner product to the tangent space $T_x(\mathcal{M})$ for each $x \in \mathcal{M}$. If this inner product is not positive-definite, we say that it is *degenerate*. We consider $\mathbb{R}^n$ with a degenerate Riemannian metric defined as follows.

Definition 3.1. *Let $x \in \mathbb{R}^n$, and let $\xi, \eta \in T_x(\mathbb{R}^n)$. We define a (degenerate) inner product on $T_x(\mathbb{R}^n)$ by*

$$\langle \xi, \eta \rangle_x \equiv (V(x) - E)_+ \langle \xi, \eta \rangle_E, \tag{3.7}$$

where $\langle \cdot, \cdot \rangle_E$ is the usual Euclidean inner product and $f(x)_+ \equiv \max\{f(x), 0\}$.

We call this the Agmon metric on $\mathbb{R}^n$, even though it may vanish at some points x. Note that it depends on V and E. We want to use this structure to induce a metric on $\mathbb{R}^n$. Let $\gamma : [0, 1] \to \mathbb{R}^n$ be a differentiable path in $\mathbb{R}^n$. For any Riemannian structure, the length of γ is

$$L(\gamma) = \int_0^1 \|\dot{\gamma}(t)\|_{\gamma(t)} dt. \tag{3.8}$$

Note that $\|\xi\|_x = \langle \xi, \xi \rangle_x^{1/2}$. In the Agmon structure (3.7), the length of the curve γ (3.8) is

$$L_A(\gamma) = \int_0^1 (V(\gamma(t)) - E)_+^{1/2} \|\dot{\gamma}(t)\|_E dt, \tag{3.9}$$

where $\|\cdot\|_E$ denotes the usual Euclidean norm. Recall that a path γ is a *geodesic* if it minimizes the energy functional $E(\gamma) \equiv \frac{1}{2} \int_0^1 \|\dot{\gamma}(t)\|^2_{\gamma(t)} dt$.

Definition 3.2. *Given a continuous potential V and energy E, the distance between $x, y \in \mathbb{R}^n$ in the Agmon metric is*

$$\rho_E(x, y) \equiv \inf_{\gamma \in P_{x,y}} L_A(\gamma), \tag{3.10}$$

where $P_{x,y} \equiv \{\gamma : [0, 1] \to \mathbb{R}^n \mid \gamma(0) = x, \gamma(1) = y, \ \gamma \in AC[0, 1]\}$. Here $AC[0, 1]$ is the space of all absolutely continuous functions on $[0, 1]$.

Hence the distance between $x, y \in \mathbb{R}^n$ with the Agmon metric is the length of the shortest geodesic connecting x to y. Note that ρ_E in (3.10) reduces to the WKB factor in (3.6) for the one-dimensional case.

Problem 3.1. Show that ρ_E satisfies the triangle inequality. (*Hint:* For any given path connecting x to y, $\rho_E(x, y) \le L_A(\gamma)$.)

Proposition 3.3.

(1) The *distance function $\rho_E(x, y)$ is locally Lipschitz continuous and hence is differentiable almost everywhere in x and in y;*

(2) *at the points where it is differentiable:* $|\nabla_y \rho_E(x, y)|^2 \le (V(y) - E)_+$.

Proof.

(1) Let $x, y, z \in \mathbb{R}^n$, and fix x. By the triangle inequality: $|\rho_E(x, y) - \rho_E(x, z)| \le \rho_E(y, z)$. Let γ_0 be a straight line from y to z: $\gamma_0(t) = (1-t)y + tz, t \in [0, 1]$, and evaluate $L_A(\gamma_0)$,

$$L_A(\gamma_0) = \int_0^1 (V(\gamma_0(t)) - E)_+^{1/2} \|z - y\| dt \; = \; \|z - y\| C_{y,z}, \tag{3.11}$$

where $C_{y,z}$ is the constant $\int_0^1 (V(\gamma_0(t)) - E)_+^{1/2} dt$. For any $R > 0$, let $B_R(y)$ be the ball of radius R centered at y and let $C_R(y) \equiv \max_{w \in B_R(y)} (V(w) - E)_+^{1/2}$. Consequently, for fixed x,

$$|\rho_E(x, w) - \rho_E(x, z)| \le \rho_E(w, z) \le C_{w,z} \|w - z\| \le C_R(y) \|w - z\|, \tag{3.12}$$

so ρ_E is locally Lipschitz in its second argument. By a theorem of analysis (Rademacher's theorem—see, for example, [EG]), and the symmetry of ρ_E, ρ_E is differentiable almost everywhere in each of its arguments.

(2) Let $z = h + y$, and compute $\limsup_{\|h\| \to 0} \|h\|^{-1} |\rho_E(x, y + h) - \rho_E(x, y)| \le \limsup_{\|h\| \to 0} \|h\|^{-1} \rho_E(y + h, y)$ which, by (3.12), is bounded above by $\lim_{\|h\| \to 0} \int_0^1 (V(\gamma_0(t)) - E_+)^{1/2} dt = \lim_{\|h\| \to 0} \int_0^1 (V(y + th) - E)_+^{1/2} dt \le (V(y) - E)_+^{1/2}$. Thus, at points y where ρ_E is differentiable, the above two results indicate that $|\nabla_y \rho_E(x, y)|^2 \le (V(y) - E)_+$, proving the proposition.

□

3.3 The Main Theorem

Theorem 3.4. *Let $H = -\Delta + V$, with V real and continuous, be a closed operator bounded below with $\sigma(H) \subset \mathbb{R}$. Suppose E is an eigenvalue of H and that $supp(E - V(x))_+$ is a compact subset of $\mathbb{R}^n$. Let $\psi \in L^2(\mathbb{R}^n)$ be an eigenfunction* *of H such that $H\psi = E\psi$. Then, for any $\epsilon > 0$ $\exists$ a constant c_ϵ, $0 < c_\epsilon < \infty$, such that*

$$\int e^{2(1-\epsilon)\rho_E(x)}|\psi(x)|^2 dx \leq c_\epsilon, \tag{3.13}$$

where $\rho_E(x) \equiv \rho_E(x, 0)$.

Remark 3.5. Result (3.13) roughly means that $\psi(x) = O(e^{-(1-\epsilon)\rho_E(x)})$ as $\|x\| \to \infty$. As $\rho_E(x) \sim V(x)^{1/2}$ for $\|x\| \to \infty$, we see that $\psi(x)$ decays exponentially as $\|x\| \to \infty$ at a rate controlled by the distance to the origin in the Agmon metric. We will show how to extract pointwise exponential bounds on the wave function in certain situations ahead.

The geometric condition in the theorem is crucial. It implies that there exists $R > 0$ such that $V(x) > E$ for all $\|x\| > R$. As we will see in Chapter 8, this condition implies that $E < \inf \sigma_{\text{ess}}(H)$. To prove Theorem 3.4, we give two preliminary lemmas. We will make use of the Sobolev spaces $H^s(\mathbb{R}^n)$, for positive integers s. These are reviewed in Appendix 4.

Lemma 3.6. *For any $\epsilon > 0$, let $f \equiv (1-\epsilon)\rho_E$. Let $\phi \in D(V) \cap H^1(\mathbb{R}^n)$ be such that supp$(\phi) \subset \mathcal{F}_{E,\delta} \equiv \{x \mid V(x) - E > \delta\}$. Then there exists $\delta_1 > 0$ such that*

$$\text{Re}\langle e^f\phi, (H-E)e^{-f}\phi\rangle \geq \delta_1\|\phi\|^2. \tag{3.14}$$

Proof.

(1) We compute the left side of (3.14). Consider the gauge transformed H : $H_f \equiv e^f H e^{-f}$. To compute this, we have for any $u \in C_0^\infty(\mathbb{R}^n)$, $-e^f\Delta e^{-f}u = -(e^f\nabla e^{-f})(e^f\nabla e^{-f})u$ and $e^f\nabla e^{-f}u = (-\nabla f + \nabla)u$. It follows that $-e^f\Delta e^{-f}u = -(\nabla - \nabla f)^2 u = (p + i\nabla f)^2 u$ (where $p \equiv -i\nabla$) and $H_f = (p + i\nabla f)^2 + V$. Writing this out, we have $H_f = p^2 - |\nabla f|^2 + (\nabla \cdot \nabla f + \nabla f \cdot \nabla) + V$. (Note: This is to be understood in the sense of quadratic forms, because f is differentiable almost everywhere.)

(2) Returning to (3.14), the left side is

$$\text{Re}\langle\phi, (H_f - E)\phi\rangle \geq \langle\phi, (V - |\nabla f|^2 - E)\phi\rangle, \tag{3.15}$$

where we used $p^2 > 0$ and the fact that $\nabla \cdot \nabla f + \nabla f \cdot \nabla$ is antisymmetric so that the real part of its contribution to the quadratic form vanishes. Now to compute (3.15), we use the definition of f, Proposition 3.3, and recall supp$(\phi) \subset \mathcal{F}_{E,\delta}$, $|\nabla f|^2 = (1-\epsilon)^2|\nabla\rho_E(x)|^2 \leq (1-\epsilon)(V-E)_+$, so $\langle\phi, (V - |\nabla f|^2 - E)\phi\rangle \geq \epsilon\delta\|\phi\|^2$ and the lemma is proved by taking $\delta_1 \equiv \epsilon\delta$. □

We mention that the left side of (3.15) might be infinite because the function f is unbounded in general. It will be convenient to introduce the notation $[A, B]$, which is called the *commutator* of the operators A and B and denotes $AB - BA$.

Lemma 3.7. *Let* $f \equiv (1-\epsilon)\rho_E$, *and let* $f_\alpha \equiv f(1+\alpha f)^{-1}$ *(which is bounded). Suppose* η *is a smooth, bounded function with* supp $|\nabla \eta|$ *compact. Set* $\phi \equiv \eta \cdot \exp(f_\alpha) \cdot \psi$, *where* $H\psi = E\psi$. *Then*

$$Re\langle e^{f_\alpha}\phi, (H-E)e^{-f_\alpha}\phi\rangle = \langle \xi e^{2f_\alpha}\psi, \psi\rangle, \tag{3.16}$$

where $\xi \equiv |\nabla\eta|^2 + 2(\nabla\eta \cdot \nabla f_\alpha)\eta$.

Proof. Observe that $e^{f_\alpha}\phi \in L^2$ as f_α is bounded by 1. We compute the left side of (3.16):

$$\langle e^{f_\alpha}\phi, (H-E)e^{-f_\alpha}\phi\rangle = \langle \eta e^{2f_\alpha}\psi, (H-E)\eta\psi\rangle = \langle \eta e^{2f_\alpha}\psi, [p^2, \eta]\psi\rangle,$$

since $(H-E)\psi = 0$. Now the commutator $[p^2, \eta] = -\Delta\eta - 2\nabla\eta \cdot \nabla$, and so by integration by parts, we compute

$$\begin{aligned} &\langle -2(\nabla\eta)\eta e^{2f_\alpha}\psi, \nabla\psi\rangle \\ &= \langle [2(\Delta\eta)\eta + 2|\nabla\eta|^2 + 4(\nabla\eta\cdot\nabla f_\alpha)\eta]e^{2f_\alpha}\psi, \psi\rangle + \langle 2(\nabla\eta)\eta e^{2f_\alpha}\nabla\psi, \psi\rangle, \end{aligned}$$

so that

$$\mathrm{Re}\langle -2(\nabla\eta)\eta e^{2f_\alpha}\psi, \nabla\psi\rangle = \langle [(\Delta\eta)\eta + |\nabla\eta|^2 + 2(\nabla\eta\cdot\nabla f_\alpha)\eta]e^{2f_\alpha}\psi, \psi\rangle.$$

Combining this with the $(-\Delta\eta)$-term from the commutator, we get

$$\mathrm{Re}\langle e^{f_\alpha}\phi, (H-E)e^{-f_\alpha}\phi\rangle = \langle [|\nabla\eta|^2 + 2(\nabla\eta\cdot\nabla f_\alpha)\eta]e^{2f_\alpha}\psi, \psi\rangle,$$

which is the result, (3.16). □

3.4 Proof of Theorem 3.4

Consider the following regions associated with a discrete eigenvalue $E \in \sigma_d(H)$:

$$\mathcal{F}_{E,2\delta} \equiv \{x \mid V(x) - E > 2\delta\}$$

and

$$\mathcal{A}_{E,\delta} \equiv \{x \mid V(x) - E < \delta\};$$

see Figure 3.3.

Let $\eta \in C^\infty$ be such that $\eta(x) = 1,\ x \in \mathcal{F}_{E,2\delta}$, and $\eta(x) = 0,\ x \in \mathcal{A}_{E,\delta}$. By our assumption on V, such an η has supp$|\nabla\eta|$ compact. Let $f = (1-\epsilon)\rho_E$ and $f_\alpha = f(1+\alpha f)^{-1}$ as before. Then $\phi \equiv \eta \cdot \exp(f_\alpha)\cdot\psi$ satisfies the hypotheses of Lemma 3.6. Consequently, $\exists \delta_1 > 0$ such that

$$\begin{aligned} \delta_1\|\phi\|^2 &\le \mathrm{Re}\,\langle e^{f_\alpha}\phi, (H-E)e^{-f_\alpha}\phi\rangle \\ &\le |\langle \xi e^{2f_\alpha}\psi, \psi\rangle| \\ &\le \left(\sup_{x \in \mathrm{supp}|\nabla\eta|} |\xi e^{2f_\alpha}(x)|\right)\|\psi\|^2, \end{aligned} \tag{3.17}$$

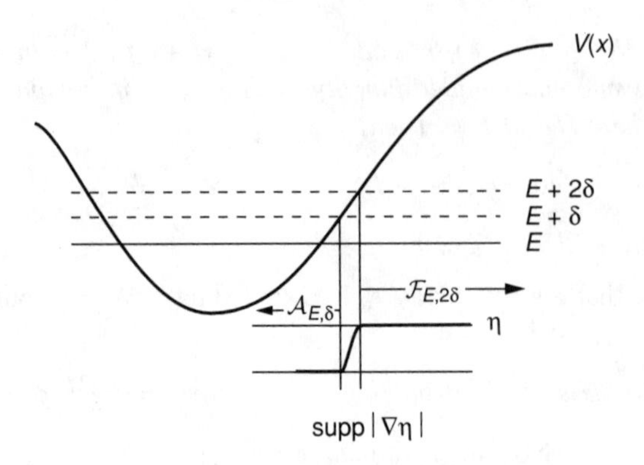

FIGURE 3.3. Classically forbidden region $\mathcal{F}_{E,2\delta}$ and allowed region $\mathcal{A}_{E,\delta}$.

where we used Lemma 3.7. The fact that $\text{supp}|\nabla\eta| = (\mathbb{R}^n \backslash \mathcal{F}_{E,2\delta}) \backslash \mathcal{A}_{E,\delta}$ is compact means that we can take $\alpha = 0$ in the right side of (3.17). Let $f_0 \equiv \sup\{|f(x)| \mid x \in supp|\nabla\eta|\}$, and take ψ normalized. Then

$$\|e^{f_\alpha}\eta\psi\|^2 \leq \tilde{c}_\epsilon, \tag{3.18}$$

where $\tilde{c}_\epsilon$ is independent of α. Hence we can take $\alpha = 0$ in the left side of (3.18). Now $\text{supp}|\nabla\eta| \cup \bar{\mathcal{A}}_{E,\delta}$ is a compact set, so $\{e^{2f(x)}|x \in \text{supp}|\nabla\eta| \cup \bar{\mathcal{A}}_{E,\delta}\}$ is bounded, and, consequently, the integral $\int_{\text{supp}|\nabla\eta|\cup\bar{\mathcal{A}}_{E,\delta}} e^{2f}|\psi|^2$ is finite. Thus, $\exists c_\epsilon$, $0 < c_\epsilon < \infty$, such that

$$\int e^{2(1-\epsilon)\rho_E(x)}|\psi(x)|^2 = \left[\int_{\{x|\eta(x)=1\}} + \int_{\text{supp}|\nabla\eta|\cup\bar{\mathcal{A}}_{E,\delta}}\right] e^{2f}|\psi|^2 \leq c_\epsilon. \qquad \square$$

3.5 Pointwise Exponential Bounds

In some cases, the L^2 exponential bound of Theorem 3.4 can be converted to a pointwise exponential bound. This is particularly true when the potential V is sufficiently regular. This is so because the regularity of V will imply the regularity of any eigenfunction of H. This is an example of what is known as *elliptic regularity*. These results are based on the Sobolev embedding theorems, one of which is given in Appendix 4, Theorem A4.6. We state a version of this needed here.

Theorem 3.8. *Let $H = -\Delta + V$, and suppose $V \in C^k(\mathbb{R}^n) \cap L^p(\mathbb{R}^n)$ for $p > n/2$, $n > 2$, and $k \geq 0$. Then if u is an eigenfunction of H, $u \in L^2(\mathbb{R}^n) \cap C^{k+2}(\mathbb{R}^n)$.*

A consequence of this theorem is that if $V \in C^k(\mathbb{R}^n) \cap L^p(\mathbb{R}^n)$, p as above, then $u \in H^s(\mathbb{R}^n)$, for $s = k + 2$. Let $C_b^k(\mathbb{R}^n)$ denote the bounded elements of $C^k(\mathbb{R}^n)$.

Lemma 3.9. *Let* $V \in C_b^k(\mathbb{R}^n) \cap L^p(\mathbb{R}^n)$, $p > n/2$, $n > 2$, *and take* $k > \min\{0, (n-4)/2\}$. *Let* $H\psi = E\psi$. *Then* $\exists C_{E,M,V}$ *depending only on* $-\infty < M < \inf \sigma(H)$, E, *and* $\sup_{x\in\mathbb{R}^n} |V^{(\alpha)}(x)|$, $|\alpha| = 0, ..., k$, *such that for any* $x_0 \in \mathbb{R}^n$,

$$\max_{x\in B(x_0,1/2)} |\psi(x)| \le C_{E,M,V} \|\psi\|_{L^2(B(x_0,1))}.$$

Proof. Let χ be a smooth characteristic function for $B(x_0, 1/2)$, that is, $\chi \in C_0^\infty(\mathbb{R}^n)$, $0 \le \chi \le 1$, $\chi | B(x_0, 1/2) = 1$. By Theorem 3.8, $\chi\psi \in L^1(\mathbb{R}^n)$ and

$$\max_{x\in B(x_0,1/2)} |\psi(x)| \le \max_{x\in B(x_0,1/2)} |(\chi\psi)(x)| \le \int |\hat{\chi} * \hat{\psi}(k)| d^n k, \tag{3.19}$$

where $\hat{f}$ is the Fourier transform of f (see Appendix 4). Note that $(|k|^2+1)^{-s/2} \in L^2$ for $s > n/2$. This condition and the fact that $s = k + 2$ imply that we must have $k > \min\{0, (n-4)/2\}$. It follows from (3.19) and the Schwarz Inequality that there exists $c_s > 0$ such that

$$\max_{x\in B(x_0,1/2)} |\psi(x)| \le c_s \left[\int |(-\Delta + 1)^{\frac{s}{2}} (\chi\psi)(x)|^2 \right]^{\frac{1}{2}}. \tag{3.20}$$

The right side is finite since $\psi\chi \in H^s(\mathbb{R}^n)$ for k as above by Theorem 3.8. Now we consider the integral in the square brackets in (3.20). Since χ has compact support, we can express this as the inner product

$$\langle \chi\psi, (-\Delta + 1)^s \chi\psi \rangle.$$

The idea is to evaluate this by successively computing the action of $(-\Delta + 1)$ on $\chi\psi$ using the eigenvalue equation. We have

$$\begin{aligned} (-\Delta + 1)(\chi\psi)(x) &= (H + 1)(\chi\psi)(x) - (V\chi\psi)(x) \\ &= (E + 1)(\chi\psi)(x) + [-\Delta, \chi]\psi(x) - (V\chi\psi)(x). \end{aligned} \tag{3.21}$$

We write the commutator in (3.21) as

$$[-\Delta, \chi]\psi = (-\Delta\chi)\psi - 2\nabla\chi \cdot \nabla\psi.$$

Let $\tilde{\chi}$ be a smooth function such that $\chi\tilde{\chi} = \chi$ and $\operatorname{supp}\tilde{\chi} \subset B(x_0, 1)$. Then

$$\begin{aligned} \nabla\chi \cdot \nabla\psi &= \nabla\chi \cdot \nabla(\tilde{\chi}\psi) \\ &= \nabla\chi \cdot \nabla(H + M)^{-1}(H + M)\tilde{\chi}\psi \\ &= \nabla\chi \cdot \nabla(H + M)^{-1}[\tilde{\chi}(E + M)\psi \\ &\quad + (-\Delta\tilde{\chi} + 2\nabla \cdot \nabla\tilde{\chi})\psi]. \end{aligned} \tag{3.22}$$

Note that $\nabla(H+M)^{-1}$ and $\nabla(H+M)^{-1}\nabla$ are bounded and that a characteristic function or its derivative sits next to ψ. Hence, $\exists \tilde{C}_{E,M,V}$ such that

$$\|\nabla\chi\cdot\nabla\psi\| \leq \tilde{C}_{E,M,V}\|\psi\|_{L^2(B(x_0,1))}.$$

Combining (3.20)–(3.22) and repeating this argument s times, it is clear that $\exists C_{E,M,V}$ such that

$$\|(-\Delta+1)^s\chi\psi\| \leq C_{E,M,V}\|\psi\|_{L^2(B(x_0,1))}.$$

The lemma follows from this and (3.20). □

Theorem 3.10. *Let* $V \in L^p(\mathbb{R}^n)\cap C_b^k(\mathbb{R}^n)$, $p > n/2$, $n > 2$, *and* $k > \min\{0, (n-4)/2\}$. *Let* E *be an eigenvalue of* H *with* $\operatorname{supp}(E - V(x))_+$ *compact, and* $\psi \in L^2(\mathbb{R}^n)$ *a corresponding eigenfunction,* $H\psi = E\psi$. *Then for each* $\epsilon > 0$, $\exists c_\epsilon > 0$ *such that*

$$|\psi(x)| \leq c_\epsilon e^{-(1-\epsilon)\rho_E(x)},$$

where $\rho_E(x) = \rho_E(x, 0)$.

Proof. By Lemma 3.9, for any $x_0 \in \mathbb{R}^n$,

$$\max_{x\in B(x_0,1/2)} |\psi(x)e^{(1-\epsilon)\rho_E(x)}| \leq C_{E,M,V}\|e^{(1-\epsilon)\rho_E}\psi\|_{L^2(B(x_0,1))}$$

$$\times\left\{\left(\sup_{y\in B(x_0,1))} e^{-(1-\epsilon)\rho_E(y)}\right)\left(\sup_{x\in B(x_0,1)} e^{(1-\epsilon)\rho_E(x)}\right)\right\}.$$

The two exponentials can be combined using the triangle inequality, $\rho_E(x) \leq \rho_E(x,y) + \rho_E(y)$, and so we get

$$\begin{aligned}\max_{x\in B(x_0,1/2)} |\psi(x)e^{(1-\epsilon)\rho_E(x)}| &\leq C_{E,M,V}\|e^{(1-\epsilon)\rho_E}\psi\|_{L^2(\mathbb{R}^n)}\\ &\quad\times\left\{\sup_{x,y\in B(x_0,1)} e^{(1-\epsilon)\rho_E(x,y)}\right\}. \qquad (3.23)\end{aligned}$$

Finally, as $\rho_E(x,y) \leq c\|x-y\|$, where c is uniform in x_0 and depends only on M, and the norm on the right side of (3.22) is bounded by Theorem 3.4, there is a $c_\epsilon > 0$ such that

$$\max_{x\in B(x_0,1/2)} |\psi(x)e^{(1-\epsilon)\rho_E(x)}| \leq c_\epsilon.$$

Since x_0 is arbitrary, this proves the result. □

Remark 3.11. The pointwise exponential bound of Theorem 3.10 can be obtained under weaker conditions on V, but this requires much more work.

3.6 Notes

There is a large amount of literature on the decay rate of eigenfunctions of second–order differential operators. The methods of this chapter are taken from the book by Agmon [Ag1], which also contains many references to earlier work. Isotropic exponential bounds on eigenfunctions for Schrödinger operators were considered before the anisotropic estimates presented here. These isotropic estimates have the form

$$|\psi(x)| \le c_\alpha e^{-\alpha\|x\|},$$

for some $\alpha > 0$. We mention the papers of Combes and Thomas [CT] and of O'Connor [OC] because their proofs employ the method of complex scaling (see Chapter 17) to prove bounds of this type. R. Carmona, P. Deift, W. Hunziker, B. Simon, and E. Vock undertook a systematic study of the decay rates of eigenfunctions for 2– and N–body Schrödinger operators in [Sim1], [Sim2], [Sim3], [DHSV], and [CS]. There is also a body of work concerning the relation between the rate of decay of an eigenfunction for eigenvalues of Schrödinger operators and the positions of the thresholds for the operator. We refer the reader to the papers by Froese, Herbst, M. Hoffmann-Ostenhof, and T. Hoffmann-Ostenhoff [FHHO1], and Froese and Herbst [FrHe] and to the book [CFKS] for an account of this theory and other references. We will mention a simple case of this in Chapter 16 for two–body operators. The theory is much richer in the N-body case because there may be many thresholds. For other results on exponential decay of wave functions, we refer to [HOAM] and references therein. Upper bounds are among the weakest estimates one can obtain on an eigenfunction. One might ask if there are lower bounds for an eigenfunction. In the case that the eigenfunction is positive, which occurs when it corresponds to the lowest eigenvalue for Schrödinger operators with reasonable potentials (see [RS4]), one can obtain isotropic lower bounds [FHHO2] and lower bounds in terms of the Agmon metric [CS]. It is still an open problem to determine in what cases the Agmon metric gives the actual rate of decay of an eigenfunction, even in the two–body case; see, for example, [He2] and the references therein.

4

Operators on Hilbert Spaces

We now continue to develop additional properties of linear operators on Hilbert spaces. The inner product structure of a Hilbert space has many consequences for the structure of operators mapping the space into itself. The most important of these is the existence of an *adjoint operator* acting on the same space. Although it is possible to define an adjoint operator corresponding to an operator on a Banach space, the operator acts on a different space in general. Because of the Riesz representation theorem, the adjoint of a Hilbert space operator can be taken to act on the same space. We conclude the chapter with a discussion of the resolvent of the Laplacian on $\mathbb{R}^3$. This important partial differential operator is another example of a Schrödinger operator and will play a central role throughout the remainder of the book.

4.1 Remarks on the Operator Norm and Graphs

Let $\mathcal{H}$ be a (separable) Hilbert space, and recall that $\mathcal{L}(\mathcal{H})$ denotes the set of bounded operators on $\mathcal{H}$. This set $\mathcal{L}(\mathcal{H})$ is a Banach space with the norm

$$\|A\| \equiv \sup\{ \|Ax\| \mid x \in \mathcal{H}, \|x\| = 1\}. \tag{4.1}$$

We first derive a useful formula for this norm.

Theorem 4.1. *For any $A \in \mathcal{L}(\mathcal{H})$,*

$$\|A\| = \sup\{ |\langle Ax, y\rangle| \mid x, y \in \mathcal{H}, \|x\| = \|y\| = 1\}. \tag{4.2}$$

In fact, the supremum can be taken over a dense subset of $\mathcal{H}$.

Proof. For fixed $x \in \mathcal{H}$, consider the map

$$y \in \mathcal{H} \mapsto \langle y, Ax \rangle.$$

It is easy to check that this is a bounded linear functional on $\mathcal{H}$. By the Riesz lemma (Theorem 2.13), the norm of this functional,

$$\sup\{ |\langle y, Ax \rangle| \mid y \in \mathcal{H}, \|y\| = 1\},$$

is equal to the norm of the defining vector, namely, Ax. Consequently,

$$\|A\| = \sup_x \{ \|Ax\| \mid \|x\| = 1\} = \sup_{x,y} \{ |\langle y, Ax \rangle| \mid \|x\| = \|y\| = 1\},$$

which proves the first part of the theorem. □

Problem 4.1. Prove the second statement in Theorem 4.1, that is, that the supremum can be taken over a dense set.

Let us recall from Appendix 3 the notion of the graph of an operator. Let A be an operator on $\mathcal{H}$ with domain $D(A)$. The *graph* of A, $\Gamma(A)$, is the subset of $\mathcal{H} \times \mathcal{H} \equiv \{(x, y) \mid x, y \in \mathcal{H}\}$ defined by

$$\Gamma(A) = \{(x, Ax) | x \in D(A)\}. \tag{4.3}$$

The set $\mathcal{H} \times \mathcal{H}$ can be made into a Hilbert space with the inner product

$$\langle (x, y), (w, z) \rangle_{\mathcal{H} \times \mathcal{H}} \equiv \langle x, w \rangle_{\mathcal{H}} + \langle y, z \rangle_{\mathcal{H}}. \tag{4.4}$$

Problem 4.2. Prove that $\mathcal{H} \times \mathcal{H}$, with the inner product given in (4.4), is a Hilbert space.

The norm induced by the inner product in (4.4) is given by

$$\|(x, y)\|_{\mathcal{H} \times \mathcal{H}} = [\|x\|_{\mathcal{H}}^2 + \|y\|_{\mathcal{H}}^2]^{\frac{1}{2}} \tag{4.5}$$

and is sometimes called the *graph norm*. Finally, let us recall the notion of a closed operator. An operator A with domain $D(A)$ is said to be *closed* if $\Gamma(A)$ is a closed subset of $\mathcal{H} \times \mathcal{H}$. If $\Gamma(A)$ is not closed but its closure, $\overline{\Gamma(A)}$, is the graph of an operator, we call this operator the closure of A. We refer the reader to Appendix 3 for a more complete discussion.

4.2 The Adjoint of an Operator

We use the inner product structure of a Hilbert space $\mathcal{H}$ to associate with each operator A another operator on $\mathcal{H}$, called its *adjoint* and denoted by A^*. The adjoint operator provides a powerful tool for the study of A and its spectrum.

Definition 4.2. *Let A be an operator on $\mathcal{H}$ with domain $D(A)$. The adjoint of A, A^*, is defined on the domain*

$$D(A^*) \equiv \{x \in \mathcal{H} | \, |\langle Ay, x\rangle| \leq C_x \|y\| \text{for some constant } C_x \text{ (independent of } y\text{) and for all } y \in D(A)\}, \tag{4.6}$$

as a map $A^* : D(A^*) \to \mathcal{H}$ satisfying

$$\langle Ay, x\rangle = \langle y, A^*x\rangle, \tag{4.7}$$

for all $y \in D(A)$ and $x \in D(A^*)$.

Note that if $A \in \mathcal{L}(\mathcal{H})$, then $D(A) = \mathcal{H}$ and

$$|\langle Ay, x\rangle| \leq \|A\| \, \|y\| \, \|x\|,$$

which implies that $D(A^*) = \mathcal{H}$, since we can take $C_x = \|A\| \, \|x\|$. The adjoint A^* can be constructed explicitly as follows. For any $x \in D(A^*)$ and $y \in D(A)$, $\langle Ay, x\rangle$ is a bounded linear functional on $D(A)$, by definition. Any bounded linear functional f can be extended to all of $\mathcal{H}$. We first extend it to $\overline{D(A)}$ as follows. If $x \in \overline{D(A)}$, let $\{x_n\} \subset D(A)$ be a sequence converging to x. We define the value of the extended linear functional on x to be $\lim_{n\to\infty} f(x_n)$. If $D(A)$ is dense, this defines a linear functional on $\mathcal{H}$. If $\overline{D(A)} \neq \mathcal{H}$, we use the projection theorem (Theorem 2.8) and write

$$H = \overline{D(A)} \oplus D(A)^{\perp}.$$

We can then define an extension of the functional to $D(A)^{\perp}$ by $f(w) = 0$, for example, for any $w \in D(A)^{\perp}$. We note that the extension is unique if $D(A)$ is dense. Otherwise, it is not unique and, consequently, the adjoint will not be uniquely defined. To avoid this problem, we will always assume that $D(A)$ is dense.

So we extend $\langle Ay, x\rangle$ to $\mathcal{H}$. By the Riesz lemma, for each $x \in D(A^*)$ there is a *unique* $z \in \mathcal{H}$ such that

$$\langle Ay, x\rangle = \langle y, z\rangle, \; y \in D(A).$$

The adjoint $A^* : D(A^*) \to \mathcal{H}$ is then given by

$$A^*x = z.$$

Problem 4.3. Show that A^* defined above is a linear operator.

We can summarize the construction of A^*, given a densely defined operator A on $D(A)$, as follows. We first define $D(A^*)$ as in (4.6). Then, for $y \in D(A)$, the map

$$x \in D(A^*) \to f_x(y) \equiv \langle Ay, x\rangle$$

can be extended by the Riesz lemma to obtain

$$\langle Ay, x\rangle = \langle y, z(x)\rangle.$$

We then define

$$A^*x = z(x).$$

By construction, A^* is uniquely defined and satisfies

$$\langle Ay, x\rangle = \langle y, z(x)\rangle = \langle y, A^*x\rangle$$

for $x \in D(A^*)$, $y \in D(A)$.

Proposition 4.3. *Let A, B be operators defined on a common, dense domain D, and let $\lambda, \mu \in \mathbb{C}$. Then*

$$(\mu A + \lambda B)^* = \bar{\mu} A^* + \bar{\lambda} B^*.$$

Problem 4.4. Prove Proposition 4.3.

Proposition 4.4. *If $A \in \mathcal{L}(\mathcal{H})$, then $A^* \in \mathcal{L}(\mathcal{H})$, $\|A^*\| = \|A\|$, and $A^{**} = A$. If $A, B \in \mathcal{L}(\mathcal{H})$, then $(AB)^* = B^*A^*$.*

Proof. As noted above, $D(A^*) = \mathcal{H}$. By the Riesz lemma, for any $x \in \mathcal{H}$:

$$\begin{aligned} \|A^*x\| &= \sup\{|\langle A^*x, y\rangle| \mid \|y\| = 1\} \\ &= \sup\{|\langle x, Ay\rangle| \mid \|y\| = 1\} \qquad (4.8) \\ &\leq \|A\|\,\|x\|, \end{aligned}$$

by (4.7) and the Schwarz inequality. Hence $A^* \in \mathcal{L}(\mathcal{H})$ and $\|A^*\| \leq \|A\|$. Next, by (4.7),

$$\langle A^*x, y\rangle = \langle x, Ay\rangle,$$

thus $(A^*)^* = A$; and by (4.8),

$$\|A\| = \|(A^*)^*\| \leq \|A^*\|,$$

whence $\|A\| = \|A^*\|$. Finally, we compute $(AB)^*$. For any $x, y \in \mathcal{H}$:

$$\begin{aligned} \langle (AB)^*x, y\rangle &= \langle x, ABy\rangle \\ &= \langle A^*x, By\rangle \\ &= \langle B^*A^*x, y\rangle, \end{aligned}$$

and so $(AB)^* = B^*A^*$. □

Remark 4.5. If $A \in \mathcal{L}(\mathcal{H})$, then we can give an equivalent definition of A^* as follows. A^* is the unique bounded operator satisfying

$$\langle A^*u, v\rangle = \langle u, Av\rangle$$

for all $u, v \in \mathcal{H}$. The existence of A^* follows from the Riesz lemma as above. We used this fact in the last part of the proof of Proposition 4.4.

Proposition 4.6. *For any densely defined linear operator A,*

$$\overline{\operatorname{Ran} A} \oplus \ker A^* = \mathcal{H}.$$

Proof. It suffices to prove that $\ker A^*$ is the orthogonal complement of $\operatorname{Ran} A$, as the result then follows by use of the projection theorem. Let $u \in \operatorname{Ran} A$ and $v \in \ker A^*$. Then there exists $f \in D(A)$ such that $u = Af$. We compute

$$\langle u, v\rangle = \langle Af, v\rangle = \langle f, A^*v\rangle = 0\,,$$

and thus $\ker A^* \subset (\operatorname{Ran} A)^\perp$. Now let $w \in (\operatorname{Ran} A)^\perp$. For any $u = Af \in \operatorname{Ran} A$, we have

$$0 = \langle u, w\rangle = \langle Af, w\rangle = \langle f, A^*w\rangle, \tag{4.9}$$

(note that $\langle Af, w\rangle = 0$ implies that $w \in D(A^*)$). As $D(A)$ is dense, it follows that $A^*w = 0$ (by Proposition 2.9), that is, $(\operatorname{Ran} A)^\perp \subset \ker A^*$. □

Proposition 4.7. *Let A be densely defined with dense range. If A has an inverse, then so does A^* and $(A^*)^{-1} = (A^{-1})^*$.*

Proof. Since $\operatorname{Ran} A$ is dense, it follows from Proposition 4.5 that $\ker A^* = \{0\}$, and so by Theorem A3.26, A^* has an inverse. Now $A^{-1} = A^{-1}A = id$, so by carefully considering domains, we get the second statement. □

Problem 4.5. Fill in the details of the proof of $(A^*)^{-1} = (A^{-1})^*$.

Corollary 4.8. *If A is invertible, then so is A^* and $(A^*)^{-1} = (A^{-1})^*$.*

It is a peculiar fact about unbounded operators that even if A is densely defined, A^* may not be. However, we can make the following proposition about A and A^* in general.

Proposition 4.9. *Let A be a densely defined operator. Then*

(1) *A^* is always closed;*

(2) *if A is closed, then $D(A^*)$ is dense (i.e., A^* is densely defined);*

(3) *if A is closed, then $A^{**} = A$.*

Proof. (1) We will prove (2). The proofs of (1) and (3) will follow by a slight extension of the method given here. Suppose A is closed but $D(A^*)$ is not dense, that is, there exists a nonzero vector x perpendicular to $D(A^*)$. Let $\Gamma(A)$ be the graph of A defined in (4.3).

Claim 1. $(x, 0) \in \Gamma(A^*)^\perp$. For any $(u, v) \in \Gamma(A^*)$, we compute

$$\langle (x, 0), (u, v)\rangle = \langle x, u\rangle = 0, \tag{4.10}$$

since $u \in D(A^*)$ and $x \perp D(A^*)$. In (4.10), the first inner product is in the Hilbert space formed from $\mathcal{H} \times \mathcal{H}$ as in Problem 4.2.

(2) Our goal is to show that $(0, x) \in \Gamma(A)$. Since A is linear, this is a contradiction. We introduce an important operator V on $\mathcal{H} \times \mathcal{H}$ by

$$V(x, y) \equiv (-y, x)\,. \tag{4.11}$$

It follows that $V^2 = -1$ and

$$V^*(x, y) = (y, -x),$$

so that $VV^* = 1 = V^*V$.

Claim 2. $(V\Gamma(A))^\perp \subset \Gamma(A^*)$. Let $x \in D(A)$, so that $(x, Ax) \in \Gamma(A)$. Then if (u, v) is orthogonal to $V\Gamma(A)$, we have

$$0 = \langle (u, v), (-Ax, x)\rangle = -\langle u, Ax\rangle + \langle v, x\rangle$$

or

$$\langle u, Ax\rangle = \langle v, x\rangle,$$

thus, by Definition 4.2, $u \in D(A^*)$ and $A^*u = v$. Hence $(u, A^*u) \in \Gamma(A^*)$, and the claim is proved.

Problem 4.6. Let V be a *unitary operator*, that is $V^* = V^{-1}$. Show that for any subspace M,

$$V(M^\perp) = (VM)^\perp. \tag{4.12}$$

(3) By applying Problem 4.6 to Claim 2, we have

$$V(\Gamma(A)^\perp) \subset \Gamma(A^*)\,,$$

and as $V^2 = -1$:

$$\Gamma(A)^\perp \subset V\Gamma(A^*)\,. \tag{4.13}$$

Now, we use the facts that $(M^\perp)^\perp = \bar{M}$ and that $M \subset N$ implies $N^\perp \subset M^\perp$, together with (4.13), to arrive at

$$(V\Gamma(A^*))^\perp \subset \Gamma(\bar{A}) = \Gamma(A),$$

since A is closed. Finally, using (4.12) again, we get

$$V(\Gamma(A^*)^\perp) \subset \Gamma(A)\,. \tag{4.14}$$

By claim 1, $(x, 0) \in \Gamma(A^*)^\perp$, $x \in D(A^*)^\perp$, and so

$$(0, x) = V(x, 0) \in V(\Gamma(A^*)^\perp) \subset \Gamma(A)$$

by (4.14). This shows that $(0, x) \in \Gamma(A)$, which is a contradiction unless $x = 0$. Hence $D(A^*)$ is dense. □

Problem 4.7. Prove that if $D(A^*)$ is dense, then A^{**} extends A. Then conclude that A is closable if and only if $D(A^*)$ is dense.

As a final aspect of the adjoint operator, we apply Proposition 4.9 to obtain a spectral relation between A and A^*.

Corollary 4.10. *Let A be closed. Then*

$$\sigma(A^*) = \overline{\sigma(A)} \text{ (complex conjugate).}$$

Proof. Since A is closed, Proposition 4.9 implies that $D(A^*)$ is dense, and so $\sigma(A^*)$ is well defined. By Corollary 4.8, if $A-\lambda$ is invertible, then $(A-\lambda)^* = A^*-\bar{\lambda}$ is also. This implies that $\overline{\rho(A)} \subset \rho(A^*)$ or, equivalently, $\sigma(A^*) \subset \overline{\sigma(A)}$. Furthermore, as A is closed, $A^{**} = A$. Again, if $A^* - \lambda$ is invertible, Corollary 4.8 shows that $(A^* - \lambda)^* = A^{**} - \bar{\lambda} = A - \bar{\lambda}$ is also. Hence, $\overline{\rho(A^*)} \subset \rho(A)$ or $\sigma(A) \subset \overline{\sigma(A^*)}$. These two relations on the spectra imply that $\overline{\sigma(A)} = \sigma(A^*)$. □

4.3 Unitary Operators

We next turn to the study of another important class of operators on a Hilbert space.

Definition 4.11. *A bounded operator U on a Hilbert space $\mathcal{H}$ is called an isometry if and only if $\|Uf\| = \|f\|$ for all $f \in \mathcal{H}$.*

Isometric transformations preserve the norms of vectors and as such play an important role in quantum mechanics. Note that by the parallelogram law (1.8), if U is an isometry, then

$$\langle Uf, Ug \rangle = \langle f, g \rangle,$$

for all $f, g \in \mathcal{H}$. Furthermore, from (4.1), $\|U\| = 1$. We could equally well define an isometry by the relation

$$U^*U = 1. \tag{4.15}$$

Problem 4.8. Prove that (1) for $U \in \mathcal{L}(\mathcal{H})$, (4.15) is equivalent to Definition 4.11; (2) if $\mathcal{H}$ is finite-dimensional, (4.15) implies that $UU^* = 1$, and so $U^{-1} = U^*$.

Definition 4.12. *A bounded operator U on a Hilbert space $\mathcal{H}$ is called a unitary operator if (a) U is an isometry and (b) Ran $U = \mathcal{H}$.*

Problem 4.9. Prove that U is unitary if and only if U is invertible and $U^* = U^{-1}$.

If U is a unitary operator and A is a closed operator, the *conjugation of A* by U, $A_U \equiv UAU^{-1}$, is often useful in the spectral analysis of A.

Proposition 4.13. *Let A be a closed operator and U a unitary operator. Let $A_U \equiv UAU^{-1}$ on $D(A_U) \equiv UD(A)$.*

(1) *A_U is closed on $D(A_U)$ and $\sigma(A_U) = \sigma(A)$;*

(2) *if A is self-adjoint, then A_U is self-adjoint.*

Proof.

(1) The closure of A_U on $D(A_U)$ is easy to check. As for the spectrum, note that $R_{A_U}(z) = UR_A(z)U^{-1}$ for any $z \in \rho(A)$ and hence

$$\|R_{A_U}(z)\| = \|R_A(z)\|.$$

Thus $\rho(A) = \rho(A_U)$, and the result follows.

(2) A_U is clearly symmetric (see Section 8.1). To prove A is self-adjoint (see Chapters 5 and 8), it suffices to show that $\text{Ran}(A_U + i) = \mathcal{H}$. But U is invertible, so $\text{Ran}(UAU^{-1} + i) = U\,\text{Ran}(A + i) = \mathcal{H}$. □

Consequently, each unitary operator generates an isospectral mapping on operators. This may be very useful for calculations of the spectrum, for many times a unitary operator can be found such that the spectral analysis of the transformed operator is very transparent. An example of this is provided by the Laplacian Δ and the unitary operator given by the Fourier transform. We use material concerning the Fourier transform summarized in Appendix 4.

With regard to spectral analysis of constant-coefficient differential operators, we note the following properties of the Fourier transform:

(1) $F(\partial f/\partial x_i)(k) = -ik_i F(f)(k)$, that is, F takes $\partial/\partial x_i$ to multiplication by $-ik_i$;

(2) $F(x_i f)(k) = i(\partial(Ff)/\partial k_i)(k)$.

We now apply these properties to the Laplacian, a second-order partial differential operator on $\mathbb{R}^n$ defined by

$$\Delta \equiv \sum_{i=1}^{n} \frac{\partial^2}{\partial x_i^2}.$$

For any $f \in \mathcal{S}(\mathbb{R}^n)$,

$$F(\Delta f)(k) = -\|k\|^2(Ff)(k), \tag{4.16}$$

where $\|k\|^2 = \sum_{i=1}^{n} k_i^2$.

Problem 4.10. Prove that the Laplacian Δ, defined in (4.16), is a closed operator on the domain $H^2(\mathbb{R}^n)$ (see Appendix 4 for a review of Sobolev spaces).

It follows from (4.16) and the comments after Proposition 4.13 that $(z - \Delta)$ is invertible for any z with $\text{Im}\, z \neq 0$. Let $R_\Delta(z) \equiv (z - \Delta)^{-1}$ be the resolvent of Δ. From (4.16) we have for any z with $\text{Im}\, z \neq 0$,

$$F(R_\Delta(z)f)(k) = (z + \|k\|^2)^{-1}(Ff)(k). \tag{4.17}$$

Our goal here is to compute an x-space representation for $R_\Delta(z)$ for dimension $n = 3$. The general case is considered in Problem 16.1. To do this, we use the inversion formula (see Appendix 4) and the following property:
Let $f * g$ be the convolution of f and g (see Section A4.3):

$$(f * g)(x) \equiv \int f(x - y)g(y)dy;$$

then

$$F(fg)(k) = (2\pi)^{\frac{n}{2}} \hat{f} * \hat{g}(k)$$

and

$$F^{-1}(fg)(k) = (2\pi)^{\frac{n}{2}} (F^{-1} f) * (F^{-1} g)(k).$$

We now see that from the convolution formula and (4.17),

$$(R_\Delta(z) f)(x) = (G_z * f)(x),$$

where

$$G_z(x) = (2\pi)^{-\frac{n}{2}} \int e^{-ik \cdot x} (z + \|k\|^2)^{-1} dk.$$

Indeed, we can compute this result as follows:

$$\begin{aligned} R_\Delta(z) f &= F^{-1} F R_\Delta(z) f \\ &= F^{-1}((z + \|k\|^2)^{-1} (Ff) \\ &= F^{-1}(z + \|k\|^2) * f. \end{aligned}$$

When $n = 3$, it follows by elementary contour integration that

$$G_z(x) = [4\pi \|x\|]^{-1} e^{-z^{\frac{1}{2}} \|x\|}, \tag{4.18}$$

where the principal branch of the square root is taken. The formula for general n is given in Problem 16.1.

Problem 4.11. Derive (4.18).
Hence, the resolvent of Δ in $\mathbb{R}^3$ is given by an integral operator:

$$(R_\Delta(z) f)(x) = \int [4\pi \|x - y\|]^{-1} e^{-z^{\frac{1}{2}} \|x-y\|} f(y) dy,$$

and this operator is bounded for all $z \in \mathbb{C} \setminus (-\infty, 0]$. We will verify in Chapter 8 that $-\Delta$ is self-adjoint (Example 8.4). This result and our calculation of $R_\Delta(z)$ show that $\sigma(-\Delta) = [0, \infty)$.

5

Self-Adjoint Operators

In this chapter, we make a preliminary study of one of the most important classes of operators on a Hilbert space, the self-adjoint operators. These operators are the infinite-dimensional analogues of symmetric matrices. They play an essential role in quantum mechanics as they determine the time evolution of quantum states. The goal of this chapter is to describe symmetric and self-adjoint operators and to understand some characteristics of the spectrum of a self-adjoint operator. In Chapter 8, we will present the fundamental criteria for self-adjointness. This will be applied in Chapters 8 and 13 to various families of Schrödinger operators.

5.1 Definitions

Let A be an operator on a Hilbert space $\mathcal{H}$ with domain $D(A)$. Unless explicitly stated to the contrary, we always assume that $D(A)$ is dense in $\mathcal{H}$. Recall that if $D(A^*)$, the domain of the adjoint of A, is dense, then A is closable. The closability of A means that the closure of the graph of A, $\Gamma(A)$, is the graph of an operator. This closed operator has an important relation to the original closable operator A. By an *extension* of an operator A with domain $D(A)$, we mean an operator $\tilde{A}$ with domain $D(\tilde{A})$, such that $D(A) \subset D(\tilde{A})$ and $\tilde{A}|D(A)$ (the *restriction* of $\tilde{A}$ to the domain of A) is equal to A. Consequently, it is not hard to check that for a closable operator A, the closure of A is the unique smallest closed extension of A.

Definition 5.1. *An operator A with domain $D(A)$ is called symmetric (or hermitian) if A^* is an extension of A or, equivalently, if $\langle Ax, y\rangle = \langle x, Ay\rangle$ for all $x, y \in D(A)$.*

Problem 5.1. Verify that the two definitions are indeed equivalent.

Example 5.2. Let $\mathcal{H} = L^2([0,1])$, and let A_1 be the operator $A_1 \equiv -d^2/dx^2$ with

$$D(A_1) = \{u \in \mathcal{H} | u, u' \in AC[0,1] \text{ and } u^{(k)}(0) = 0 = u^{(k)}(1), \quad \text{for } k = 0, 1\}. \tag{5.1}$$

Recall that $AC[0,1]$ are precisely those functions that are antiderivatives, and so they can be integrated by parts. Then $D(A_1)$ is dense and A_1 on $D(A_1)$ is symmetric. To show that $D(A_1)$ is dense, note that $D(A)$ contains $C_0^\infty[0,1]$, which is dense in $\mathcal{H}$. The symmetry of A_1 is easily verified by two integrations by parts. Let $u, v \in D(A_1)$; then

$$\begin{aligned} \langle u, A_1 v\rangle &= -\int_0^1 u(x)\overline{\frac{d^2 v}{dx^2}}(x) = -u(x)\overline{\frac{dv}{dx}}(x)|_0^1 + \int_0^1 \frac{du}{dx}(x)\overline{\frac{dv}{dx}}(x) \\ &= \frac{du}{dx}(x)\bar{v}(x)|_0^1 - \int_0^1 \frac{d^2u}{dx^2}(x)\bar{v}(x) \\ &= \langle A_1 u, v\rangle. \end{aligned} \tag{5.2}$$

Definition 5.3. *An operator A is called self-adjoint if* $A = A^*$, *that is, if (a) A is symmetric and (b)* $D(A) = D(A^*)$.

Note that the property of self-adjointness depends both on the form of the operator A (i.e., what A does to a permissible vector) and on the domain $D(A)$. The same symbol may define a self-adjoint operator on a domain D_1, but it might not represent a self-adjoint operator on a domain D_2. As the adjoint A^{**} is always closed, a self-adjoint operator is closed. On the other hand, a symmetric operator need not be closed, and a closed, symmetric operator need not be self-adjoint.

Problem 5.2. Let A be symmetric on $D(A)$. Prove that A^{***} is also symmetric.

Example 5.4. We continue Example 5.2. The operator A_1 defined there is not self-adjoint on $D(A_1)$. To see this, note that $u \in D(A_1^*)$ if and only if $u, u' \in AC[0,1]$ and for all $v \in D(A_1)$

$$-u(x)\frac{dv}{dx}(x)|_0^1 + \frac{du}{dx}(x)v(x)|_0^1 = 0. \tag{5.3}$$

The conditions on v mean that (5.3) is satisfied provided $u, u' \in AC[0,1]$ with no other conditions. Hence

$$D(A_1^*) = \{v \in \mathcal{H} | v, v' \in AC[0,1]\}.$$

Clearly, $D(A_1^*) \supset D(A_1)$. The problem is that A_1 is "too small" for self-adjointness; as a consequence, $D(A_1^*)$ is much bigger than $D(A_1)$. We can search for an extension of A_1 that is self-adjoint by relaxing the requirements on the domain while preserving (5.3).

Problem 5.3. First, show that if $u \in D(A_1^*)$, then $u, u' \in AC[0,1]$. Second, relaxing the constraints on the domain, find other dense domains on which the symbol $A \equiv -d^2/dx^2$ is symmetric. Try to find a domain on which the operator is self-adjoint. (*Hint*: Consider linear boundary conditions of the form $u(0)+\alpha u(1) = 0$.)

As Example 5.4 shows, a symmetric operator need not be self-adjoint, but it follows from the definition that every self-adjoint operator is symmetric.

5.2 General Properties of Self-Adjoint Operators

The property of self-adjointness is strong enough to enable us to make some rather specific statements about the spectrum and to obtain some bounds on the resolvent. These resolvent bounds are crucial for the study of the discrete and essential spectrum. In later chapters, we will see how to extend some of these bounds to various closed, non–self-adjoint operators. Although we will not discuss it in this book, the property of self-adjointness of the Schrödinger operator is necessary and sufficient for the existence of quantum dynamics; see [RS1].

Let us recall from Chapter 2 the definition of the residual spectrum of an operator A:

$$\sigma_{\text{res}}(A) = \{\lambda \in \mathbb{C} |\ \ker\,(A - \lambda) = \{0\} \text{ and } \operatorname{Ran}(A - \lambda) \text{ is not dense}\}.$$

Our first theorem presents some results about the spectrum of self-adjoint operators.

Theorem 5.5. *Let A be self-adjoint. Then*

(1) $\sigma(A) \subset \mathbb{R}$;

(2) $\sigma_{\text{res}}(A) = \phi$;

(3) *eigenvectors corresponding to distinct eigenvalues are orthogonal.*

Proof.

(1) We will prove this after we obtain a preliminary estimate on the resolvent in Theorem 5.6.

(2) Let λ be such that $\ker(A - \lambda) = \{0\}$. From Chapter 4, we have

$$\operatorname{Ran}(A - \lambda)^{\perp} = \ker(A^* - \lambda) = \ker(A - \lambda) = \{0\}, \tag{5.4}$$

so that $\operatorname{Ran}(A - \lambda)$ is dense in $\mathcal{H}$ and $\lambda \notin \sigma_{res}(A)$.

(3) Let ψ, ϕ be eigenvectors of A such that

$$A\phi = \lambda\phi, \ \ A\psi = \mu\psi, \ \text{ and } \mu \neq \lambda.$$

Then, from the eigenvalue equation,

$$\begin{aligned}\lambda\langle\phi,\psi\rangle &= \langle A\phi,\psi\rangle = \langle\phi, A\psi\rangle \\ &= \bar{\mu}\langle\phi,\psi\rangle,\end{aligned}$$

and the fact that $\bar{\mu} = \mu$, which follows from (1), we obtain

$$(\lambda - \mu)\langle\phi,\psi\rangle = 0.$$

As $\lambda \neq \mu$, we get $\langle\phi,\psi\rangle = 0$. □

Theorem 5.6. *Let A be self-adjoint. If for some $M > 0$ and for all $u \in D(A)$,*

$$\|(\lambda - A)u\| \geq M\|u\|, \tag{5.5}$$

then $\lambda \in \rho(A)$. Moreover, we have

$$\{z \subset \mathbb{C} \mid |z - \lambda| < M\} \subset \rho(A).$$

Proof.

(1) We first show that equation (5.5) implies

(a) $\ker(A - \lambda) = \{0\}$;

(b) $\text{Ran}(A - \lambda) = \mathcal{H}$.

As for statement (a), if $A\psi = \lambda\psi$, then inserting ψ into (5.5) implies that $\|\psi\| = 0$, so $\psi = 0$. To prove (b), we note by (5.4) that $\text{Ran}(A - \lambda)$ is dense, and so it suffices to show that it is closed. Let $x_n \in \text{Ran}(A - \lambda)$ form a Cauchy sequence. Then there exists $\{y_n\}$ in $D(A)$ such that $x_n = (A - \lambda)y_n$. We claim $\{y_n\}$ is Cauchy. Indeed, by (5.5),

$$\|y_n - y_m\| \leq M^{-1}\|(A - \lambda)(y_n - y_m)\| = M^{-1}\|x_n - x_m\|,$$

so that as $\{x_n\}$ is Cauchy, the sequence $\{y_n\}$ is also. Hence there is $y \in \mathcal{H}$ such that $y = \lim_{n\to\infty} y_n$. Since $y_n \to y$ and $x_n = (A - \lambda)y_n$ is a Cauchy sequence, it follows from the fact that A is closed that $x_n \to (A - \lambda)y \in \text{Ran}(A - \lambda)$; thus $\text{Ran}(A - \lambda)$ is closed.

(2) It is a consequence of (1) that $(A - \lambda)$ has an everywhere-defined inverse $(A - \lambda)^{-1}$, and we must show that it is bounded. Let $x \in \mathcal{H}$ and $y \equiv (A - \lambda)^{-1}x \in D(A)$. Then $x = (A - \lambda)y$ and

$$\|x\| = \|(A - \lambda)y\| \geq M^{-1}\|y\| = M^{-1}\|(A - \lambda)^{-1}x\|,$$

so for all $x \in \mathcal{H}$,

$$\|(A - \lambda)^{-1}x\| \leq M\|x\|, \tag{5.6}$$

which shows that $(A - \lambda)^{-1} \in \mathcal{L}(\mathcal{H})$ and hence that $\lambda \in \rho(A)$. We leave the proof of the remaining part of the theorem as a problem. □

Problem 5.4. Prove the remaining part of Theorem 5.6. (*Hint*: Write $A - z = (\lambda - z) + (A - \lambda)$, and use the second resolvent identity, Proposition 1.9.)

Proof of part (1) of Theorem 5.5.
Let $z = \lambda + i\mu$ with $\mu \neq 0$. Then, by the self-adjointness of A,

$$\begin{aligned} \|(A-z)u\|^2 &= \langle (A-z)u, (A-z)u \rangle \\ &= \|(A-\lambda)u\|^2 + |\mu|^2 \|u\|^2 \\ &\geq \mu^2 \|u\|^2, \end{aligned} \tag{5.7}$$

for any $u \in D(A)$. By Theorem 5.6, this implies that $z \in \rho(A)$, and so $\sigma(A) \subset \mathbb{R}$. □

Corollary 5.7. *Let A be self-adjoint and $z \in \mathbb{C}$, $Im\, z \neq 0$. Then*

$$\|R_A(z)\| \leq |\text{Im}\, z|^{-1}. \tag{5.8}$$

Proof. This follows directly from (5.6) and (5.7). □

We can sharpen Corollary 5.7 in the following form.

Theorem 5.8. *Let A be self-adjoint, and let $\lambda \in \rho(A)$. Then*

$$\|R_A(\lambda)\| \leq [\text{dist}(\lambda, \sigma(A))]^{-1}. \tag{5.9}$$

We will not prove this here, although we will use the result in this book. The standard proof of this theorem uses the functional calculus; see [RS1]. We leave it as an exercise to prove the following simple extension of Corollary 5.7. Suppose that A is self-adjoint and $\sigma(A) \subset [\lambda_0, \infty)$. Then for $\lambda \in \mathbb{R}$ and $\lambda < \lambda_0$, we have

$$\|R_A(\lambda)\| \leq |\lambda - \lambda_0|^{-1}.$$

We want to emphasize that, despite the simplicity of condition (5.5), the notion that lower bounds on $\|(A-z)u\|$, for u in various subsets, leads to spectral estimates is a very powerful one. We will encounter variations on this idea in our discussion of spectral stability in Chapter 19 and in the discussion of nontrapping estimates in Chapter 21.

5.3 Determining the Spectrum of Self-Adjoint Operators

The estimates on the resolvent of a self-adjoint operator A given in the preceding section allow us to obtain some detailed information about the location of $\sigma(A)$ in $\mathbb{R}$. We give two such results here.

Theorem 5.9. *Let A be self-adjoint. If for some $\epsilon > 0$ there exists some $u \in D(A)$ such that*

$$\|(A - \lambda)u\| \leq \epsilon\|u\|, \tag{5.10}$$

then $\sigma(A) \cap [\lambda - \epsilon, \lambda + \epsilon] \neq \phi$, *that is, A has spectrum inside* $[\lambda - \epsilon, \lambda + \epsilon]$.

Proof. Assume to the contrary that $[\lambda - \epsilon, \lambda + \epsilon] \subset \rho(A)$. By Theorem 5.8,

$$\|R_A(\lambda)\| < \epsilon^{-1}, \tag{5.11}$$

since $\rho(A)$ is open, and therefore $\text{dist}(\lambda, \sigma(A)) > \epsilon$. For any $v \in \mathcal{H}$, set $u = R_A(\lambda)v$. Then, $u \in D(A)$, and (5.11) implies

$$\|u\| < \epsilon^{-1}\|v\|,$$

$$\text{so that} \quad \|u\| < \epsilon^{-1}\|(A - \lambda)u\|,$$

for all $u \in D(A)$. This contradicts (5.10). □

We have seen that, in general, the spectrum of an operator on an infinite-dimensional space consists of much more than eigenvalues (see the end of Chapter 4). In fact, in many cases there are no eigenvalues at all. This is in stark contrast to the case of operators between finite-dimensional spaces. One may ask for a characterization of the elements of the spectrum which are not eigenvalues. We present one such characterization here for self-adjoint operators, and we will return to a similar characterization when we discuss the essential spectrum, in general, in Chapter 7. The following theorem, which states a version of Weyl's criterion for the spectrum of self-adjoint operators, may be interpreted as stating that any $\lambda \in \sigma(A)$ is an *approximate eigenvalue*: Given any $\epsilon > 0$, there exists $u \in D(A)$ such that $\|(A - \lambda)u\| < \epsilon\|u\|$.

Theorem 5.10. *Let A be self-adjoint. Then* $\lambda \in \sigma(A)$ *if and only if there exists a sequence* $\{u_n\}$, $u_n \in D(A)$, *such that* $\|u_n\| = 1$ *and* $\|(A-\lambda)u_n\| \to 0$ *as* $n \to \infty$.

Proof.

(1) Let $\lambda \in \sigma(A)$. Two cases arise:

(a) $\ker(A - \lambda) \neq \{0\}$ (i.e., λ is an eigenvalue). Then let $u_n = f$ for any $f \in \ker(A - \lambda)$ with $\|f\| = 1$.

(b) $\ker(A - \lambda) = \{0\}$. Then $\text{Ran}(A - \lambda)$ is dense but not equal to $\mathcal{H}$, so $(A - \lambda)^{-1}$ exists but is unbounded. Consequently, there exists a sequence $\{v_n\}$, $v_n \in D((A - \lambda)^{-1})$, $\|v_n\| = 1$ such that

$$\|(A - \lambda)^{-1}v_n\| \to \infty.$$

Define $u_n \equiv [(A - \lambda)^{-1}v_n]\|(A - \lambda)^{-1}v_n\|^{-1}$. Then $u_n \in D(A)$, $\|u_n\| = 1$, and

$$\|(A - \lambda)u_n\| = \|v_n\| \ \|(A - \lambda)^{-1}v_n\|^{-1} \to 0,$$

as $n \to \infty$.

(2) Conversely, let $\lambda \in \rho(A)$. Then there exists $M > 0$ such that for any $u \in \mathcal{H}$,

$$\|R_A(\lambda)u\| \leq M\|u\|.$$

Let $u = R_A(\lambda)^{-1}v$ for $v \in D(A)$ so that

$$\|v\| \leq M\|(A - \lambda)v\|,$$

and thus no sequence having the properties described can exist.

(Remark: We proved the contrapositive of the "if" part of the theorem.) □

We finish this section with two results about specific types of self-adjoint operators.

Definition 5.11. *An operator A is positive, $A \geq 0$, if $\langle u, Au\rangle \geq 0$ for all $u \in D(A)$.*

Proposition 5.12. *Let A be a self-adjoint operator. Then $A \geq 0$ if and only if $\sigma(A) \subset [0, \infty)$.*

Problem 5.5. Prove the "only if" part of Proposition 5.11.

Completion of proof of Proposition 5.12.

If $\sigma(A) \subset [0, \infty)$, then for any $a > 0$, $-a \in \rho(A)$ and dist $(-a, \sigma(A)) \geq a$. Hence, by Theorem 5.8,

$$\|(A + a)u\| \geq a\|u\|.$$

This implies that

$$a^2\|u\|^2 \leq \|(A + a)u\|^2 = \|Au\|^2 + 2a\langle Au, u\rangle + a^2\|u\|^2, \tag{5.12}$$

or

$$\langle Au, u\rangle \geq -(2a)^{-1}\|Au\|^2. \tag{5.13}$$

In deriving (5.12), we used the symmetry of A. The result, $\langle Au, u\rangle \geq 0$, follows from (5.13) since $a > 0$ is arbitrary. □

For bounded self-adjoint operators, the norm of the operator can be expressed in terms of the "size" of the spectrum.

Definition 5.13. *Let A be a bounded, self-adjoint operator. The spectral radius of A, $r(A)$, is defined by*

$$r(A) \equiv \sup\{|\lambda| \mid \lambda \in \sigma(A)\}.$$

Theorem 5.14. *Let A be a bounded, self-adjoint operator. Then*

$$\|A\| = r(A) = \sup\{|\lambda| \mid \lambda \in \sigma(A)\}.$$

Proof.
(1) We first show that $r(A) \leq \|A\|$. If $|z| > \|A\|$, then

$$\|(A - z)u\| \geq (|z| - \|A\|)\|u\|,$$

and so $z \in \rho(A)$ and $\sigma(A) \subset \{\lambda | \, |\lambda| \leq \|A\|\}$. Thus, $r(A) \leq \|A\|$.
(2) Conversely, let $\lambda_0 = \|A\|$. By an extension of (4.2) valid for self-adjoint operators, there exists a sequence $\{u_n\}$, $\|u_n\| = 1$, such that $\lim_{n\to\infty}\langle Au_n, u_n\rangle = \lambda_0$. We then have

$$\begin{aligned} \|(A - \lambda_0)u_n\|^2 &= \lambda_0^2 + \|Au_n\|^2 - 2\lambda_0\langle Au_n, u_n\rangle \\ &\leq 2\lambda_0^2 - 2\lambda_0\langle Au_n, u_n\rangle \to 0, \end{aligned}$$

and so $\|(A - \lambda_0)u_n\| \to 0$. By Weyl's criterion, Theorem 5.10, $\lambda_0 \in \sigma(A)$. Hence, $\|A\| \leq r(A)$. □

5.4 Projections

Definition 5.15. *A bounded operator P is called a projection if $P^2 = P$. If, in addition, $P^* = P$ (i.e., P is self-adjoint), then P is called an orthogonal projection.*

Projections are the building blocks of self-adjoint operators. The spectral theorem (which we will not discuss in these chapters) associates with each self-adjoint operator a family of projections that completely determines the operator (see, for example, [RS1]). More basically, projections are intimately related to the subspaces of a Hilbert space, as we describe in the next theorem. Let us note that $u \in \text{Ran}\, P$, a projection P, if and only if $Pu = u$.

Proposition 5.16. *Let P be an orthogonal projection.*

(1) Ran P *is a closed subspace of* $\mathcal{H}$, ker P *is orthogonal to* Ran P, *and*

$$\mathcal{H} = \ker P \oplus \text{Ran}\, P.$$

(2) *If M is a closed subspace of $\mathcal{H}$, then there exists a unique orthogonal projection P such that* Ran $P = M$.

Thus, there is a one-to-one correspondence between closed subspaces of $\mathcal{H}$ and orthogonal projections. The geometry of the family of closed subspaces can be stated in terms of properties of orthogonal projections.

Proof of Proposition 5.16.

(1) To show that Ran P is closed, suppose $\{y_n = Px_n\}$ is a sequence in Ran P and $y_n \to y$. Then by the remark preceding the proposition, $Py_n = y_n$, and so $Py = y$ and $y \in \text{Ran}\, P$. The rest of the statement follows from the result $\text{Ran}\, A \oplus \ker A^* = H$, but it is also easy to prove directly using the facts that if $\bar{P} \equiv 1 - P$, then $\bar{P}$ is an orthogonal projection, $\bar{P} + P = 1$, and $\bar{P}P = 0$.

(2) If M is a closed subspace of $\mathcal{H}$, let $M^{\perp}$ be its orthogonal complement so that $M \oplus M^{\perp} = \mathcal{H}$. Hence any $x \in \mathcal{H}$ has a unique representation $x = x_1 + x_2$ with $x_1 \in M,\ x \in M^{\perp}$. Define an operator P by

$$Px = x_1.$$

Then P is bounded as $\|Px\| = \|x_1\|$, and $P^2 = P$ is obvious. To show that P is symmetric, let $x = x_1 + x_2$, let $y = y_1 + y_2$, and compute

$$\langle Px, y\rangle = \langle x_1, y_1 + y_2\rangle = \langle x_1, y_1\rangle = \langle x_1, Py\rangle = \langle x, Py\rangle,$$

using the orthogonality of x_1, y_1 with x_2, y_2. □

6

Riesz Projections and Isolated Points of the Spectrum

In this and the following chapter, we will discuss the spectrum of a closed operator A. We know that we can form a disjoint decomposition of $\sigma(A)$ as $\sigma_d(A) \cup \sigma_{\text{ess}}(A)$. The discrete spectrum of A, $\sigma_d(A)$, consists of isolated eigenvalues of finite algebraic multiplicity, and $\sigma_{\text{ess}}(A)$, the essential spectrum of A, is the remaining part of the spectrum. We will study $\sigma_d(A)$ in this chapter through the projection operators that can be obtained from the resolvent of A for each distinct eigenvalue in $\sigma_d(A)$. We develop the basic theory of these projections, called *Riesz projections*. These operators provide a powerful tool for the study of the discrete spectrum of closed operators.

The Riesz projections have the additional property that when the closed operator A is self-adjoint, they are orthogonal and project onto the subspace spanned by the eigenfunctions of A for a given discrete eigenvalue. In the last section of this chapter, we discuss the projection onto the eigenspace of a self-adjoint operator corresponding to an eigenvalue embedded in the essential spectrum. Although a Riesz projection cannot be constructed in this case, we still obtain a representation of the projection in terms of the resolvent. We will discuss $\sigma_{\text{ess}}(A)$ in the next chapter.

6.1 Riesz Projections

Let A be a closed operator on a Banach space. Let $\lambda_0 \in \sigma(A)$ be an isolated point of the spectrum, and let Γ_{λ_0} be a simple closed contour around λ_0 such that the closure of the region bounded by Γ_{λ_0} and containing λ_0 intersects $\sigma(A)$ only at λ_0 (see Figure 6.1). We refer to such a contour as *admissible* for λ_0 and A.

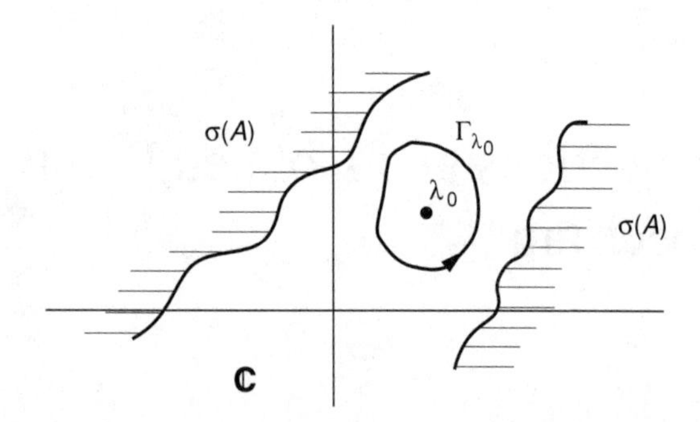

FIGURE 6.1. An admissible contour Γ_{λ_0} for a Riesz projection.

Consider the following contour integral:

$$P_{\lambda_0} \equiv \frac{1}{2\pi i} \oint_{\Gamma_{\lambda_0}} R_A(\lambda) d\lambda. \tag{6.1}$$

Since $R_A(\lambda)$ is analytic in a neighborhood of Γ_{λ_0}, the integral exists as a uniform limit of Riemann sums. Approximating the contour by a union of straight-line segments Δ_i so that $\cup_{i=1}^n \Delta_i$ is a closed polygonal contour containing λ_0, the Nth Riemann sum is

$$P_{\lambda_0}^N \equiv (2\pi i)^{-1} \sum_{i=1}^{N} R_A(\lambda_i) \Delta_i,$$

where $\lambda_i \in \Delta_i$. One can then show that $n - \lim_{N\to\infty} P_{\lambda_0}^N$ exists as a bounded operator.

Problem 6.1. Define the Riemann sums $P_{\lambda_0}^N$ corresponding to the contour integral (6.1) and show that $\lim_{N\to\infty} P_{\lambda_0}^N$ exists. (*Hint*: Show that $\{P_{\lambda_0}^N\}$ is a norm Cauchy sequence.)

Alternately, we can use the analyticity of $R_A(\lambda)$ to first define the integral in the weak sense. For any $u \in X$ and $l \in X^*$, define

$$l(P_{\lambda_0} u) \equiv (2\pi i)^{-1} \oint_{\Gamma_{\lambda_0}} l(R_A(\lambda)u) d\lambda. \tag{6.2}$$

Note that $l(R_A(\lambda)u)$ is analytic on the resolvent set of A. By a simple calculation, there exists a constant $M > 0$ such that

$$|l(P_{\lambda_0} u)| \leq M \|l\|_{X^*} \|u\|_X. \tag{6.3}$$

Consequently, since $u \to l(P_{\lambda_0} u)$ is linear and bounded, the integral defines a bounded map from X^* to X^* (the adjoint). By a standard argument, this induces a bounded linear transformation on X.

Problem 6.2. Repeat the above argument, filling in all the details, when X is a Hilbert space, taking advantage of the fact that we can identify X with X^*.

Let us now study some of the properties of the bounded linear transformation P_{λ_0}.

Lemma 6.1. *P_{λ_0} is independent of the contour Γ_{λ_0} provided that the contour is admissible for λ_0 and A, that is, it lies in $\rho(A)$ and contains no other part of $\sigma(A)$ besides λ_0.*

Proof. We will work with the integrals understood in the uniform sense. By the Cauchy theorem (see, for example, [Ma]), if C is a contour in $\rho(A)$ with no spectrum inside C,

$$\oint_C R_A(\lambda)d\lambda = 0. \tag{6.4}$$

Consider two admissible contours Γ_{λ_0} and $\tilde{\Gamma}_{\lambda_0}$, with Γ_{λ_0} containing $\tilde{\Gamma}_{\lambda_0}$ in its interior. We add two segments C_1 and C_2, connecting the two contours, and call the piece of $\tilde{\Gamma}_{\lambda_0}$ delimited by them C_3, and the piece of Γ_{λ_0} delimited by them C_4 (see Figure 6.2). We consider closed contours $\Gamma_1 \equiv \cup_{i=1}^4 C_i$, $\Gamma'_{\lambda_0} \equiv \{\tilde{\Gamma}_{\lambda_0} \cup C_1 \cup C_4 \cup C_2\} - C_3$, and the simple closed contour Γ_2 formed from $\tilde{\Gamma}_{\lambda_0}$, Γ_{λ_0}, C_1, and C_2. By (6.4),

$$\oint_{\Gamma_i} R_\lambda(A)d\lambda = 0, \quad i = 1, 2. \tag{6.5}$$

Using this result for $i = 1$ and the cancellation of integrals along the same path with opposite orientation, we have

$$\oint_{\Gamma_{\lambda_0}} R_\lambda(A)d\lambda = \oint_{\Gamma'_{\lambda_0}} R_\lambda(A)d\lambda,$$

where Γ'_{λ_0} is the intermediate contour defined above. Again, by (6.5) for $i = 2$,

$$\begin{aligned}\oint_{\Gamma_{\lambda_0}} R_\lambda(A)d\lambda &= \oint_{\Gamma'_{\lambda_0}} R_\lambda(A)d\lambda + \oint_{\Gamma_2} R_\lambda(A)d\lambda \\ &= \oint_{\tilde{\Gamma}_{\lambda_0}} R_\lambda(A)d\lambda,\end{aligned}$$

since the contribution from the overlap contours vanishes. In the case that the two contours Γ_{λ_0} and $\tilde{\Gamma}_{\lambda_0}$ overlap, one uses the same two facts, that is, Cauchy's theorem (6.4) and the fact that integrals along the same contour in opposite directions cancel, to verify the result. □

Definition 6.2. *Let A be a closed operator on X and λ_0 be an isolated point of $\sigma(A)$. For an admissible contour Γ_{λ_0} as described above,*

$$P_{\lambda_0} \equiv (2\pi i)^{-1} \oint_{\Gamma_{\lambda_0}} R_\lambda(A)d\lambda,$$

is called the Riesz integral for A and λ_0.

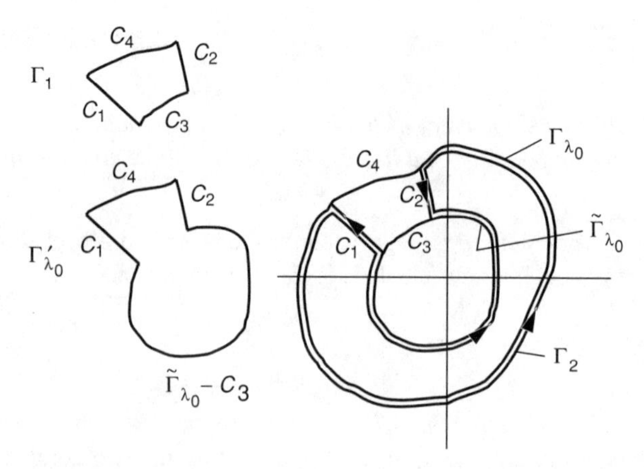

FIGURE 6.2. Principal contours Γ_{λ_0} and $\tilde{\Gamma}_{\lambda_0}$, and the intermediate contours Γ_1, Γ_2, and Γ'_{λ_0}.

Riesz integrals are extremely useful in studying $\sigma(A)$. Their main properties are stated in the next proposition.

Proposition 6.3. *Let P_{λ_0} be the Riesz integral for A and λ_0.*

(1) *P_{λ_0} is a projection.*

(2) Ran $P_{\lambda_0} \supset \ker(A - \lambda_0)$.

(3) *If X is a Hilbert space and A is self-adjoint, then P_{λ_0} is the orthogonal projection onto* $\ker(A - \lambda_0)$.

Proof. (1) Let Γ_{λ_0} and $\tilde{\Gamma}_{\lambda_0}$ be two admissible contours for defining P_{λ_0}. Let us suppose that Γ_{λ_0} is contained in the interior of the region bounded by $\tilde{\Gamma}_{\lambda_0}$ (see Figure 6.2). By the first resolvent equation, Proposition 1.6, we obtain

$$\begin{aligned} P_{\lambda_0}^2 &= (2\pi i)^{-2} \oint_{\Gamma_{\lambda_0}} d\lambda \oint_{\tilde{\Gamma}_{\lambda_0}} d\mu R_A(\lambda) R_A(\mu). \\ &= (2\pi i)^{-2} \oint_{\Gamma_{\lambda_0}} d\lambda \oint_{\tilde{\Gamma}_{\lambda_0}} d\mu (\mu - \lambda)^{-1} [R_A(\lambda) - R_A(\mu)]. \end{aligned} \tag{6.6}$$

We apply the residue theorem [Ma] to the first integral in (6.6) to obtain

$$\oint_{\Gamma_{\lambda_0}} d\lambda \oint_{\tilde{\Gamma}_{\lambda_0}} d\mu (\mu - \lambda)^{-1} R_A(\lambda) = 2\pi i \oint_{\Gamma_{\lambda_0}} d\lambda R_A(\lambda). \tag{6.7}$$

For the second integral, we have

$$\oint_{\Gamma_{\lambda_0}} d\lambda \oint_{\tilde{\Gamma}_{\lambda_0}} d\mu (\mu - \lambda)^{-1} R_A(\mu)$$

$$= \oint_{\tilde{\Gamma}_{\lambda_0}} d\mu R_A(\mu) \oint_{\Gamma_{\lambda_0}} d\lambda(\mu - \lambda)^{-1} = 0,$$

since the integrals are absolutely convergent and $(\mu - \lambda)^{-1}$ is analytic on and inside Γ_{λ_0}. It follows from (6.7) that $P_{\lambda_0}^2 = P_{\lambda_0}$.
(2) Let $f \in \ker(A - \lambda_0)$. Then for $\lambda \neq \lambda_0$,

$$(A - \lambda)^{-1} f = (\lambda_0 - \lambda)^{-1} f. \tag{6.8}$$

We show that $P_{\lambda_0} f = f$, so $f \in \text{Ran } P_{\lambda_0}$. By the definition of P_{λ_0} and (6.8),

$$\begin{aligned} P_{\lambda_0} f &= (2\pi i)^{-1} \oint_{\Gamma_{\lambda_0}} (A - \lambda)^{-1} f d\lambda \\ &= (2\pi i)^{-1} \oint_{\Gamma_{\lambda_0}} (\lambda_0 - \lambda)^{-1} f d\lambda = f. \end{aligned}$$

(3) Let X be a Hilbert space and suppose that $A = A^*$. We first show that $P_{\lambda_0} = P_{\lambda_0}^*$, that is, P_{λ_0} is orthogonal. By self-adjointness, $R_A(\lambda)^* = R_A(\bar{\lambda})$.
Problem 6.3. Prove that if $A = A^*$, then $R_A(\lambda)^* = R_A(\bar{\lambda})$.
Choose $r > 0$ such that $\Gamma_{\lambda_0} \equiv \{\lambda \mid |\lambda - \lambda_0| = r\}$ is an admissible contour, and take $\lambda = \lambda_0 + re^{i\theta}$. Then

$$P_{\lambda_0} = (2\pi)^{-1} \int_{-\pi}^{\pi} R_A(\lambda_0 + re^{i\theta}) r d\theta$$

and

$$P_{\lambda_0}^* = (2\pi)^{-1} \int_{-\pi}^{\pi} R_A(\lambda_0 + re^{-i\theta}) r d\theta. \tag{6.9}$$

Upon reparametrizing (6.9) with $\theta \to -\theta$, we easily find that $P_{\lambda_0} = P_{\lambda_0}^*$. Finally, we must show that $\text{Ran } P_{\lambda_0} = \ker(A - \lambda_0)$, which, by part (b) requires that we show $\text{Ran } P_{\lambda_0} \subset \ker(A - \lambda_0)$. We compute

$$(A - \lambda_0) P_{\lambda_0} = (2\pi i)^{-1} \oint_{\Gamma_{\lambda_0}} (A - \lambda_0)(A - \lambda)^{-1} d\lambda$$

$$= (2\pi i)^{-1} \oint_{\Gamma_{\lambda_0}} (\lambda - \lambda_0)(A - \lambda)^{-1} d\lambda. \tag{6.10}$$

Let U_{λ_0} denote the interior of Γ_{λ_0}. On $U_{\lambda_0} \setminus \{\lambda_0\}$, the operator $(\lambda - \lambda_0)(A - \lambda)^{-1}$ is an analytic, operator-valued function and satisfies the bound

$$|\lambda_0 - \lambda| \|(A - \lambda)^{-1}\| \le |\lambda_0 - \lambda| d(\lambda, \sigma(A))^{-1},$$

where $d(x, y)$ is the distance from x to y. Now we take the diameter of Γ_{λ_0} small enough so that λ_0 is the closest point of $\sigma(A)$ to Γ_{λ_0}. Consequently, $|\lambda_0 - \lambda| \, \|(A - \lambda)^{-1}\| < 1$ and this function is uniformly bounded on $U_{\lambda_0} \setminus \{\lambda_0\}$. It follows from standard results that $(\lambda_0-\lambda)(A-\lambda)^{-1}$ extends to an analytic function on U_{λ_0}, and hence, by Cauchy's theorem, the integral in (6.10) vanishes. This establishes that Ran $P_{\lambda_0} \subset \ker(A-\lambda_0)$ as well as the result. □

6.2 Isolated Points of the Spectrum

We now wish to consider an isolated point $\lambda \in \sigma(A)$ for a self-adjoint operator A on a Hilbert space $\mathcal{H}$. Recall that in this case we have the equality of $\ker(A-\lambda)$ and the Ran P_λ. We first consider two properties of an isolated point $\lambda \in \sigma(A)$, $A = A^*$.

Question 1. If $\lambda \in \sigma(A)$ is isolated and $A = A^*$, is λ an eigenvalue of A?

Proposition 6.4. *The isolated points of* $\sigma(A)$, $A = A^*$, *are eigenvalues of* A.

Proof. Let λ be an isolated point of $\sigma(A)$. Suppose that $\ker(A-\lambda) = \{0\}$, that is, λ is not an eigenvalue. Then by Proposition 6.3(3), Ran $P_\lambda = \{0\}$ (where P_λ is the Riesz projection for λ), so $P_\lambda = (2\pi i)^{-1} \oint_\Gamma R_A(z)dz = 0$, where Γ is any admissible contour. But by Morera's theorem [Ma], this implies that $R_A(z)$ is analytic on a neighborhood of $z = \lambda$, and so $\lambda \in \rho(A)$, a contradiction. □

Remark 6.5. Note that we had to use the self-adjointness of A to conclude $P_\lambda = 0$ from $\ker(A-\lambda) = \{0\}$. For a general, closed operator A we only know that Ran $P_\lambda \supset \ker(A-\lambda)$. In this case, any $\psi \in$ Ran P_λ is a *generalized eigenvector* of A in the sense that there exists some $n \in \mathbb{Z}_+$ such that $(A-\lambda)^n\psi = 0$. The largest such n for which this holds is equal to the *algebraic multiplicity* of λ. This phenomenon occurs in matrix theory and is encountered in discussions of Jordan canonical form. It is not difficult to see that the algebraic multiplicity of an eigenvalue λ is equal to the dimension of the range of P_λ. The *geometric multiplicity* of the eigenvalue λ is equal to the dimension of $\ker(A-\lambda)$. It follows from this proposition that the geometric multiplicity of an eigenvalue is always less than or equal to its algebraic multiplicity.

Question 2. If $\lambda \in \sigma(A)$ is isolated and $A = A^*$, can "other parts" of $\sigma(A)$ be located at λ?

What we are asking is the following: If we remove from $\mathcal{H}$ the subspace $\ker(A-\lambda)$ and study $A|[\ker(A-\lambda)]^\perp$, will λ occur in the spectrum of this reduced operator?

Proposition 6.6. *Let* λ *be a isolated point of* $\sigma(A)$ *and* A *be self-adjoint. Then* $(A-\lambda)$ *restricted to* $[\ker(A-\lambda)]^\perp$ *has a bounded inverse, that is,* $\lambda \notin \sigma(A|[\ker(A-\lambda)]^\perp)$.

Proof. Recall that we have a direct sum decomposition of $\mathcal{H}$,

$$\mathcal{H} = \ker(A-\lambda) \oplus [\ker(A-\lambda)]^\perp, \tag{6.11}$$

and that each subspace is closed and A-invariant. Let $A_1 \equiv A|[\ker(A-\lambda)]^\perp \cap D(A)$. Recall that A_1 is self-adjoint on $[\ker(A-\lambda)]^\perp \cap D(A)$ and that $\ker(A_1 -$

$\lambda) = \{0\}$. Since λ is isolated from $\sigma(A_1)$, we conclude from Proposition 6.4 that $\lambda \in \rho(A_1)$. Hence $(A_1 - \lambda)$ has a bounded inverse on $[\ker(A - \lambda)]^{\perp}$. □

We now can combine these two propositions to give a characterization of the discrete spectrum of a self-adjoint operator.

Theorem 6.7. *Let A be self-adjoint, and let $\lambda \in \sigma(A)$. Then $\lambda \in \sigma_d(A)$ if and only if* $\ker(A - \lambda)$ *is finite-dimensional and* $(A - \lambda)|[\ker(A - \lambda)]^{\perp} \cap D(A)$ *has a bounded inverse.*

Proof.

(1) $\Rightarrow$ If $\lambda \in \sigma_d(A)$, then by Proposition 6.6, $(A - \lambda)$ acting on $[\ker(A - \lambda)]^{\perp}$ has a bounded inverse. As λ is an isolated point in $\sigma(A)$, by Proposition 6.4, λ is an eigenvalue, and so $\ker(A - \lambda)$ contains more than the zero function. The finite dimensionality of $\ker(A - \lambda)$ follows from the definition of $\sigma_d(A)$.

(2) $\Leftarrow$ The subspace $\ker(A - \lambda)$ is a closed subspace of $\mathcal{H}$ because it is finite-dimensional. Let P be the orthogonal projection onto $\ker(A - \lambda)$, and let $A_\lambda \equiv (A - \lambda)|[\ker(A - \lambda)]^{\perp}$. Then A_λ has a bounded inverse (on $[\ker(A - \lambda)]^{\perp}$) if and only if $0 \in \rho(A_\lambda)$. Since the resolvent set is open, there exists a neighborhood W of 0 such that $W \subset \rho(A_\lambda)$. For $z \neq \lambda$ but close to λ such that $\lambda - z \in W$, $(A_\lambda - (\lambda - z))$ has a bounded inverse. Hence we define an operator on $\mathcal{H}$, via the decomposition (6.11), by

$$R(z) \equiv (\lambda - z)^{-1} P \oplus (A_\lambda + (z - \lambda))^{-1}, \tag{6.12}$$

where the first component acts on $\ker\,(A - \lambda)$, and the second component acts on $[\ker\,(A - \lambda)]^{\perp}$. Note that $R(z)$ is a bounded operator. We claim that $R(z)$ is an inverse for $(A - z)$. Once we establish this, the theorem will be proved. To prove this claim, we compute

$$\begin{aligned} (A - z)R(z) &= (\lambda - z)^{-1}(A - z)P \oplus (A - z)(A_\lambda + z - \lambda)^{-1} \\ &= P \oplus (A - z)(A - z)^{-1}(1 - P) \\ &= P \oplus (1 - P) = 1, \end{aligned}$$

where we used the invariance of $P\mathcal{H}$ and the definition of P. We conclude from this that $z \in \rho(A)$, so $(W + \lambda) \setminus \{\lambda\} \subset \rho(A)$. This means that λ is an isolated point of $\sigma(A)$. By Proposition 6.4, λ is an eigenvalue of A, and so $\lambda \in \sigma_d(A)$. □

Corollary 6.8. *Let* $\ker(A - \lambda)$ *be finite-dimensional. Then $\lambda \in \sigma_{\text{ess}}(A)$ if and only if* $(A - \lambda)|[\ker(A - \lambda)]^{\perp}$ *has an unbounded inverse.*

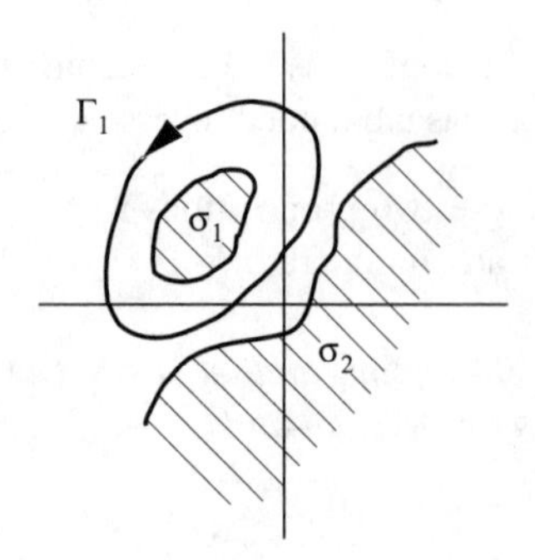

FIGURE 6.3. Contour for P_{σ_1} where $\sigma(A) = \sigma_1 \cup \sigma_2$.

6.3 More Properties of Riesz Projections

We now develop some more properties of the Riesz integrals defined earlier. Let us suppose that for a closed operator A, $\sigma(A)$ decomposes into two disconnected components σ_1 and σ_2 (both nonempty). Let us suppose that σ_1 is bounded. Because these are closed and disconnected, we can enclose σ_1, say, by a simple closed curve Γ_1 with positive orientation. Hence we can associate a Riesz integral with σ_1 :

$$P_{\sigma_1} \equiv (2\pi i)^{-1} \oint_{\Gamma_1} R_A(z) dz \tag{6.13}$$

(see Figure 6.3).

Proposition 6.9. *Let A be closed and as described above. Let P_{σ_1} be as defined in (6.13). Then*

(1) P_{σ_1} *commutes with A on $D(A)$;*

(2) $\sigma(AP_{\sigma_1}) = \sigma_1$ *and* $\sigma(A(1 - P_{\sigma_1})) = \sigma(A) \setminus \sigma_1$.

Proof. The first statement follows from the fact that A and $R_A(z)$ commute on $D(A)$. To prove the second, we first show that $\sigma(AP_{\sigma_1}) \subset \sigma_1$. Since AP_{σ_1} is a restriction of A to an A-invariant subspace, $\rho(AP_{\sigma_1}) \supset \rho(A)$. To see this, note that the restriction of $R_A(z)$ to Ran P_{σ_1} is the resolvent of the restricted operator AP_{σ_1}, and so the inclusion follows. Now to get a better estimate on $\rho(AP_{\sigma_1})$, take any z outside Γ_1 (see Figure 6.3) (z may be in σ_2) and compute

$$R_A(z)P_{\sigma_1} = (2\pi i)^{-1} \oint_{\Gamma_1} R_A(z)R_A(w)dw$$

$$= (2\pi i)^{-1} \left[\oint_{\Gamma_1} (z - w)^{-1} R_A(z) dw - \oint_{\Gamma_1} (z - w)^{-1} R_A(w) dw \right], \tag{6.14}$$

where we used the first resolvent identity. The first integral in (6.14) vanishes since z is outside of Γ_1. It is easy to check that the second integral is analytic in z outside of Γ_1. Hence $R_A(z)P_{\sigma_1} = (z - AP_{\sigma_1})^{-1}$ is analytic on $\mathbb{C} \setminus \overline{(\text{Int}\,\Gamma_1)}$ (the closure of the region bounded by Γ_1). By shrinking Γ_1 close to $\partial\sigma_1$, the boundary of σ_1, and using the openness of the resolvent set, we see that $R_A(z)P_{\sigma_1}$ is analytic on $\mathbb{C} \setminus \sigma_1$. In a similar way one shows that $\sigma(A(1 - P_{\sigma_1})) \subset \sigma(A) \setminus \sigma_1$.

Problem 6.4. Complete the proof of Proposition 6.9. □

6.4 Embedded Eigenvalues of Self-Adjoint Operators

Let A be a self-adjoint operator on a Hilbert space $\mathcal{H}$. An eigenvalue λ of A with multiplicity $m_\lambda \geq 1$ is called an *embedded eigenvalue* of A if λ is not isolated in the spectrum of A. In this case, the projection P_λ for the m_λ-dimensional eigenspace $\mathcal{E}_\lambda$, spanned by the eigenvectors $\{\psi_i \mid i = 1, \ldots, m_\lambda\}$ for A and λ, cannot be obtained as a Riesz projection as discussed above for isolated eigenvalues. This is because there is no simple closed contour $\Gamma \subset \rho(A)$ such that the only point of the spectrum of A in the interior of Γ is λ. We can, however, obtain an expression for the projection onto the eigenspace in terms of the resolvent of A. We will use this representation in Chapter 16 when we discuss the Aguilar–Balslev–Combes–Simon theory of resonances.

As we will see in Chapters 16–23, embedded eigenvalues play an important role in the theory of quantum resonances. Unlike isolated eigenvalues, which are relatively stable under perturbations, as we will see in Chapter 15, embedded eigenvalues generically disappear from the spectrum of a self-adjoint operator under perturbations (see [FrHe] and [AgHeSk] for additional discussion and references). Actually, they do not disappear but, as we will see, they become resonances of the operator.

Since the operator A is self-adjoint, the projection P_λ, corresponding to an embedded eigenvalue of A, is an orthogonal projection. We will denote the projection orthogonal to P_λ by $Q_\lambda \equiv 1 - P_\lambda$.

Theorem 6.10. *Let A be a self-adjoint operator on the Hilbert space $\mathcal{H}$ with an embedded eigenvalue λ. The projection P_λ onto the eigenspace $\mathcal{E}_\lambda$ is given by*

$$P_\lambda = s - \lim_{\epsilon \to 0^\pm} (-i\epsilon)(A - \lambda - i\epsilon)^{-1}. \tag{6.15}$$

Proof. Let us define $P_\epsilon \equiv (-i\epsilon)(A - \lambda - i\epsilon)^{-1}$. By Corollary 5.7, this is a bounded operator as long as $\epsilon \neq 0$ and,

$$\|P_\epsilon\| \leq 1.$$

For any $\epsilon \neq 0$, we have

$$P_\epsilon \cdot P_\lambda = P_\lambda, \tag{6.16}$$

since $\mathcal{E}_\lambda$ is an A-invariant subspace and $A \mid \mathcal{E}_\lambda = \lambda$. For any $\psi \in \mathcal{H}$, a simple calculation, using the self-adjointness of A, shows that

$$\begin{aligned} \|P_\epsilon Q_\lambda \psi\|^2 &= \langle Q_\lambda \psi, \epsilon^2[(A-\lambda)^2+\epsilon^2]^{-1} Q_\lambda \psi\rangle \\ &= \|Q_\lambda \psi\|^2 - \langle Q_\lambda \psi, (A-\lambda)^2[(A-\lambda)^2+\epsilon^2]^{-1} Q_\lambda \psi\rangle. \end{aligned} \tag{6.17}$$

Since $\ker(A-\lambda) \mid Q_\lambda \mathcal{H}$ is empty, we know that for all $\psi \notin \mathcal{E}_\lambda$, $(A-\lambda)Q_\lambda \psi \neq 0$ (i.e., $(A-\lambda) \mid Q_\lambda \mathcal{H}$ is invertible). It is clear that the vector $(A-\lambda) \mid Q_\lambda \psi \in D(((A-\lambda) \mid Q_\lambda \mathcal{H})^{-1})$, and so it follows that

$$\lim_{\epsilon \to 0} \|(A-\lambda-i\epsilon)(A-\lambda)Q_\lambda \psi\| = \|Q_\lambda \psi\|, \tag{6.18}$$

for any $\psi \notin \mathcal{E}_\lambda$. Thus, equations (6.18) and (6.17) imply that

$$\lim_{\epsilon \to 0} \|P_\epsilon Q_\lambda \psi\| = 0, \tag{6.19}$$

for any $\psi \in \mathcal{H}$. By the orthogonality of the projections P_λ and Q_λ, it follows from (6.16) and (6.19) that

$$s - \lim_{\epsilon \to 0} P_\epsilon = \begin{cases} 1 & \text{on } \mathcal{E}_\lambda, \\ 0 & \text{on } Q_\lambda \mathcal{H}. \end{cases} \tag{6.20}$$

This shows that the strong limit is an orthogonal projection and equal to P_λ. □

Problem 6.5. Show that the formula for P_λ is independent of the sign of ϵ.

7

The Essential Spectrum: Weyl's Criterion

We continue to discuss the general properties of the spectrum of a closed operator A. Here we will require that A act on a Hilbert space $\mathcal{H}$. In the general Banach space setting, recall that we decomposed $\sigma(A)$ into two disjoint parts:

$\sigma_d(A) \equiv$ the *"discrete spectrum"* of A, which is the set of all isolated eigenvalues of A with finite algebraic multiplicity;

$\sigma_{\text{ess}}(A) \equiv$ the *"essential spectrum"* of A, which is simply given by $\sigma(A) \backslash \sigma_d(A)$. Properties of $\sigma_d(A)$ were studied in Chapter 6. The set $\sigma(A) \setminus \sigma_d(A)$ is called the essential spectrum because, unlike other subsets of $\sigma(A)$, it is stable under relatively compact perturbations of A. This result, called Weyl's theorem, is discussed in Chapter 14. The main goal of this chapter is to develop a convenient characterization of $\sigma_{\text{ess}}(A)$. In particular, given $\lambda \in \sigma(A)$, we want to know whether $\lambda \in \sigma_{\text{ess}}(A)$. A clue for such a characterization in the self-adjoint case can be found in Theorem 5.10. Elements in $\sigma(A)$ are approximate eigenvalues and can be characterized by the existence of sequences of approximating eigenfunctions.

7.1 The Weyl Criterion

Let us recall the following characterization of points in $\sigma(A)$, where A is a self-adjoint operator on a Hilbert space $\mathcal{H}$:

(a) $\lambda \in \sigma(A)$ if and only if $\exists \{u_n\} \subset D(A)$, $\|u_n\| = 1$, such that $\lim_{n\to\infty} \|(A - \lambda)u_n\| = 0$;

(b) λ is an eigenvalue of A with finite multiplicity if and only if there exists a finite number of linearly independent functions $u_i \in D(A)$ such that $(A - \lambda)u_i = 0$;

(c) λ is an eigenvalue of A with infinite multiplicity if and only if there exists a sequence of linearly independent functions $\{u_i\} \subset D(A)$ such that $(A - \lambda)u_i = 0$.

Of course, both (b) and (c) are of the form in (a). In (b), if we form an infinite sequence, the functions will not be linearly independent. We will use this fact to separate the $\sigma_d(A)$ from $\sigma(A)$. In case (c) it is possible, by the Gram–Schmidt procedure, to choose the elements of the sequence to be orthonormal:

$$\langle u_i, u_j \rangle = \delta_{ij}.$$

Consequently, if $v \in \mathcal{H}$, then Bessel's inequality implies that

$$\sum_{i=1}^{\infty} |\langle v, u_i \rangle|^2 < \infty$$

and hence

$$\lim_{i \to \infty} |\langle v, u_i \rangle| = 0. \tag{7.1}$$

Statement (7.1) implies that the sequence $\{u_i\}$ converges weakly to zero.

Problem 7.1. In case (b), use the Gram–Schmidt procedure to prove that the elements of the sequence can be chosen to be orthonormal. Verify that an orthonormal sequence converges weakly to zero, hence proving that any infinite sequence of linearly independent vectors converges weakly to zero.

Recall that in a Hilbert space a sequence $\{u_i\}$ converges *weakly* to $u \in \mathcal{H}$ if for each $v \in \mathcal{H}$, $\langle u_i, v \rangle \to \langle u, v \rangle$, and we write $u_n \overset{w}{\to} u$. A sequence converges *strongly* to u if $\|u_i - u\| \to 0$ as $i \to \infty$, and we write $u_n \overset{s}{\to} u$ (see Appendix 1, Section 2, for a discussion of these points).

It is now clear that the eigenvalues of A with finite multiplicity (because A is self-adjoint, the geometric and algebraic multiplicities are equal) are characterized by the fact that if we construct a sequence satisfying (a) the terms will not weakly converge to zero. Consequently, if the terms do weakly converge to zero, then we suspect that $\lambda \in \sigma_{\text{ess}}(A)$. This is the idea of Weyl's criterion.

Definition 7.1. *A sequence $\{u_n\}$ is called a Weyl sequence for A and λ if $\exists\ \{u_n\} \subset D(A)$, such that $\|u_n\| = 1$, $u_n \overset{w}{\to} 0$ and $(A - \lambda)u_n \overset{s}{\to} 0$.*

Theorem 7.2 (Weyl's criterion). *Let A be self-adjoint. Then $\lambda \in \sigma_{\text{ess}}(A)$ if and only if there exists a Weyl sequence for A and λ.*

We give the proof of this and an application in the next sections. We emphasize that the difference between the Weyl criterion and condition (a) lies in the weak convergence of the sequence to zero.

We mention that this criterion can be extended (with some technical conditions) to non–self-adjoint operators. We discuss this in Chapter 10. In the case that $\lambda \in \sigma_{\text{ess}}(A)$ and λ is not an eigenvalue, we can think of the elements u_i of a Weyl sequence for A and λ as approximate eigenfunctions. For any $\epsilon > 0$, we can find an $N \in \mathbb{Z}_+$ such that for all $n \geq N$, $\|(A - \lambda)u_n\| < \epsilon$. This idea is familiar from quantum mechanics. The operator $p \equiv -i(d/dx)$ on $L^2(\mathbb{R})$ represents momentum and is self-adjoint (we will show this later). Physicists commonly say that the plane wave e^{ikx}, $k \in \mathbb{R}$, is an *eigenfunction* of p with eigenvalue k since we have the eigenvalue equation $pe^{ikx} = ke^{ikx}$. However, the function $e^{ikx} \notin L^2(\mathbb{R})$. We can form wave packets $\psi_n(x)$ from functions $f_n(q) \in L^2(\mathbb{R})$ which are supported in intervals about k of decreasing length using the Fourier transform:

$$\psi_n(x) \equiv \int_{-\infty}^{\infty} e^{iqx} f_n(q) dq. \tag{7.2}$$

These functions are in $L^2(\mathbb{R})$ and approximate e^{ikx} as $n \to \infty$. In fact, it is not too hard to show that these form a Weyl sequence for p and k. One might wonder about functions of the form e^{izx} for $z \in \mathbb{C}$, $\text{Im}\, z \neq 0$, since $pe^{izx} = ze^{izx}$ also. But p is self-adjoint, so $z \notin \sigma(p)$. Luckily, $e^{izx} \notin L^2(\mathbb{R})$ and, because of the exponential increase of these functions, it is impossible to construct a Weyl sequence as in (7.2). In fact, no Weyl sequence exists for p and z, with $\text{Im}\, z \neq 0$.

7.2 Proof of Weyl's Criterion: First Part

($\Rightarrow$) Let $\lambda \in \sigma_{\text{ess}}(A)$. If $\ker(A - \lambda)$ is infinite-dimensional, then form a Weyl sequence from any infinite orthonormal basis for $\ker(A - \lambda)$. By Remark 7.1, any infinite orthonormal set converges weakly to zero. If $\ker(A - \lambda)$ is finite-dimensional (including the case that $\ker(A - \lambda) = \{0\}$), we consider $[\ker(A - \lambda)]^\perp$ and use Theorem 6.7, which characterizes $\sigma_d(A)$. First we need a lemma.

Lemma 7.3. *Let A be self-adjoint. Then*

(i) $(\ker A)^\perp$ *is an A-invariant subspace, that is,* $A : (\ker A)^\perp \cap D(A) \to (\ker A)^\perp$;

(ii) *the restriction of A to* $(\ker A)^\perp$ *has trivial kernel* $\{0\}$ *and is self-adjoint.*

Problem 7.2. Prove Lemma 7.3. For part (ii), use the fundamental criterion for self-adjointness: K, a closed, symmetric operator, is self-adjoint if and only if $\text{Ran}(K \pm i) = \mathcal{H}$. (See Chapter 8, Theorem 8.3.)

Now let A_λ denote $(A - \lambda)|[\ker(A - \lambda)]^\perp$. The operator A_λ is self-adjoint because λ is real, and hence it has a trivial kernel by Lemma 7.3. By the results of Section 4 of Appendix 3, the operator $(A - \lambda)$ has an inverse. It follows from Theorem 6.7 that if $\ker(A - \lambda)$ is finite-dimensional, then $\lambda \in \sigma_{\text{ess}}(A)$ if A_λ^{-1} is unbounded. Hence (using $\Rightarrow$ part) there is a sequence $\{v_n\} \subset D(A_\lambda^{-1})$ such that $\|v_n\| = 1$ and $\|A_\lambda^{-1} v_n\| \to \infty$ as $n \to \infty$. Set $u_n \equiv A_\lambda^{-1} v_n \|A_\lambda^{-1} v_n\|^{-1}$; then $\|u_n\| = 1$ and

$$\|A_\lambda u_n\| = \|v_n\| \, \|A_\lambda^{-1} v_n\|^{-1} \to 0.$$

It remains to show that $u_n \stackrel{w}{\to} 0$. Now let $f \in \mathcal{H}$; then we have the decomposition $f = f_1 \oplus f_2$, with $f_1 \in \ker(A-\lambda)$ and $f_2 \in [\ker(A-\lambda)]^{\perp}$. Since $A_\lambda^{-1} : \operatorname{Ran} A_\lambda \to D(A_\lambda) \cap [\ker\,(A-\lambda)]^{\perp}$, we have

$$\langle f, u_n \rangle = \langle f_2, u_n \rangle,$$

and so it suffices to take $f \in [\ker(A-\lambda)]^{\perp}$. Moreover, it is sufficient to prove that $\langle f, u_n \rangle \to 0$ as $n \to \infty$ for f in a dense subset of $[\ker\,(A-\lambda)]^{\perp}$.

Problem 7.3. Prove that if $\langle f, u_n \rangle \to 0$ for all f in a dense set, then $u_n \stackrel{w}{\to} 0$.

We claim that $D((A_\lambda^{-1})^*)$ is dense in $[\ker(A-\lambda)]^{\perp}$. Clearly, any element of the form $A_\lambda f$, $f \in D(A_\lambda)$, belongs to $D((A_\lambda^{-1})^*)$ since, for any $u \in D(A_\lambda^{-1})$,

$$\langle A_\lambda f, A_\lambda^{-1} u \rangle = \langle f, u \rangle.$$

Now $\operatorname{Ran} A_\lambda$ is dense, for suppose $\exists u \in [\ker(A-\lambda)]^{\perp}$ such that for all $v \in D(A_\lambda)$

$$\langle u, A_\lambda v \rangle = 0.$$

This implies that $u \in D(A_\lambda)$ (since A_λ is self-adjoint) and that $A_\lambda u = 0$. (This fact also follows from the fundamental criterion for self-adjointness, Theorem 8.3.) But by Lemma 7.3(b), $\ker\ A_\lambda = \{0\}$, so $u = 0$. Hence $\operatorname{Ran}\ A_\lambda$, and therefore $D((A_\lambda^{-1})^*)$, is dense. So, letting $f \in D((A_\lambda^{-1})^*)$, we compute

$$\begin{aligned} |\langle f, u_n \rangle| &= |\langle (A_\lambda^{-1})^* f, v_n \rangle| \|A_\lambda^{-1} v_n\|^{-1} \\ &\leq \|(A_\lambda^{-1})^* f\| \ \|A_\lambda^{-1} v_n\|^{-1}, \end{aligned}$$

and the right side converges to zero as $n \to \infty$. Hence, by Problem 7.3, $u_n \stackrel{w}{\to} 0$. □

7.3 Proof of Weyl's Criterion: Second Part

($\Leftarrow$) Let $\{u_n\}$ be a Weyl sequence for A and λ. By Proposition 5.10, $\lambda \in \sigma(A)$. It remains to show that λ is not an isolated eigenvalue of finite multiplicity. If $\ker(A-\lambda)$ is infinite-dimensional, then $\lambda \in \sigma_{\text{ess}}(A)$. Hence, we assume that $\dim\,[\ker(A-\lambda)] < \infty$ (if there are no eigenfunctions, then we are done). By Theorem 6.7, it suffices to show that the restriction of $(A-\lambda)$ to $[\ker(A-\lambda)]^{\perp}$, called A_λ, does not have a bounded inverse. Let $\{\phi_i\}_{i=1}^k$ be a finite orthonormal basis for $\ker(A-\lambda)$, and let $P_\lambda \equiv$ orthogonal projection onto $\ker(A-\lambda)$. Then, as the vectors form an orthonormal basis for $\operatorname{Ran} P_\lambda$, we get that

$$\|P_\lambda u_n\|^2 = \sum_{i=1}^{k} |\langle u_n, \phi_i \rangle|^2,$$

and

$$\lim_{n\to\infty} \|P_\lambda u_n\|^2 = 0, \tag{7.3}$$

since $u_n \xrightarrow{w} 0$. It follows that

$$\|(1 - P_\lambda)u_n\|^2 = 1 - \|P_\lambda u_n\|^2 \to 1. \tag{7.4}$$

Let $\bar{P}_\lambda \equiv 1 - P_\lambda$ be the projection orthogonal to P_λ. We define $v_n \equiv \bar{P}_\lambda u_n \|\bar{P}_\lambda u_n\|^{-1}$, so $\|v_n\| = 1$ and the denominator remains bounded by (7.4). Since $v_n \in D(A)$, we have

$$\|(A - \lambda)v_n\| = \|(A - \lambda)u_n\| \, \|\bar{P}_\lambda u_n\|^{-1} \to 0 \tag{7.5}$$

as $n \to \infty$ since $\{u_n\}$ is a Weyl sequence. Our result now follows from (7.5) and the following lemma.

Lemma 7.4. *Suppose that a closed operator A has an inverse. Then A^{-1} is unbounded if and only if $\exists \; \{u_n\} \subset D(A)$, $\|u_n\| = 1$, such that $\|Au_n\| \to 0$ as $n \to \infty$.*

Problem 7.4. Prove Lemma 7.4. (*Hint*: Consider A^{-1} on the sequence $v_n \equiv Au_n\|Au_n\|^{-1}$.)

To conclude the proof, it follows from Lemma 7.4 that A_λ^{-1} is unbounded. Hence, by Theorem 6.7, $\lambda \in \sigma_{\text{ess}}(A)$. □

Example 7.5. *Laplacian on $\mathbb{R}^n$.*

Let Δ denote the operator $\sum_{i=1}^n \partial^2/\partial x_i^2$ with domain $D(\Delta) = H^2(\mathbb{R}^n) \subset L^2(\mathbb{R}^n)$, where $H^s(\mathbb{R}^n)$ is the Sobolev space of order s defined in Appendix 4. If $f \in H^2(\mathbb{R}^n)$, then $\|k\|^2 \hat{f} \in L^2(\mathbb{R}^n)$, where $\hat{f}$ denotes the Fourier transform of f. We discussed some aspects of the Laplacian and the relationship between the Fourier transform and differential operators in Section 4.3. There we showed that $(\Delta f)^\wedge(k) = -\|k\|^2 \hat{f}(k)$. Since the Fourier transform is a unitary operator on $L^2(\mathbb{R}^n)$, it follows that $f \in D(\Delta)$ if and only if $f \in H^2(\mathbb{R}^n)$. Moreover, the subspace $H^2(\mathbb{R}^n)$ is dense in $L^2(\mathbb{R}^n)$. It follows from these observations and the definition of a closed operator, that Δ is closed on the domain $H^2(\mathbb{R}^n)$. It is also easy to check that Δ on $H^2(\mathbb{R}^n)$ is symmetric by using the Fourier transform. Hence, the Laplacian is a closed, symmetric operator with domain $H^2(\mathbb{R}^n)$. In Chapter 8, Example 8.4, we will prove that Δ on $H^2(\mathbb{R}^n)$ is self-adjoint.

Theorem 7.6. *The spectrum of the self-adjoint operator $-\Delta$ on $H^2(\mathbb{R}^n)$ is $\sigma(-\Delta) = \sigma_{\text{ess}}(-\Delta) = [0, \infty)$.*

Proof.

(1) Let $u \in D(-\Delta) = H^2(\mathbb{R}^n)$. Then by the Plancherel theorem, Theorem A4.2,

$$-\langle u, \Delta u \rangle = \int \|k\|^2 |\hat{u}(k)| dk \geq 0,$$

and so $\sigma(-\Delta) \subset [0, \infty)$ by Proposition 5.12. We use Weyl's criterion to show that if $\lambda > 0$, then $\lambda \in \sigma_{\text{ess}}(-\Delta)$. To this end, we construct a sequence of approximate eigenfunctions, called Gaussian wave packets, concentrated at λ. This construction is similar to the construction of approximate eigenfunctions discussed in Section 1 of this chapter. For example, we consider functions u_m defined by

$$u_m(x) \equiv (2\pi)^{-\frac{n}{2}} \int \hat{u}_m(k) e^{ik\cdot x} dk, \tag{7.6}$$

where $k \cdot x = \sum_{i=1}^n k_i x_i$, and

$$\hat{u}_m(k) \equiv (2\pi m)^{\frac{n}{2}} e^{-m^2|k-k_0|^2}, \qquad \|k_0\|^2 = \lambda. \tag{7.7}$$

This integral is absolutely convergent and defines a function in $\mathcal{S}(\mathbb{R}^n)$. Notice that $\hat{u}_m(k)$ is a Gaussian function strongly localized around k_0 and that the root mean width decreases as $m^{-1/2}$.

(2) We claim that $\{u_m\}$ in (7.6) and (7.7) is a Weyl sequence for $-\Delta$ and λ. We compute the normalization:

$$\begin{aligned} \|u_m\|^2 &= \|\hat{u}_m\|^2 = (2\pi m)^n \int e^{-2m^2\|k-k_0\|^2} dk \\ &= 2^{\frac{n}{2}} \pi^n \Omega_n \int_0^\infty e^{-u^2} u^{n-1} du = 1, \end{aligned}$$

where Ω_n is the area of a unit sphere in n dimensions. Next, we show that $u_n \xrightarrow{w} 0$. Let $f \in \mathcal{S}(\mathbb{R}^n)$, the dense set of Schwarz functions. Then by the Plancherel theorem, Theorem A4.2, and (7.7), we have

$$\begin{aligned} \langle u_m, f \rangle &= \langle \hat{u}_m, \hat{f} \rangle = (2\pi m)^{\frac{n}{2}} \int e^{-m^2\|k-k_0\|^2} \hat{f}(k) dk \\ &= \left(\frac{2\pi}{m}\right)^{\frac{n}{2}} \int e^{-\|k\|^2} \hat{f}\left(\frac{k}{m} + k_0\right) dk, \end{aligned}$$

where we changed variables from k to $m(k - k_0)$. Now $\hat{f} \in \mathcal{S}$ also (see Appendix 4), so $\lim_{m\to\infty} \hat{f}(k/m + k_0) = \hat{f}(k_0)$. Consequently, we find

$$|\langle u_m, f \rangle| \leq c m^{-\frac{n}{2}} |\hat{f}(k_0)| \to 0,$$

for some $c > 0$, independent of f. By Problem 7.3, $\{u_m\}$ converges weakly to zero.

(3) Finally, we show that $(-\Delta - \lambda)u_m \xrightarrow{s} 0$. By the Fourier transform and the smoothness of u_m,

$$((-\Delta - \lambda)u_m)(x) = (2\pi)^{-\frac{n}{2}} \int (\|k\|^2 - \lambda)\hat{u}_m(k)e^{ik\cdot x}dk$$

and by (7.7):

$$\begin{aligned} \|(-\Delta - \lambda)u_m\|^2 &= \|(\|k\|^2 - \lambda)\hat{u}_m\|^2 \\ &= (2\pi m)^n \int (\|k\|^2 - \lambda)^2 e^{-2m^2\|k-k_0\|^2} dk \\ &= \left(2^{\frac{1}{2}}\pi\right)^n \int e^{-\|k\|^2} \left(|(2^{\frac{1}{2}}m)^{-1} k + k_0|^2 - \lambda\right)^2 dk \\ &= \left(2^{\frac{1}{2}}\pi\right)^n \int e^{-\|k\|^2}((2m^2)^{-1}\|k\|^2 \\ &\quad +2^{\frac{1}{2}}m^{-1}k\cdot k_0)^2 dk, \end{aligned} \tag{7.8}$$

where we used the new variable $\tilde{k} \equiv 2^{1/2}m(k - k_0)$ and the fact that $\lambda = k_0^2$. It follows directly by the superexponential convergence of the integral in (7.8), that the integral in (7.8) converges to zero like $1/m$ as $m \to \infty$. Hence, $\{u_m\}$ is a Weyl sequence for $-\Delta$ and λ, and by the Weyl criterion, $\lambda \in \sigma_{\text{ess}}(-\Delta)$. Thus $(0, \infty) \subset \sigma_{\text{ess}}(-\Delta)$, so as the spectrum is a closed set, $[0, \infty) \subset \sigma(-\Delta)$. These imply that

$$\sigma(-\Delta) = \sigma_{\text{ess}}(-\Delta) = [0, \infty),$$

because zero is not an isolated point of the spectrum. □

Remark 7.7. It follows from the previous calculations that the Laplacian, as a self-adjoint operator on the Hilbert space $L^2(\mathbb{R}^n)$, has no eigenvalues. Suppose that $-\Delta$ has an eigenvalue at $\lambda \geq 0$. Any corresponding eigenfunction ψ has to satisfy

$$(\|k\|^2 - \lambda)\hat{\psi} = 0.$$

This means that supp $\hat{\psi} = \{k \mid \|k\| = \lambda\}$, the sphere of radius λ. But, $\hat{\psi} \in L^2(\mathbb{R}^n)$, and so $\hat{\psi} = 0$. This follows from the following simple version of elliptic regularity (see Theorem 3.8).

Proposition 7.8. *If $u \in L^2(\mathbb{R}^n)$ satisfies $-\Delta u = \lambda u$, for some $\lambda \geq 0$, then u is infinitely differentiable.*

Proof. It follows from the eigenvalue equation that if $u \in H^2(\mathbb{R}^n)$, then $u \in H^4(\mathbb{R}^n)$. Iterating in this manner, it follows that u is in any positive indexed Sobolev space. So, by Theorem A4.6, the function u is infinitely differentiable. □

Problem 7.5. Show that Proposition 7.8 and the preceding analysis imply that any eigenfunction ψ of the Laplacian must be zero.

8

Self-Adjointness: Part 1. The Kato Inequality

We will now concentrate on the class of self-adjoint operators in a Hilbert space $\mathcal{H}$. Our first task will be to develop criteria that will allow us to determine which operators occurring in applications are self-adjoint. Then we will apply this to prove that Schrödinger operators with positive potentials are self-adjoint. After discussing in Chapters 11 and 12 the semiclassical analysis of eigenvalues for Schrödinger operators with positive, growing potentials, we will return to the question of self-adjointness in Chapter 13 and present the Kato–Rellich theory.

8.1 Symmetric Operators

Let us recall from Chapter 5 that an operator is self-adjoint if it is equal to its adjoint. A necessary condition is that the operator be symmetric. In Definition 5.1, we gave two characterizations of a symmetric operator. We recall here the most computationally convenient one.

Definition 8.1. *An operator A on a dense domain $D(A)$ is symmetric if for all $u, v \in D(A)$,*

$$\langle Au, v\rangle = \langle u, Av\rangle.$$

It is usually easy to check whether a given operator is symmetric.

Examples 8.2.

(1) *Multiplication operators*. Let $V \in L^2(\mathbb{R}^n) + L^\infty(\mathbb{R}^n)$ and be real, that is, V can be decomposed as $V = V_1 + V_2$, with $V_1 \in L^2$ and $V_2 \in L^\infty$. We

denote by the same letter V the linear operator "multiplication by V," that is, $(Vf)(x) \equiv V(x)f(x)$. Let $C_0(\mathbb{R}^n)$ be the space of all continuous functions on $\mathbb{R}^n$ with compact (bounded) support. We define an operator V on the dense domain $C_0(\mathbb{R}^n)$ in $L^2(\mathbb{R}^n)$ by

$$V : f \in C_0(\mathbb{R}^n) \to Vf. \tag{8.1}$$

Note that $Vf \in L^2(\mathbb{R}^n)$ as

$$\begin{aligned} \|Vf\|_2 &\leq \|V_1 f\|_2 + \|V_2 f\|_2 \\ &\leq \|f\|_\infty^2 \|V_1\|_2 + \|V_2\|_\infty \|f\|_2 < \infty, \end{aligned} \tag{8.2}$$

where $\|g\|_\infty \equiv \sup_{x\in\mathbb{R}^n} |g(x)|$.

Claim. V is symmetric.

Formally, we have for $u,\ v \in C_0(\mathbb{R}^n)$,

$$\begin{aligned} \langle Vu, v\rangle &= \int V(x)u(x)\bar{v}(x)dx = \int u(x)V(x)\bar{v}(x)dx \\ &= \langle u, Vv\rangle. \end{aligned} \tag{8.3}$$

We have to check that $\int V(x)u(x)\bar{v}(x)dx$ converges. However, by the Schwarz inequality, Theorem 1.14,

$$|\langle Vu, v\rangle| \leq \|v\|_2 \|Vu\|_2,$$

and by (8.2), $\|Vu\|_2 < \infty$ for $u \in C_0(\mathbb{R}^n)$, so the integral converges.

(2) *The Laplacian*. We recall that $\Delta \equiv \sum_{i=1}^n \partial^2/\partial x_i^2$ on $L^2(\mathbb{R}^n)$ and that Δ is naturally defined on the Sobolev space of order 2, $H^2(\mathbb{R}^n)$. The spectrum of the Laplacian was discussed in Example 7.5. If $f \in H^2(\mathbb{R}^n)$, then

$$(\hat{\Delta f})(k) = -\|k\|^2 \hat{f}(k) \tag{8.4}$$

and $\|k\|^2 \hat{f} \in L^2$. We claim that Δ on $H^2(\mathbb{R}^n)$ is symmetric. By (8.4) and the Plancherel theorem (Theorem A4.2), for any $f,\ g \in H^2(\mathbb{R}^n)$:

$$\begin{aligned} \langle \Delta f, g\rangle &= -\int \|k\|^2 \hat{f}(k)\overline{\hat{g}(k)}dk \\ &= -\int \hat{f}(k)\overline{\|k\|^2 \hat{g}(\bar{k})}dk \\ &= \langle f, \Delta g\rangle. \end{aligned}$$

We remark that this can also be established as follows. Consider Δ on the domain $\mathcal{S}(\mathbb{R}^n)$, the Schwarz functions. For $n = 1$, we use integration by parts and the vanishing of these functions at infinity. For any $u, w \in \mathcal{S}(\mathbb{R})$,

$$\begin{aligned}
\langle \Delta u, w \rangle &= \lim_{R\to\infty} \int_{-R}^{R} u'' \bar{w} \\
&= \lim_{R\to\infty} \left[u'(x)\bar{w}(x)|_{-R}^{R} - u(x)\bar{w}'(x)|_{-R}^{R} \right] + \langle u, \Delta w \rangle \\
&= \langle u, \Delta w \rangle. \qquad (8.5)
\end{aligned}$$

For $n > 1$, one uses Green's theorem. Now $\mathcal{S}(\mathbb{R}^n)$ is dense in $H^2(\mathbb{R}^n)$ in the H^2-norm (i.e., if $f \in H^2$, then $\|f\|_{H^2}^2 \equiv \int (1 + \|k\|^2)^2 |\hat{f}(k)|^2 dk$), so it follows from (8.5) that Δ is symmetric on $H^2(\mathbb{R}^n)$. This example also shows that the same operator (i.e., symbol) can be symmetric on several different dense domains: Δ is symmetric on $H^2(\mathbb{R}^n)$ and on $\mathcal{S}(\mathbb{R}^n)$ (of course, Δ is not closed on $\mathcal{S}(\mathbb{R}^n)$).

Problem 8.1. Prove that Δ is closed on $H^2(\mathbb{R}^n)$, that it is not closed on $C_0^\infty(\mathbb{R}^n)$, but that its closure is $(\Delta, H^2(\mathbb{R}^n))$. (*Hint*: Consider (8.4).)

8.2 Fundamental Criteria for Self-Adjointness

Because it is easy to prove that an operator is symmetric, the consequences of being symmetric are rather weak. The property of self-adjointness is much more powerful. As it is difficult to compute $D(A^*)$ exactly and show, for a symmetric operator A, that $D(A) = D(A^*)$, we develop alternate criteria for establishing the self-adjointness of a closed symmetric operator.

Theorem 8.3 (Basic Criteria for Self-Adjointness). *Let A be a closed symmetric operator. Then the following statements are equivalent:*

(a) *A is self-adjoint*;

(b) $\ker(A^* \pm i) = \{0\}$;

(c) $\operatorname{Ran}(A \pm i) = \mathcal{H}$.

Proof.

(1) Statements (b) and (c) are equivalent by Proposition 4.6, which states that $\ker A^* \oplus \overline{\operatorname{Ran} A} = \mathcal{H}$, and the fact that $\operatorname{Ran}(A \pm i)$ is closed.

Problem 8.2. Prove that if A is a closed, symmetric operator, then Ran $(A \pm i)$ is closed. (*Hint*: Use the facts that for a symmetric operator, $\|(A \pm i)y\|^2 = \|y\|^2 + \|Ay\|^2$ and that A is closed).

(2) Clearly (a) implies (b) since $\sigma(A) \subset \mathbb{R}$. We must show that (b) and (c) imply (a). Since $D(A) \subset D(A^*)$, we must show that $D(A^*) \subset D(A)$. Let $f \in D(A^*)$, and define $\phi \equiv (A^* + i)f$. By (c), there exists a $g \in D(A)$ such that $(A + i)g = \phi$. Moreover, since A is symmetric, $Ag = A^*g$. Thus we have

$$(A^* + i)f = \phi = (A^* + i)g \quad \text{or } (A^* + i)(f - g) = 0.$$

By (b), $f = g$ and $f \in D(A)$. Thus $D(A) = D(A^*)$, and A is self-adjoint.

□

Example 8.4. The Laplacian Δ on $H^2(\mathbb{R}^n)$ is *self-adjoint.*

Proof. We prove that $\text{Ran}(\Delta + i) = L^2(\mathbb{R}^n)$. By Example 8.2, the Laplacian Δ is symmetric on $H^2(\mathbb{R}^n)$. It is also closed on that domain, for suppose $\{f_n\} \subset H^2(\mathbb{R}^n)$ is an L^2-Cauchy sequence such that Δf_n is also Cauchy. This means that $f = \lim_n f_n \in H^2(\mathbb{R}^n)$ and therefore that Δf_n converges to Δf. Now from Theorem 7.6, we see that $\sigma(\Delta) \subset (-\infty, 0]$, so $-i \in \rho(\Delta)$. This means that $(\Delta + i)$ has bounded inverse, and this is only possible if $\text{Ran}(\Delta + i) = L^2(\mathbb{R}^n)$.

□

We can also consider a symmetric operator A with a domain $D(A)$, which is not necessarily closed. Our question then becomes: What are some sufficient conditions so that the closure of A is self-adjoint? Note that since A^{**} is also symmetric, such criteria follow from Theorem 8.3.

Corollary 8.5. Let A with domain $D(A)$ be a symmetric operator. Then the closure of A, A^{**}, is self-adjoint if and only if $\ker(A^* \pm i) = 0$.

Problem 8.3. Prove Corollary 8.5.

Problem 8.4. Prove the following extension of Corollary 8.5: The closure of a symmetric operator A, A^{**}, is self-adjoint if and only if $\overline{\text{Ran}(A \pm i)} = \mathcal{H}$.

A symmetric operator A with domain $D(A)$, having the property that its closure is self-adjoint, is called *essentially self-adjoint*. Any domain with this property is called a *core* for the corresponding self-adjoint operator. Since it is sometimes difficult to prove that a symmetric operator on a natural domain is closed on that domain, Corollary 8.5 and the result of Problem 8.4 can be extremely useful.

We now combine the two types of operators discussed in Examples 8.2. A linear operator of the form $-\Delta + V \equiv H$ is called a *Schrödinger operator* with potential V. All the information about a quantum mechanical system (atoms, molecules, nuclei, solids, etc.) is contained in the Schrödinger operator for the system. A linear operator $H = -\Delta + V$ is also called the *Hamiltonian* for the quantum system described by the potential V. We will devote the remainder of this book to a detailed study of the spectra of Schrödinger operators.

Our first application of the fundamental criteria will be to Schrödinger operators $H = -\Delta + V$, where V is a real, positive potential function. We give this application first because we want to study the eigenvalues for such a Schrödinger operator H.

Later, we will return to the Kato–Rellich theorem, which provides the second basic result on the self-adjointness of Schrödinger operators. These two results on the self-adjointness of Schrödinger operators are different in the following sense. For Schrödinger operators, we can consider H as a *perturbation* of $-\Delta$ by the potential V. Our first problem in perturbation theory is to consider which potentials V preserve the self-adjointness of $-\Delta$. The Kato inequality allows us to say that for *positive* V with some regularity, the resulting Schrödinger operator H is self-adjoint. The Kato–Rellich theorem says that for any sufficiently regular V that is small relative to $-\Delta$ in a certain sense, the resulting Schrödinger operator is again self-adjoint.

8.3 The Kato Inequality for Smooth Functions

Let us recall the following function spaces:

- $C^\infty(\mathbb{R}^n) \equiv$ all continuously differentiable functions on $\mathbb{R}^n$;
- $C_0^\infty(\mathbb{R}^n) \equiv$ all continuously differentiable functions with bounded support in $\mathbb{R}^n$.

Definition 8.6. *For any function u, define $(sgn\ u)$ by*

$$(sgn\ u)(x) = \begin{cases} 0, & u(x) = 0, \\ \bar{u}(x)|u(x)|^{-1}, & u(x) \neq 0, \end{cases} \tag{8.6}$$

and, for any $\epsilon > 0$, define a regularized absolute value of u by

$$u_\epsilon(x) \equiv [|u(x)|^2 + \epsilon^2]^{\frac{1}{2}}. \tag{8.7}$$

Note that $\lim_{\epsilon \to 0} u_\epsilon(x) = |u(x)|$ pointwise and that $|u(x)| = \text{sgn}(u) \cdot u$. We now give Kato's inequality in the smooth case.

Lemma 8.7. *For any $u \in C^\infty(\mathbb{R}^n)$, define $|u| \equiv [\bar{u}u]^{\frac{1}{2}}$. We then have*

$$\Delta|u| \geq Re[(sgn\ u)\Delta u], \tag{8.8}$$

except where $|u|$ is not differentiable.

Proof. Observe that $u_\epsilon \geq |u|$. Then from (8.7), if we differentiate $u_\epsilon^2 = |u|^2 + \epsilon^2$, we get

$$u_\epsilon \nabla u_\epsilon = \text{Re}\ \bar{u} \nabla u. \tag{8.9}$$

Squaring this and using the observation gives

$$|\nabla u_\epsilon| \leq u_\epsilon^{-1}|u|\ |\nabla u| \leq |\nabla u|. \tag{8.10}$$

Next, take the divergence of (8.9), to obtain

$$|\nabla u_\epsilon|^2 + u_\epsilon \Delta u_\epsilon = |\nabla u|^2 + \mathrm{Re}\, \bar{u} \Delta u.$$

By (8.10), this is equivalent to

$$u_\epsilon \Delta u_\epsilon \geq \mathrm{Re}\, \bar{u} \Delta u. \tag{8.11}$$

Let $\mathrm{sgn}_\epsilon u \equiv \bar{u} u_\epsilon^{-1}$, so that (8.11) is

$$\Delta u_\epsilon \geq \mathrm{Re}[(\mathrm{sgn}_\epsilon u) \Delta u]. \tag{8.12}$$

Since $\Delta u_\epsilon \to \Delta |u|$ pointwise and $\mathrm{sgn}_\epsilon u \to \mathrm{sgn}\, u$ pointwise, we obtain the result by taking $\epsilon \to 0$ in (8.12). □

Problem 8.5. Verify the calculations in the proof of Lemma 8.7, especially the claims concerning the pointwise convergence at the end of the proof.

We remark that one can also prove (8.8) by straightforward differentiation of $|u|$ and by using the positivity of resulting terms to get a lower bound. Our next goal is to extend (8.8) to a more general class of functions. For this we need a few technical tools.

8.4 Technical Approximation Tools

Let us recall some notions from Appendix 4 about operations in the distributional sense. For this we need to introduce some new function spaces.

Definition 8.8. *For any $1 \leq p < \infty$, we define a local L^p-space by*

$$L^p_{\mathrm{loc}}(\mathbb{R}^n) = \left\{ f \mid \int_\Omega |f(x)|^p dx < \infty, \text{ for any bounded } \Omega \subset \mathbb{R}^n \right\}.$$

We have $L^p(\mathbb{R}^n) \subset L^p_{\mathrm{loc}}(\mathbb{R}^n)$. We do not need any other structure on $L^p_{\mathrm{loc}}(\mathbb{R}^n)$. Note that if $f \in L^1_{\mathrm{loc}}(\mathbb{R}^n)$ and $g \in C_0^\infty(\mathbb{R}^n)$, then $\int f\bar{g}$ converges by the Hölder inequality, Theorem A2.4.

Definition 8.9. *Let $\phi \in L^1_{\mathrm{loc}}(\mathbb{R}^n)$, and let $\langle f, g \rangle \equiv \int f\bar{g}$.*

(1) *A function $\psi \in L^1_{\mathrm{loc}}$ is the distributional derivative of ϕ with respect to x_i (formally, $\psi = \partial\phi/\partial x_i$) if*

$$\langle \psi, f \rangle = -\left\langle \phi, \left(\frac{\partial f}{\partial x_i} \right) \right\rangle$$

for all $f \in C_0^\infty(\mathbb{R}^n)$.

(2) *Let* $\phi_n, \phi \in L^1_{loc}(\mathbb{R}^n)$. *Then* ϕ_n *converges to* ϕ *in the distributional sense if* $\langle \phi_n, f\rangle \to \langle \phi, f\rangle$ *for all* $f \in C_0^\infty(\mathbb{R}^n)$.

(3) *Let* $\phi, \psi \in L^1_{loc}(\mathbb{R}^n)$. *Then* $\phi \geq \psi$ *in the distributional sense if* $\langle \phi, f\rangle \geq \langle \psi, f\rangle$ *for all positive* $f \in C_0^\infty(\mathbb{R}^n)$.

Problem 8.6. Prove that distributional derivatives and distributional limits are unique.

Next we need the notion of an approximate identity.

Definition 8.10. *Let* $\omega \in C^\infty(\mathbb{R}^n)$, $\omega \geq 0$, *and* $\int \omega(x)dx = 1$. *For* $\delta > 0$, *we define* $\omega_\delta(x) \equiv \delta^{-n}\omega(\delta^{-1}x)$. *Note that then* $\int \omega_\delta(x)dx = 1$. *We define a map* I_δ *by*

$$I_\delta u \equiv \omega_\delta * u, \tag{8.13}$$

whenever the right side exists, and where $(f * g)(x) \equiv \int f(x-y)g(y)dy$ *is the convolution of* g *and* f. *The map* I_δ *is called an approximation of the identity, or simply, an approximate identity.*

Lemma 8.11. *Let* I_δ *be an approximation of the identity.*

(i) *If* $u \in L^1_{loc}(\mathbb{R}^n)$, *then* $I_\delta u \in C^\infty(\mathbb{R}^n)$.

(ii) *For any differentiable function* u,

$$\left(\frac{\partial}{\partial x_i}\right) I_\delta u = I_\delta\left(\frac{\partial u}{\partial x_i}\right),$$

that is, the map I_δ *commutes with* $\partial/\partial x_i$.

(iii) *The map* $I_\delta : L^p(\mathbb{R}^n) \to L^p(\mathbb{R}^n)$ *is bounded, and* $\|I_\delta\| \leq 1$.

(iv) *For any* $u \in L^p(\mathbb{R}^n)$, $\lim_{\delta\to 0} \|I_\delta u - u\|_p = 0$.

(v) *For any* $u \in L^1_{loc}(\mathbb{R}^n)$, $I_\delta u \to u$ *in the distributional sense as* $\delta \to 0$.

Problem 8.7. Prove (i)–(iii). (*Hint*: Use Young's inequality, Theorem A4.7, for (iii).)

Proof of (iv) and (v). We will show that $\sup_{x\in\mathbb{R}^n} |(I_\delta u)(x) - u(x)| \to 0$ as $\delta \to 0$ for any $u \in C_0^\infty(\mathbb{R}^n)$. Statement (iv) then follows from (iii), the density of $C_0^\infty(\mathbb{R}^n)$ in $L^p(\mathbb{R}^n)$, and the inequality

$$\|I_\delta u - u\|_p \leq (\sup_x |(I_\delta u)(x) - u(x)|)V(u),$$

where $V(u) \equiv \int_{\text{supp}\,(u)} d^n x$. Furthermore, result (v) follows from the identity

$$\langle f, I_\delta u\rangle = (-1)^n \langle I_{-\delta} f, u\rangle$$

and the bound

$$|\langle f - (-1)^n I_{-\delta} f, u\rangle| \le (\sup_x |f(x) - (-1)^n I_{-\delta} f(x)|) C(u, f),$$

where $C(u, f) = \int_{\text{supp}(f)} |u(x)| dx$, for any $f \in C_0^\infty(\mathbb{R}^n)$. Now, turning to the proof, since $\int \omega(y) dy = 1$, we have for any $u \in C_0^\infty(\mathbb{R}^n)$,

$$(I_\delta u)(x) - u(x) = \int \omega_\delta(x-y)[u(y) - u(x)] dy. \tag{8.14}$$

Divide the region of integration in (8.14) into two parts: $\|x - y\| < \sqrt{\delta}$ and $\|x - y\| > \sqrt{\delta}$. For the first, we have

$$\int_{\|x-y\|<\sqrt{\delta}} \omega_\delta(x-y)|u(x) - u(y)| dy \le c_1 \sqrt{\delta} \int \omega_\delta(x-y) dy = c_1 \sqrt{\delta}, \tag{8.15}$$

for some $c_1 > 0$, by the mean value theorem. For the second region, we have

$$\int_{\|x-y\|>\sqrt{\delta}} \omega_\delta(x-y)|u(x) - u(y)| \le 2\|u\|_\infty \int_{\|x\|>\sqrt{\delta}} \omega_\delta(x) dx. \tag{8.16}$$

To treat the last integral, we change variables,

$$\int_{\|x\|>\sqrt{\delta}} \omega_\delta(x) dx = \int_{u>\delta^{-1}} \omega(u) du \equiv W(\delta),$$

and this latter integral $W(\delta)$ vanishes as $\delta \to 0$. In light of (8.15) and (8.16), (8.14) yields

$$\|I_\delta u - u\|_\infty \le c_1 \sqrt{\delta} + 2\|u\|_\infty W(\delta) \to 0,$$

as $\delta \to 0$. □

Problem 8.8. Check that $\omega(x) \equiv [\pi(1 + x^2)]^{-1}$ generates an approximate identity I_δ.

8.5 The Kato Inequality

Theorem 8.12. *Let $u \in L^1_{\text{loc}}(\mathbb{R}^n)$, and suppose that the distributional Laplacian $\Delta u \in L^1_{\text{loc}}(\mathbb{R}^n)$. Then*

$$\Delta |u| \ge \text{Re}\,[(\text{sgn}\, u) \Delta u] \tag{8.17}$$

in the distributional sense.

Proof. Let $u \in L^1_{\text{loc}}(\mathbb{R}^n)$. By Lemma 8.11, $I_\delta u$ is smooth for any $\delta > 0$. Inserting $I_\delta u$ into (8.12) in place of u, we obtain for any $\epsilon > 0$,

$$\Delta(I_\delta u)_\epsilon \geq \text{Re}[\text{sgn}_\epsilon\,(I_\delta u)\Delta(I_\delta u)]. \tag{8.18}$$

We must remove the two cut-offs in equation (8.18). We leave it as Problem 8.9 to show that there exists a subsequence of $\text{sgn}_\epsilon(I_\delta u)\Delta(I_\delta u)$ which converges pointwise to $\text{sgn}_\epsilon(u)\Delta u$, except possibly on a set of measure zero.

Since $\Delta u \in L^1_{\text{loc}}(\mathbb{R}^n)$, it follows from Lemma 8.11 that the limit in $L^1_{\text{loc}}(\mathbb{R}^n)$ as $\delta \to 0^+$ of $\Delta(I_\delta u)$ is Δu. It follows from this and the boundedness of $\text{sgn}_\epsilon(I_\delta u)$ that the distributional limit as $\delta \to 0^+$ of $\text{sgn}_\epsilon(I_\delta u)[\Delta(I_\delta u) - \Delta u]$ is zero.

These two facts, and the Lebesgue dominated convergence theorem, Theorem A2.7, suffice to prove that there is a subsequence such that $\text{sgn}_\epsilon\,(I_\delta u)\Delta(I_\delta u)$ converges to $\text{sgn}_\epsilon\, u\Delta u$ in the distributional sense. Taking this subsequential limit in (8.18) yields

$$\Delta u_\epsilon \geq \text{Re}[(\text{sgn}_\epsilon\, u)\Delta u]. \tag{8.19}$$

Finally, we take $\epsilon \to 0$ in this equation to obtain the result. □

Problem 8.9. Use the $L^1_{\text{loc}}(\mathbb{R}^n)$ properties of Δu and Theorem A2.8 to prove the assertion in the proof of Theorem 8.12.

8.6 Application to Positive Potentials

Let $V \in L^2_{\text{loc}}(\mathbb{R}^n)$ and be real. We define $H = -\Delta + V$ on $D(H) \equiv D(\Delta) \cap D(V)$, where $D(\Delta) = H^2(\mathbb{R}^n)$ and

$$D(V) = \{f \in L^2(\mathbb{R}^n) | \int |Vf|^2 < \infty\}.$$

Note that $C_0^\infty(\mathbb{R}^n) \subset D(H)$, so H is densely defined. The Schrödinger operator H is symmetric on this domain:

$$\langle Hu, v\rangle = \langle u, Hv\rangle,$$

as is easily checked (see Example 8.2). Hence, we have that $D(H) \subset D(H^*)$. Moreover, if $V \geq 0$, then $H \geq 0$ as

$$\langle Hu, u\rangle = \|\nabla u\|^2 + \langle Vu, u\rangle \geq 0,$$

for any $u \in D(H)$. The fact that $\langle -\Delta u, u\rangle = \|\nabla u\|^2$ can be checked by using Fourier transform methods or integration by parts.

Example 8.13. Suppose the real function $V(x) \in L^2_{\text{loc}}(\mathbb{R}^n)$ satisfies the bound $\lim_{\|x\|\to\infty} |V(x)|(1 + \|x\|^M)^{-1} < C_0$, for some constant C_0. Then, $D(V) = \{f \in L^2(\mathbb{R}^n) | \int |f(x)|^2(1 + \|x\|^M)^2 dx < \infty\}$. Perhaps the most well known example is $V(x) \equiv \langle x, Ax \rangle$, where A is a positive, definite $n \times n$ matrix. This is a *harmonic oscillator* potential. This model will be studied in Chapter 11 and plays an important role in the semiclassical analysis of Schrödinger operators.

Theorem 8.14. *Let $V \in L^2_{\rm loc}$ and $V \geq 0$. Then, the Schrödinger operator $H = -\Delta + V$ is essentially self-adjoint on $C_0^\infty(\mathbb{R}^n)$.*

Proof. We will prove in Lemma 8.15 a slight extension of the fundamental criteria, Theorem 8.3, by which it is sufficient to show that ker $(H^* + 1) = \{0\}$. Since $D(H^*) \subset L^2$, the triviality of the kernel is implied by the statement

$$\text{If} \quad -\Delta u + Vu + u = 0, u \in L^2, \text{ then } u = 0. \tag{8.20}$$

We prove (8.20) by Kato's inequality. We note that $u \in L^2$ and $V \in L^2_{\rm loc}$ imply, by the Schwarz inequality, that $uV \in L^1_{\rm loc}$. Since we have the inclusion,

$$L^2 \subset L^2_{\rm loc} \subset L^1_{\rm loc},$$

which follows from the estimate

$$\int_\Omega |u(x)| \cdot 1 \leq V(\Omega)[\int_\Omega |u(x)|^2]^{\frac{1}{2}},$$

where $V(\Omega)$ is the volume of Ω, we have $u \in L^1_{\rm loc}$. Hence by (8.20), $\Delta u \in L^1_{\rm loc}$, where the derivative is a distributional derivative. From (8.17), we obtain

$$\begin{aligned} \Delta|u| &\geq \text{Re}\,[(\text{sgn}\, u)\Delta u] \\ &\geq \text{Re}\,[(\text{sgn}\, u)(V+1)u] \\ &= |u|(V+1) \geq 0. \end{aligned} \tag{8.21}$$

Hence, the function $\Delta|u| \geq 0$, and by Lemma 8.10,

$$\Delta I_\delta |u| = I_\delta \Delta |u| \geq 0. \tag{8.22}$$

On the other hand, $I_\delta|u| \in D(\Delta)$, and therefore

$$\langle \Delta(I_\delta|u|), (I_\delta|u|)\rangle = -\|\nabla(I_\delta|u|)\|^2 \leq 0. \tag{8.23}$$

By (8.22), the left side of (8.23) is nonnegative, and so $\nabla(I_\delta|u|) = 0$ (in the L^2-sense) and hence $I_\delta|u| = c \geq 0$. But $|u| \in L^2$ and $I_\delta|u| \to |u|$ in the L^2-sense (by Lemma 8.10), and so $c = 0$. Hence, $I_\delta|u| = 0$, so $|u| = 0$ and $u = 0$. □

Lemma 8.15. *Let H be a closed, positive, symmetric operator. Then H is self-adjoint if and only if ker $(H^* + b) = \{0\}$ for some $b > 0$. Similarly, if H is a positive, symmetric operator, then the closure of H, H^{**}, is self-adjoint if and only if this condition holds.*

Proof.

(1) We can, without loss of generality, take $b = 1$. We repeat the same arguments as in the proof of Theorem 8.3.

(a) $\text{Ran}(H+1)$ is closed. Let $u_n \in \text{Ran}(H+1)$ form a convergent sequence. There exists a sequence $\{f_n\} \subset D(H)$ such that $u_n = (H+1)f_n$. Then

$$\langle f_n, u_n \rangle = \langle f_n, Hf_n \rangle + \|f_n\|^2 \ \geq \ \|f_n\|^2,$$

and so by the Schwarz inequality,

$$\|f_n\| \ \leq \ \|u_n\|. \tag{8.24}$$

Since $u_n \to u$, the set $\{u_n\}$ is uniformly bounded (i.e. $\sup_n \|u_n\| \ < \infty$) and so (8.24) implies that $\sup_n \|f_n\| \ < \infty$. Now, by positivity,

$$\begin{aligned} \|f_n - f_m\|^2 &\leq \langle (f_n - f_m), (H+1)(f_n - f_m) \rangle \\ &\leq (\|f_n\| + \|f_m\|)\|u_n - u_m\| \\ &\leq C \ \|u_n - u_m\|, \end{aligned}$$

and the sequence $\{f_n\}$ is Cauchy. Since H is closed, there exists $f = s - \lim_n \ f_n \in D(H)$ such that $(H+1)f = u$. Hence, Ran $(H+1)$ is closed.

(b) Since $\ker \ B^* \oplus \overline{\text{Ran } B} = \mathcal{H}$, the fact that $\ker \, (H^* + 1) = \{0\}$ implies that $\text{Ran}(H+1) = \mathcal{H}$.

(c) We show $D(H^*) \subset D(H)$. Let $f \in D(H^*)$, and set $\phi = (H^* + 1)f$. Then there exists $g \in D(H)$, by (b), such that

$$(H+1)g = (H^*+1)g = \phi = (H^*+1)f, \tag{8.25}$$

since $D(H) \subset D(H^*)$. From (8.24), $(H^* + 1)(f - g) = 0$, and so $f = g$.

(2) The converse is trivial since if $H = H^*$ and $H \geq 0$, then $\ker(H + b) = \{0\}$, for any $b > 0$, since $\sigma(H) \subset \mathbb{R}_+$; see Proposition 5.11. □

9

Compact Operators

A compact operator is, loosely speaking, an operator that behaves as if it were almost an operator on a finite-dimensional space. These operators form a very important class of operators, and a great deal of spectral analysis is based on them. We shall see that compact operators have a very transparent canonical form. Consequently, the spectral properties of a self-adjoint compact operator mimic those of a symmetric matrix as closely as possible. Moreover, the Fredholm alternative, Theorem 9.12, is a fundamental tool in the solution of partial differential equations and in solving operator equations. Compact operators also enter the theory of Schrödinger operators and, more generally, partial differential operators, through the notions of *local compactness* and *relative compactness*. These important tools are discussed in Chapters 10 and 14, respectively.

9.1 Compact and Finite-Rank Operators

We will use a streamlined notion of a compact operator. The standard definition is that an operator K on a Banach space X is compact if for every bounded set $\mathcal{N} \subset X$, $K\mathcal{N}$ has compact closure. We refer the reader to Kato [K] for more details. The following definition is equivalent to the usual one for reflexive Banach spaces and, in particular, for Hilbert spaces.

Definition 9.1. *A bounded linear operator K on a reflexive Banach space X is called compact if it maps any weakly convergent sequence into a strongly convergent sequence.*

This means that if $\{x_n\}$ is a sequence and $x_n \xrightarrow{w} 0$, then $Kx_n \xrightarrow{s} 0$, that is, $\|Kx_n\| \to 0$. Recall that $x_n \xrightarrow{w} 0$ if for all bounded linear functionals f on X, $f(x_n) \to 0$.

Example 9.2. An *integral operator* K defined on $\mathcal{C}([0,1])$ is determined by a kernel function $K(x,y)$ by

$$(Kf)(x) \equiv \int_0^1 K(x,y)f(y)dy,$$

for all f such that the integral exists. This operator can be extended to a bounded operator on $L^2[0,1]$ when $K(x,y)$ is continuous in both variables. This follows from the boundedness of the function K and the Schwarz inequality.

Proposition 9.3. *If the kernel $K(x,y)$ is continuous on $R \equiv [0,1] \times [0,1]$, then K is compact on $L^2[0,1]$.*

Proof. Let $f_n \xrightarrow{w} 0$. Then, by continuity of the kernel,

$$\mathcal{I}_n(x) \equiv (Kf_n)(x) = \int_0^1 K(x,y)f_n(y)dy \to 0, \tag{9.1}$$

that is, it converges pointwise to zero on $[0,1]$, since, for each x, the integral defines a bounded linear functional on $L^2[0,1]$. Since $[0,1]$ is compact and K is continuous, it is uniformly continuous in x. Thus, given $\epsilon > 0$, there exists $\delta > 0$ such that

$$\max_{y\in[0,1]} |K(x,y) - K(x',y)| < \epsilon, \tag{9.2}$$

whenever $|x - x'| < \delta$. Next, divide $[0,1]$ into $[\delta^{-1}]+1$ (greatest integer less than δ^{-1}) intervals Δ_i of length $\leq \delta$. Let x_i be the middle of Δ_i. Because of (9.1), we can choose n such that $\max_i |\mathcal{I}_n(x_i)| < \epsilon$. Then,

$$\begin{aligned}
\sup|\mathcal{I}_n(x)| &< \max_i \left(\sup_{x\in\Delta_i} |\mathcal{I}_n(x)| \right) \\
&\leq \max_i \left(\sup_{\Delta_i} |\mathcal{I}_n(x) - \mathcal{I}_n(x_i)| \right) + \max_i |\mathcal{I}_n(x_i)| \\
&\leq \max_i \left(\sup_{\Delta_i} \max_{y\in[0,1]} |K(x,y) - K(x_i,y)| \right) \|f_n\| + \epsilon \\
&\leq \epsilon(\|f_n\| + 1),
\end{aligned} \tag{9.3}$$

where we used (9.2) and the definition of $\mathcal{I}_n$ in (9.1). Since $f_n \xrightarrow{w} 0$, the sequence is uniformly bounded, that is, there exists $M > 0$ such that $\|f_n\| < M < \infty$. Then, from (9.3),

$$\|\mathcal{I}_n\|_\infty \leq \epsilon(M+1).$$

This easily implies that $\mathcal{I}_n \xrightarrow{s} 0$. □

We next turn to a special class of compact operators that do behave as if they are operators on a finite-dimensional space.

Definition 9.4. *A bounded operator F on a Hilbert space $\mathcal{H}$ is called a finite-rank operator if and only if there exist two sets $\{\phi_i\}_{i=1}^n$ and $\{\psi_i\}_{i=1}^n$ of $n < \infty$ linearly independent vectors such that for any $f \in \mathcal{H}$,*

$$Ff = \sum_{i=1}^{n} \langle f, \phi_i \rangle \psi_i. \tag{9.4}$$

Again, we have opted to take this definition for a finite-rank operator rather than the more general one. It is standard to show that Definition 9.4 is equivalent to the usual one.

Problem 9.1. An operator F on a Hilbert space $\mathcal{H}$ is *finite-rank* if Ran F is finite-dimensional. Prove that this definition is equivalent to Definition 9.4. (*Hint*: Use the Riesz representation theorem, Theorem 2.13.)

Problem 9.2. Prove that a finite-rank operator is compact.

Note that for a finite-rank operator F, if $f \in \mathcal{H}$ is orthogonal to the span of $\{\phi_i\}_{i=1}^n$, then $Ff = 0$. Furthermore, $\text{Ran}(F) = \text{span}\{\psi_i\}$ is a finite-dimensional subspace of $\mathcal{H}$. Consequently, we can consider F as acting on the $\leq 2n$-dimensional subspace of $\mathcal{H}$ spanned by $\{\phi_i, \psi_i\}_{i=1}^n$, since F acts as the zero operator on the orthogonal complement. Since this subspace is finite-dimensional, F has a matrix representation on it relative to any basis.

9.2 The Structure of the Set of Compact Operators

Let $\mathcal{K}(\mathcal{H}, \mathcal{H}')$ denote the set of all compact operators from a Hilbert space $\mathcal{H}$ to a Hilbert space $\mathcal{H}'$. This set has much structure, as we will now describe.

Theorem 9.5. *Let $K,\ L \in \mathcal{K}(\mathcal{H},\ \mathcal{H}')$ and let $B \in \mathcal{L}(\mathcal{H})$ and $B' \in \mathcal{L}(\mathcal{H}')$. Then we have the following relations:*

(a) *$aK + bL$ is compact, $a, b \in \mathbb{C}$;*

(b) *$B'K$ and KB are compact;*

(c) *K^* is compact.*

Proof of (a) and (b). Part (a) is obvious from the definition. This shows that $\mathcal{K}(\mathcal{H}, \mathcal{H}')$ is a linear vector space. To prove (b), note that if $u_n \xrightarrow{w} 0$, then $\|B'Ku_n\| \leq \|B'\|\,\|Ku_n\| \to 0$, and similarly for KB since $Bu_n \xrightarrow{w} 0$. □

To prove part (c), we need a technical device, called the *polar decomposition*, which is extremely important in its own right. We refer the reader to the standard literature, for example [RS1], for the proof. Recall from Chapter 5 the notion of a positive operator:

Definition 5.11. *An operator A is positive* ($A \geq 0$) *if* $\langle Af, f\rangle \geq 0$ *for all* $f \in D(A)$.

In connection with positive operators, we have the next definition.

Definition 9.6. *A positive operator B is called the square root of a positive operator A if* $B^2 = A$. *In this case, we write* $B = A^{1/2}$. *If A is bounded, the absolute value of A is defined by* $|A| \equiv (A^*A)^{1/2}$.

The main result we need is the following.

Theorem 9.7 (The polar decomposition). *For any bounded operator A, there exists a bounded operator U such that* $A = U|A|$. *The operator U is a partial isometry, that is, U is an isometry when restricted to* $(\ker A)^{\perp}$.

Proof of Theorem 9.5 (c). We prove that K is compact if and only if $|K|$ is compact. To see this, note that $\||K|f\| = \|Kf\|$. Thus if K is compact, then $|K|$ is compact. Let $K = U|K|$ be the polar decomposition of K. Taking the adjoint of this equation, we get $K^* = |K|U^*$. It follows from part (b) that K^* is compact since U^* is bounded and $|K|$ is compact. □

We remark that there is a form of the polar decomposition theorem for a closed operator. It follows from Theorem 9.5 that $\mathcal{K}(\mathcal{H})$ is a two-sided ideal (over $\mathbb{C}$) in the algebra $\mathcal{L}(\mathcal{H})$ and that it is closed under the operation of taking the adjoint. Next, we will show that this ideal is, in fact, norm-closed.

Theorem 9.8. *The norm limit of a convergent sequence of compact operators in* $\mathcal{L}(\mathcal{H}, \mathcal{H}')$ *is compact.*

Proof. Let $\{K_n\}$ be a norm-convergent sequence of compact operators with limit $K \in \mathcal{L}(\mathcal{H}, \mathcal{H}')$. Let $\{x_n\}$ be a sequence weakly convergent to zero. We have

$$\|Kx_n\| \leq \|K - K_m\| \, \|x_n\| + \|K_m x_n\|. \tag{9.5}$$

Given $\epsilon > 0$, choose $M > 0$ such that $m > M$ implies $\|K - K_m\| < \epsilon$. Fix some $m_0 > M$. Now $K_{m_0}x_n \xrightarrow{s} 0$, so choose $N > 0$ such that $n > N$ implies $\|K_{m_0}x_n\| < \epsilon$. Thus, for $n > N$ and $m = m_0$ in (9.5), we get

$$\|Kx_n\| \leq \epsilon(\sup_n \|x_n\| + 1), \tag{9.6}$$

and since $x_n \xrightarrow{w} 0$, the sequence $\{\|x_n\|\}$ is uniformly bounded and so the result follows from (9.6). □

Corollary 9.9. *The set of all compact operators* $\mathcal{K}(\mathcal{H})$ *is a norm-closed ideal in* $\mathcal{L}(\mathcal{H})$ *which is closed under the taking of the adjoint* $K \to K^*$.

We remark that $\mathcal{K}(\mathcal{H})$ is an example of C^*-algebra without identity since $1 \notin \mathcal{K}(\mathcal{H})$ if $\dim \mathcal{H} = \infty$. C^*-algebras play an important role in many areas of mathematical physics, notably in quantum field theory and statistical mechanics. One might wonder what role the finite-rank operators play in determining $\mathcal{K}(\mathcal{H}, \mathcal{H}')$. We will see that they actually determine $\mathcal{K}(\mathcal{H}, \mathcal{H}')$ (Theorem 9.15), but to prove this we must first develop some spectral properties of compact operators.

9.3 Spectral Theory of Compact Operators

The spectral theory for compact operators reflects the fact that these operators behave almost as if they were acting on a finite-dimensional space. The following theorem is true for an arbitrary compact operator, but we prove it here for self-adjoint compact operators.

Theorem 9.10 (Riesz–Schauder theorem). *The spectrum of a compact operator consists of nonzero isolated eigenvalues of finite multiplicity with the only possible accumulation point at zero, and, possibly, the point zero (which may have infinite multiplicity).*

Proof (Self-adjoint case). We prove that $\sigma_{ess}(K) \subset \{0\}$. The theorem then follows from the disjoint decomposition $\sigma(K) = \sigma_d(K) \cup \sigma_{ess}(K)$ and the following observation. If $\lambda_0 \neq 0$ is an accumulation point of eigenvalues of K, then $\lambda_0 \in \sigma_{ess}(K)$. Let $\lambda \in \sigma_{ess}(K)$. Then by Weyl's criterion, Theorem 7.2, there is a Weyl sequence $\{u_n\}$ for K and λ, that is, $u_n \xrightarrow{w} 0$, $\|u_n\| = 1$, and $(K - \lambda)u_n \xrightarrow{s} 0$. Since K is compact, $Ku_n \xrightarrow{s} 0$, and so $\lambda u_n \xrightarrow{s} 0$. But $\|u_n\| = 1$, and so $\lambda = 0$ and $\sigma_{ess}(K) \subset \{0\}$. □

Corollary 9.11. *If K is compact and $0 \notin \sigma(K)$, then K is a finite-rank operator.*

Problem 9.3. Prove Corollary 9.11. (*Hint*: Use the boundedness of K^{-1} to prove that K has finitely many eigenvalues.)

Compact operators have the useful property that one can state precisely when equations of the form

$$f + Kf = g$$

have a unique solution. We now state and prove the simplest of many variants of what is known as the *Fredholm alternative*.

Theorem 9.12 (Fredholm Alternative). *Let K be a compact operator.*

(i) *The equation $f + Kf = g$ has a unique solution for every $g \in \mathcal{H}$ if and only if $-1 \notin \sigma(K)$ (i.e., if and only if $f + Kf = 0$, or $f + K^* f = 0$, has no nontrivial solutions).*

(ii) *If $-1 \in \sigma(K)$, then $f + Kf = g$ has a unique solution if and only if $g \in [\ker(1 + K^*)]^\perp$.*

Problem 9.4. Prove part (i) of Theorem 9.12.

Proof of Theorem 9.12(ii). Suppose $-1 \in \sigma(K)$.

(1) Suppose that $g \in [\ker(K^* + 1)]^\perp$. Since $\sigma(K)$ is discrete (except, possibly at 0), the restriction $(K + 1)|[\ker(K + 1)]^\perp \equiv A$ has a bounded inverse by Theorem 6.7. Now a simple calculation shows that Ran A is closed and that $\text{Ran}\, A = [\ker(1+K^*)]^\perp$. Consequently, the operator $A^{-1} : [\ker(1+K^*)]^\perp \to [\ker(K + 1)]^\perp$ is a bijection. Thus, if $g \in [\ker(1 + K^*)]^\perp$, then $f \equiv A^{-1}g$ is the unique solution of $f + Kf = g$.

(2) Conversely, if $g \in \ker(1 + K^*)$, then from $f + Kf = g$, we have

$$\|g\|^2 = \langle (1 + K)f, g\rangle = 0,$$

which is a contradiction, so $f + Kf = g$ cannot have a solution. □

There is an important class of integral operators that are compact and that afford a calculable characterization. Recall that an integral operator is determined by a kernel $K(x, y)$. We may think of this as a function on $\mathbb{R}^{2n}$.

Definition 9.13. *An integral operator K with kernel $K(x, y)$ is in the Hilbert–Schmidt class if and only if*

$$\left[\int |K(x, y)|^2 dxdy\right]^{\frac{1}{2}} < \infty. \tag{9.7}$$

We remark that there is a more general characterization of a Hilbert–Schmidt operator using the notion of trace. The family of all Hilbert–Schmidt operators forms a closed, two-sided ideal in $\mathcal{L}(\mathcal{H})$ like the compact operators. The Hilbert–Schmidt operators are one example of what are known as *trace ideals*. We refer to the excellent book by Simon [Sim4] for a discussion of these ideals and their many applications. All Hilbert–Schmidt operators are compact. We prove this for integral operators of the Hilbert–Schmidt type.

Theorem 9.14. *Let K be an integral operator on $L^2(\mathbb{R}^n)$ with kernel $K(x, y) \in L^2(\mathbb{R}^{2n})$ (i.e., satisfying (9.7)). Then K is compact and*

$$\|K\| \leq \left[\int |K(x, y)|^2 dxdy\right]^{\frac{1}{2}}. \tag{9.8}$$

Proof.

(1) By the Schwarz inequality for any x and f,

$$\int |K(x, y)f(y)|dy \leq \left[\int |K(x, s)|^2 ds\right]^{\frac{1}{2}} \|f\|,$$

from which (9.8) easily follows. This proves that K is bounded.

(2) To prove that K is compact, we show that K is the norm limit of a sequence of finite-rank operators. Let $\{\phi_i\}$ be an orthonormal basis for $\mathcal{H}$. Then for any $f \in L^2(\mathbb{R}^n)$,

$$f = \sum_i \langle f, \phi_i \rangle \phi_i \tag{9.9}$$

and

$$Kf = \sum_i \langle f, \phi_i \rangle K\phi_i . \tag{9.10}$$

Let $\psi_i \equiv K\phi_i$. Then we claim that $\sum_{i=1}^{\infty} \|\psi_i\|^2 < \infty$ for

$$\begin{aligned}
\sum_i \|\psi_i\|^2 &= \sum_i \left[\int dx |K\phi_i(x)|^2 \right] \\
&= \sum_i \int dx |\langle K(x, \cdot), \phi_i \rangle|^2 \\
&= \int dx \left(\sum_i |\langle K(x, \cdot), \phi_i \rangle|^2 \right) \\
&\leq \int dx \int dy |K(x, y)|^2 < \infty,
\end{aligned}$$

where we used the Plancherel theorem in the last step. Now we introduce a family of finite-rank operators K_N by

$$K_N f = \sum_{i=1}^{N} \langle f, \phi_i \rangle \psi_i .$$

We claim that $\lim_{N \to \infty} K_N = K$ in the uniform sense for

$$\|Kf - K_N f\| \leq \sum_{i \geq N+1} |\langle f, \phi_i \rangle| \|\psi_i\|$$

$$\leq \left[\sum_{i \geq N+1} |\langle f, \phi_i \rangle|^2 \right]^{\frac{1}{2}} \left[\sum_{i \geq N+1} \|\psi_i\|^2 \right]^{\frac{1}{2}},$$

by the Schwarz inequality. Since by the Plancherel theorem

$$\|f\|^2 = \sum_{i=1}^{\infty} |\langle f, \phi_i \rangle|^2,$$

we get

$$\|Kf - K_N f\| \leq \|f\| \left[\sum_{i \geq N+1} \|\psi_i\|^2 \right]^{\frac{1}{2}} .$$

Now, as $N \to \infty$, $\sum_{i=1}^{N} \|\psi_i\|^2$ converges, and so the sum $\sum_{i \geq N+1} \|\psi_i\|^2 \to 0$ as $N \to \infty$. Consequently, given $\epsilon > 0$, there exists N_ϵ such that $N > N_\epsilon \Rightarrow$

$$\|K - K_N\| = \sup_{f \neq 0} \|(K - K_N)f\| \, \|f\|^{-1} \leq \epsilon,$$

and so $\lim_{N \to \infty} \|K - K_N\| = 0$. By Theorem 9.8, K is compact. □

9.4 Applications of the General Theory

We now apply the results obtained in the previous sections and prove some properties of compact operators. The first theorem states that the finite-rank operators are norm-dense in the set of compact operators, a result anticipated at the end of Section 9.2.

Theorem 9.15. *The finite-rank operators form a norm-dense subset of the compact operators (i.e., any compact operator can be approximated arbitrarily closely by a finite-rank operator).*

Proof in the self-adjoint case. Let K be a self-adjoint compact operator. If $0 \notin \sigma(K)$, K is finite-rank by Corollary 9.11. Hence we assume $0 \in \sigma(K)$. Fix any $\epsilon > 0$. Let $\sigma_\epsilon \equiv \sigma(K) \cap \{\lambda \mid |\lambda| \geq \epsilon\}$. By the Riesz–Schauder theorem, Theorem 9.10, σ_ϵ consists of a finite number of isolated eigenvalues of K with finite multiplicities; see Figure 9.1. Let P_ϵ be the Riesz projection onto the eigenspace corresponding to σ_ϵ,

$$P_\epsilon \equiv (2\pi i)^{-1} \oint_{\Gamma_\epsilon} R_K(z) dz \, ,$$

where Γ_ϵ is the union of two simple closed curves shown in Figure 9.1. Since K is self-adjoint, it follows from Theorem 6.3 that Ran P_ϵ is the span of the eigenspaces of K corresponding to all the eigenvalues in σ_ϵ. This space is finite-dimensional, and thus P_ϵ is a finite-rank operator.

Now $K P_\epsilon = P_\epsilon K$ is a finite-rank operator since it has finite range. We write $K = P_\epsilon K + (1 - P_\epsilon)K$ and show that $\|(1 - P_\epsilon)K\| < \epsilon$. But, by Theorem 5.14,

$$\|(1 - P_\epsilon)K\| \leq \text{spectral radius } (1 - P_\epsilon)K \leq \epsilon.$$

Thus, K is approximated to within ϵ by the finite-rank operator $P_\epsilon K$. □

By Definition 9.4, the finite-rank operators have a particularly transparent form. In light of Theorem 9.15, compact operators are almost finite-rank, so one may ask if there is a canonical form for compact operators similar to that for finite-rank operators. Although we do not need this result for our later work, we include it for completeness.

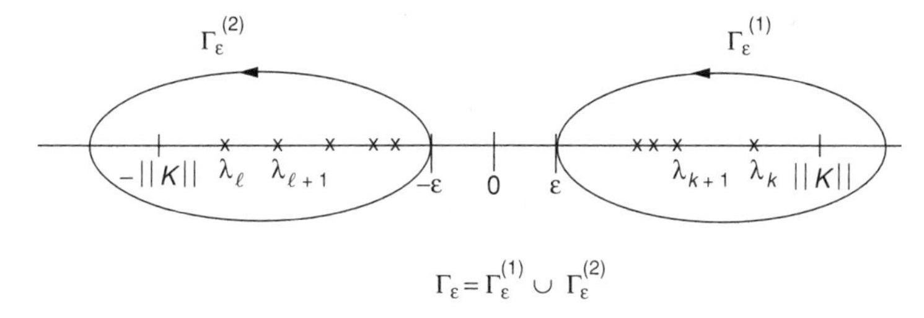

FIGURE 9.1. Spectrum of K in Theorem 9.15.

Theorem 9.16. *Let K be a compact, self-adjoint operator. Then there exists a complete orthonormal basis $\{\phi_n\}$ of $\mathcal{H}$ consisting of eigenvectors of K, that is, $K\phi_n = \lambda_n\phi_n$.*

Problem 9.5. Prove Theorem 9.16. (*Hint*: Write $\mathcal{H} = \mathcal{H}_1 \oplus \mathcal{H}_2$, where $\mathcal{H}_1$ is the span of all the eigenspaces of K and $\mathcal{H}_2 \equiv \mathcal{H}_1^\perp$.)

Theorem 9.17 (Canonical form). *Let K be a compact operator on a Hilbert space $\mathcal{H}$. Then there exist two orthonormal families $\{\psi_n\}$, $\{\phi_n\}$ of vectors in $\mathcal{H}$ and positive real numbers $\{\mu_j\}$, with $\mu_j \to 0$, such that for any $f \in \mathcal{H}$,*

$$Kf = \sum_{j=1}^{\infty} \mu_j \langle f, \psi_j \rangle \phi_j,$$

and the sum converges in norm.

Proof. Since K is compact, so is K^*K by Theorem 9.5. The operator K^*K is a nonnegative, self-adjoint, compact operator, and so by Theorem 9.16, there exists a complete orthonormal basis $\{\psi_j\}$ for $\mathcal{H}$ consisting of eigenvectors for K^*K such that

$$K^*K\psi_j = \lambda_j\psi_j,$$

with $\lambda_j \geq 0$ and $\lim_{j\to\infty}\lambda_j = 0$. For each nonzero λ_j, set $\mu_j = \lambda_j^{1/2}$ and $\phi_j \equiv \lambda_j^{-1/2}K\psi_j$. Then, for any $f = \sum_i \langle f, \psi_i \rangle \psi_i \in \mathcal{H}$, we have

$$\begin{aligned} Kf &= \sum_i \langle f, \psi_i \rangle K\psi_i \\ &= \sum_i \mu_i \langle f, \psi_i \rangle \phi_i, \end{aligned}$$

since $K\psi_j = 0$ for any j such that $K^*K\psi_j = 0$. □

The numbers $\{\mu_j\}$ appearing in Theorem 9.17 are called the *singular values* of K.

The final result we wish to discuss is the regularizing effect compact operators have upon the convergence of sequences of bounded operators. Let us recall that the family $\mathcal{L}(X)$ of all bounded operators on a Banach space X has three primary notions of convergence for sequences of operators.

Definition 9.18. *Let $\{B_n\}$ be a sequence of bounded operators on a Banach space X. Then*

(1) $B_n \overset{n}{\to} B$, *norm or uniform convergence, if* $\|B_n - B\| \to 0$;

(2) $B_n \overset{s}{\to} B$, *strong convergence, if for each* $f \in X$, $\|B_n f - Bf\| \to 0$;

(3) $B_n \overset{w}{\to} B$, *weak convergence, if for each* $l \in X^*$ *(linear functionals on* X*) and each* $f \in X$, $|l(B_n f) - l(Bf)| \to 0$.

Problem 9.6. Prove that uniform convergence implies strong convergence and that these both imply weak convergence.

Theorem 9.19. *Let K be compact and let $\{B_n\}$ be a sequence of bounded operators with $B_n \overset{s}{\to} B$. Then $KB_n \overset{n}{\to} KB$ and $B_n K \overset{n}{\to} BK$.*

Proof. For any $\epsilon > 0$, we decompose K as $K = F + L_\epsilon$, where F is finite-rank and $\|L_\epsilon\| < \epsilon$. For any u, let

$$Fu = \sum_{i=1}^{N} \langle u, \phi_i \rangle \psi_i$$

be the canonical form of F. We first show that $\|B_n F - BF\| \to 0$. To see this, we write

$$\|B_n F - BF\| = \sup_{\|u\|=1} \|(B_n F - BF)u\| \leq \sum_{i=1}^{N} |\langle u, \phi_i \rangle| \|(B_n - B)\psi_i\|,$$

and note that we can choose n large enough to make the right side as small as required, say less than ϵ, since $B_n \overset{s}{\to} B$ and the sum is finite. Then, we have

$$\|B_n K - BK\| \leq \|B_n F - BF\| + \epsilon \|B_n - B\|.$$

Since $B_n \overset{s}{\to} B$, the principle of uniform boundedness (see [RS1]) implies that there exists $0 < M < \infty$ such that $\|B_n\| \leq M$, for all n. Consequently, for all n sufficiently large,

$$\|B_n K - BK\| \leq \epsilon(1 + \|B\| + M),$$

which implies that $B_n K \overset{n}{\to} BK$. Now the same argument implies that $KB_n \overset{n}{\to} KB$, and so that theorem is proved. □

10

Locally Compact Operators and Their Application to Schrödinger Operators

10.1 Locally Compact Operators

The resolvent $R_H(z)$, Im $z \neq 0$, of a Schrödinger operator $H = -\Delta + V$ on $L^2(\mathbb{R}^n)$ is typically not compact (however, it usually is on $L^2(X)$, when X is compact). If $R_H(z)$ is compact, then $\sigma(R_H(z))$ is discrete with zero the only possible point in the essential spectrum. Hence, one would expect that H has discrete spectrum with the only possible accumulation point at infinity (i.e., $\sigma_{\text{ess}}(H)) = \emptyset$). In this way, the $\sigma(H)$ reflects the compactness of $R_H(z)$. It turns out that these properties are basically preserved if, instead of $R_H(z)$ being compact, it is compact only when restricted to any compact subset of $\mathbb{R}^n$. This is the notion of *local compactness*. From an analysis of this notion we will see that the discrete spectrum of H is determined by the behavior of H on bounded subsets of $\mathbb{R}^n$ and the essential spectrum of H is determined by the behavior of H (in particular, V) in a neighborhood of infinity.

Definition 10.1. *Let A be a closed operator on $L^2(\mathbb{R}^n)$ with $\rho(A) \neq \phi$, and let χ_B be the characteristic function for a set $B \subset \mathbb{R}^n$. Then A is locally compact if for each bounded set B, $\chi_B(A-z)^{-1}$ is compact for some (and hence all) $z \in \rho(A)$.*

Problem 10.1. Prove that if $\chi_B(A-z)^{-1}$ is compact for some $z \in \rho(A)$, then it is compact for all $z \in \rho(A)$.

Examples 10.2.

(1) Δ is locally compact on $L^2(\mathbb{R}^3)$. Note that $\chi_B(1-\Delta)^{-1}$ has kernel

$$\chi_B(x)[4\pi\|x-y\|]^{-1}e^{-\|x-y\|},$$

which follows from (4.18) and is easily seen to belong to $L^2(\mathbb{R}^3 \times \mathbb{R}^3)$. By Theorem 9.14, it follows that $\chi_B(1-\Delta)^{-1}$ is compact. We mention that the same compactness result holds in n dimensions (see [RS3] and [Sim4]).

Problem 10.2. Use the Fourier transform to prove that $(1-\Delta)^{-1}$ is not compact. (*Hint*: Compute $\sigma((1-\Delta)^{-1})$.)

(2) $(-\Delta)^{1/2}$, the positive square root of $-\Delta \geq 0$, is locally compact. We offer two proofs of this fact.

Proof 1. Note that it suffices to show that $\chi_B(i+(-\Delta)^{1/2})^{-1} \equiv A^*$ is compact. As $A = (-i+(-\Delta)^{1/2})^{-1}\chi_B$, we have

$$A^*A = \chi_B(1-\Delta)^{-1}\chi_B,$$

and by (1) above, A^*A is compact. Now we claim that this implies that A is compact, for if $u_n \xrightarrow{w} 0$,

$$\|Au_n\|^2 = \langle u_n, A^*Au_n\rangle \leq \|u_n\| \, \|A^*Au_n\|,$$

and as the sequence $\{\|u_n\|\}$ is uniformly bounded and $A^*Au_n \xrightarrow{s} 0$, we have $Au_n \xrightarrow{s} 0$. Hence, A is compact. □

Proof 2. Let $\chi_R(r)$ be 1 on the interval $[0, R]$ and $\operatorname{supp}(\chi_R) \subset \{r \mid 0 \leq r \leq R+1\}$ with $\chi_R \geq 0$. Define the bounded operator $\chi_R((-\Delta)^{1/2})$ through the Fourier transform by

$$(F\chi_R((-\Delta)^{1/2})f)(k) = \chi_R(\|k\|)(Ff)(k), \tag{10.1}$$

where F is the Fourier transform. First, note that

$$\| \left(1-\chi_R\left((-\Delta)^{\frac{1}{2}}\right)\right)\left(1+(-\Delta)^{\frac{1}{2}}\right)^{-1} \| \leq R^{-1}. \tag{10.2}$$

This follows since in the Fourier representation (as in (10.1)) the operator on the left side of (10.2) is multiplication by the function

$$g_R(k) \equiv (1-\chi_R(\|k\|))(1+\|k\|)^{-1},$$

and $\|g_R\|_\infty \leq R^{-1}$. On the other hand, $\chi_B(1+(-\Delta)^{1/2})^{-1}\chi_R((-\Delta)^{1/2})$ is compact because it can be written as

$$\chi_B(1-\Delta)^{-1}(1+\Delta)\left(1+(-\Delta)^{\frac{1}{2}}\right)^{-1}\chi_R\left((-\Delta)^{\frac{1}{2}}\right), \tag{10.3}$$

where the first factor $\chi_B(1-\Delta)^{-1}$ is compact by (1) and the remaining factors form a bounded operator, as is easily checked in the Fourier representation. Hence, it follows from (10.2) and (10.3) that $\chi_B(1+(-\Delta)^{1/2})^{-1}$ is uniformly approximated by the compact operators $\chi_B(1+(-\Delta)^{1/2})^{-1}\ \chi_R((-\Delta)^{1/2})$, and so, by Theorem 9.8, it is compact. □

We now show that certain classes of Schrödinger operators $H = -\Delta + V$ are locally compact.

Theorem 10.3. *Let V be continuous (or $V \in L^2_{\text{loc}}(\mathbb{R}^n)$), $V \geq 0$, and $V \to \infty$ as $\|x\| \to \infty$. Then $H = -\Delta + V$ is locally compact.*

Proof. Note that H is self-adjoint by the Kato inequality, Theorem 8.12, and that $H \geq 0$. We first make the following claim:

$$(-\Delta)^{\frac{1}{2}} \text{ is } H^{\frac{1}{2}}\text{-bounded and } \left((-\Delta)^{\frac{1}{2}} + 1\right)(H+1)^{-\frac{1}{2}} \text{ is bounded.} \tag{10.4}$$

Given the claim, we have

$$\chi_B(1+H)^{-\frac{1}{2}} = \chi_B\left(1+(-\Delta)^{\frac{1}{2}}\right)^{-1}\left(1+(-\Delta)^{\frac{1}{2}}\right)(H+1)^{-\frac{1}{2}}, \tag{10.5}$$

and, by Example 10.2 (2), the first factor on the right in (10.5) is compact, the second is bounded, and so $\chi_B(1+H)^{-1/2}$ is compact. To prove the theorem, simply write

$$\chi_B(1+H)^{-1} = \chi_B(1+H)^{-\frac{1}{2}}(1+H)^{-\frac{1}{2}},$$

and observe that the right side is the product of a compact and a bounded operator and is hence compact. □

Proof of the claim. Since $-\Delta \geq 0$ and $H \geq 0$, all of the operators $(-\Delta)^{1/2}$, $H^{1/2}$, and $(H+1)^{1/2}$ are well defined. We have a simple estimate for any $u \in C_0^\infty(\mathbb{R}^n)$,

$$\begin{aligned} \|(-\Delta)^{\frac{1}{2}}u\|^2 &= \langle u, -\Delta u\rangle \leq \langle u, Hu\rangle \leq \langle u, (H+1)u\rangle \\ &\leq \|(H+1)^{\frac{1}{2}}u\|^2. \end{aligned} \tag{10.6}$$

This estimate extends to all $u \in D(H^{1/2})$. Consequently, relation (10.6) shows that $(-\Delta)^{1/2}$ is $(H+1)^{1/2}$-bounded. Also, as we have $\langle u, Hu\rangle \leq \|H^{1/2}u\|^2$, which follows from the Schwarz inequality, it follows from this and the third term of (10.6) that $(-\Delta)^{1/2}$ is $H^{1/2}$-bounded. □

10.2 Spectral Properties of Locally Compact Operators

We now come to the main properties of locally compact operators. Roughly speaking, this family of operators enjoys the spectral properties described in the Introduction: The essential spectrum is determined by the action of the operator on states supported in a neighborhood of infinity. This idea provides an extremely useful tool for the calculation of the essential spectrum. We already know from Chapter 7 that σ_{ess} of a self-adjoint operator is determined by Weyl sequences. We now introduce a specific family of sequences, called *Zhislin sequences*, which will allow us to characterize the σ_{ess} of locally compact, self-adjoint operators.

Definition 10.4. *Let* $B_k \equiv \{x \in \mathbb{R}^n \mid \|x\| \leq k\ ,\ k \in \mathbb{N}\}$. *A sequence* $\{u_n\}$ *is a Zhislin sequence for a closed operator* A *and* $\lambda \in \mathbb{C}$ *if* $u_n \in D(A)$,

$$\|u_n\| = 1,\ \text{supp}\ u_n \subset \{x \mid x \in \mathbb{R}^n \setminus B_n\},\ \textit{and}\ \|(A-\lambda)u_n\| \to 0\ \textit{as}\ n \to \infty.$$

If $\{u_n\}$ is a Zhislin sequence, then u_n is supported on the complement of B_n. Note that because of this, $u_n \xrightarrow{w} 0$. We call these sequences Zhislin sequences after Zhislin [Z], who was one of the first mathematicians to introduce them into the study of Schrödinger operators. By Weyl's criterion, Theorem 7.2, it is clear that if A is self-adjoint and there exists a Zhislin sequence for A and λ, then $\lambda \in \sigma_{\text{ess}}(A)$.

Definition 10.5. *Let* A *be a closed operator. The set of all* $\lambda \in \mathbb{C}$ *such that there exists a Zhislin sequence for* A *and* λ *is called the Zhislin spectrum of* A*, which we denote by* $Z(A)$.

Let us recall from Chapter 3 that the *commutator* of two linear operators A and B is defined formally by $[A, B] \equiv AB - BA$. As earlier, let $B_R(x)$ denote the ball of radius R centered at the point $x \in \mathbb{R}^n$. Our main theorem states that the essential spectrum is equal to the Zhislin spectrum of a self-adjoint, locally compact operator that is also *local* in the sense of (10.7) ahead.

Theorem 10.6. *Let* A *be a self-adjoint and locally compact operator on* $L^2(\mathbb{R}^n)$. *Suppose that* A *also satisfies*

$$\|[A, \phi_n(x)](A-i)^{-1}\| \to 0\ \textit{as}\ n \to \infty, \tag{10.7}$$

where $\phi_n(x) = \phi(x/n)$ *for some* $\phi \in C_0^\infty(\mathbb{R}^n)$, $\text{supp}\ \phi \subset B_2(0)$, $\phi \geq 0$, *and* $\phi|B_1(0) = 1$. *Then* $\sigma_{\text{ess}}(A) = Z(A)$.

Proof.

(1) It is immediate that $Z(A) \subset \sigma_{\text{ess}}(A)$, by Weyl's criterion. To prove the converse, suppose $\lambda \in \sigma_{\text{ess}}(A)$. Then there exists a Weyl sequence $\{u_n\}$ for A and λ: $\|u_n\| = 1$, $u_n \xrightarrow{w} 0$, and $\|(A-\lambda)u_n\| \to 0$. Let ϕ_n be as in the statement of the theorem, and let $\bar{\phi}_n \equiv 1 - \phi_n$. We first observe that $(i - A)u_n \xrightarrow{w} 0$, because

$$(i-A)u_n = (\lambda - A)u_n + (i-\lambda)u_n, \tag{10.8}$$

and the first term goes strongly to zero whereas the second goes weakly to zero. Next, note that by local compactness, for any fixed n, $\phi_n u_m \xrightarrow{s} 0$ as $m \to \infty$. This can be seen by writing

$$\phi_n u_m = \phi_n(i-A)^{-1}(i-A)u_m, \tag{10.9}$$

and noting that by (10.8), $(i - A)u_m \xrightarrow{w} 0$ and $\phi_n(i - A)^{-1}$ is compact, and so the result follows by Theorem 9.19. Consequently, $\|\phi_n u_m\| \to 0$ and $\|\bar{\phi}_n u_m\| \to 1$ for any fixed n as $m \to \infty$.

(2) We want to construct a Zhislin sequence from $\bar{\phi}_n u_m$. To this end, it remains to consider

$$\|(\lambda - A)\bar{\phi}_n u_m\| \leq \|\bar{\phi}_n\| \, \|(\lambda - A)u_m\| + \|[A, \phi_n]u_m\|. \tag{10.10}$$

The commutator term is analyzed using (10.7):

$$\|[A, \phi_n]u_m\| \leq \|[A, \phi_n](i - A)^{-1}\|(\|(\lambda - A)u_m\| + |i - \lambda|),$$

since $\|u_m\| = 1$. This converges to zero as $n \to \infty$ uniformly in m because the sequence $\{(\lambda - A)u_m\}$ is uniformly bounded, say by M, so

$$\|[A, \phi_n]u_m\| \leq \|[A, \phi_n](i - A)^{-1}\|(M + |i - \lambda|) \to 0,$$

as $n \to \infty$.

(3) To construct the sequence, it follows from (10.10) that for each k there exists $n(k)$ and $m(k)$ such that $n(k) \to \infty$ and $m(k) \to \infty$, as $k \to \infty$, and

$$\|\bar{\phi}_{n(k)} u_{m(k)}\| \geq 1 - k^{-1} \tag{10.11}$$

and

$$\|(\lambda - A)\bar{\phi}_{n(k)} u_{m(k)}\| \leq k^{-1}, \tag{10.12}$$

as $k \to \infty$. We define $v_k \equiv \bar{\phi}_{n(k)} u_{m(k)} \|\bar{\phi}_{n(k)} u_{m(k)}\|^{-1}$. It then follows that $\{v_k\}$ is a Zhislin sequence for A and λ by (10.11)–(10.12) and the fact that $\operatorname{supp} v_k \subset \mathbb{R}^n \setminus B_{2k}$. Hence, $\lambda \in Z(A)$ and $\sigma_{\text{ess}}(A) \subset Z(A)$. □

Problem 10.3. Verify in detail the existence of functions $n(k)$ and $m(k)$ with the properties stated in the proof.

Theorem 10.6 is our main result about locally compact, self-adjoint operators. It states that if A is locally compact and *local* in the sense that (10.7) holds, then $\sigma_{\text{ess}}(A)$ is determined by the behavior of A in a neighborhood of infinity.

Problem 10.4. Prove that the Laplacian $-\Delta$ on $L^2(\mathbb{R}^n)$ satisfies the assumptions of Theorem 10.6. Compute the Zhislin spectrum $Z(-\Delta)$, and conclude that $\sigma_{\text{ess}}(-\Delta) = [0, \infty)$.

We will now apply these ideas to compute $\sigma_{\text{ess}}(H)$ of the locally compact Schrödinger operators $H = -\Delta + V$ studied in Theorem 10.3. We expect that if $V \to \infty$ as $\|x\| \to \infty$, then $\sigma_{\text{ess}}(H) = \{\infty\}$, that is, is empty. We will consider the complementary case, $V \to 0$ as $\|x\| \to \infty$, in Chapter 14.

Theorem 10.7. *Assume that $V \geq 0$, V is continuous (or $V \in L^2_{\text{loc}}(\mathbb{R}^n)$), and $V(x) \to \infty$ as $\|x\| \to \infty$. Then $H = -\Delta + V$ has purely discrete spectrum.*

Proof. By Theorem 10.3, the self-adjoint Schrödinger operator H is locally compact. Let ϕ_n be as in Theorem 10.6. We must verify (10.7). A simple calculation gives

$$[H, \phi_n] = \frac{2}{n} \phi_n' \cdot \nabla - \frac{1}{n^2} \phi_n'', \tag{10.13}$$

where ϕ_n' and ϕ_n'' are uniformly bounded in n. For any $u \in D(H)$, it follows as in (10.6) that

$$\|\nabla u\|^2 \leq \langle u, -\Delta u \rangle \leq \langle u, (H+1)u \rangle,$$

by the positivity of V. Taking $u = (H+1)^{-1} v$, for any $v \in L^2(\mathbb{R}^n)$, it follows that $\nabla(H+1)^{-1}$ and, consequently, $\nabla(H-i)^{-1}$ are bounded. This result and (10.13) verify (10.7).

Hence, it follows by Theorem 10.6 that $Z(H) = \sigma_{ess}(H)$. We show that $Z(H) = \{\infty\}$. If $\lambda \in Z(H)$, then there exists a Zhislin sequence $\{u_n\}$ for H and λ. By the Schwarz inequality, we compute a lower bound,

$$\begin{aligned} \|(\lambda - H)u_n\| \geq |\langle u_n, (\lambda - H)u_n \rangle| &\geq \|\nabla u_n\|^2 + \langle u_n, V u_n \rangle - |\lambda| \\ &\geq \left[\begin{array}{c} \inf V(x) \\ x \in \mathbb{R}^n \setminus B_n(0) \end{array} \right] - |\lambda|. \end{aligned} \tag{10.14}$$

As $n \to \infty$, the left side of (10.14) converges to zero whereas the right side diverges to $+\infty$ unless $\lambda = +\infty$. □

We will offer another proof of this theorem after we have discussed relatively compact operators in Chapter 13.

Theorem 10.6 is the starting point of a group of techniques for studying the essential spectra of operators which are known collectively as "geometric spectral analysis"; see [Ag1] and [DHSV]. It has become a powerful tool for studying elliptic and degenerate elliptic operators and can be extended to include the geometry of configuration and momentum space (i.e., phase space). We will give another application of these ideas in Chapter 14, where we discuss Persson's formula for the bottom of the essential spectrum of a Schrödinger operator. In the next section here, we show that the definition of $Z(A)$ applies to closed, non–self-adjoint operators and that, under additional technical assumptions, one can show that $\sigma_{\text{ess}}(A) = Z(A)$. This will be important for applications to spectral deformation in Chapters 17 and 18.

10.3 Essential Spectrum and Weyl's Criterion for Certain Closed Operators

In this section, we extend the results of Sections 10.1 and 10.2 to certain closed operators. This material is based on [DHSV], to which we refer for further information. For the first part of the discussion, concerning the Weyl spectrum, we will work in a general (separable) Hilbert space $\mathcal{H}$. Then, we will limit ourselves to

$\mathcal{H} = L^2(\mathbb{R}^n)$ when we discuss the Zhislin spectrum. We will need these results in Chapters 17 and 18 when we discuss spectral deformation. Let us recall that the decomposition of $\sigma(A)$, the spectrum of A, introduced in Definition 1.4, is valid for any closed operator.

Definition 10.8. *Let A be a closed operator on a Hilbert space $\mathcal{H}$. The discrete spectrum of A, $\sigma_d(A)$, consists of all isolated eigenvalues of A with finite algebraic multiplicity. The essential spectrum of A, $\sigma_{\text{ess}}(A)$, is the complement of $\sigma_d(A)$ in $\sigma(A)$: $\sigma_{\text{ess}}(A) = \sigma(A) \setminus \sigma_d(A)$.*

As in Definition 7.1, we define Weyl sequences and a Weyl spectrum for A by

$$W(A) \equiv \{\lambda \in \mathbb{C} \mid \exists \{u_n\} \in D(A),\ \|u_n\| = 1,\ u_n \xrightarrow{w} 0, \text{ and } \|(A - \lambda)u_n\| \to 0\}.$$

Problem 10.5. Prove that $\sigma_{\text{ess}}(A)$ and $W(A)$ are closed subsets of $\mathbb{C}$. (*Hint*: To prove that $W(A)$ contains all its accumulation points, construct a new Weyl sequence using a diagonal argument.)

Lemma 10.9. *For a closed operator A, $W(A) \subset \sigma(A)$.*

Proof. Suppose $\lambda \in W(A)$. Let $\{u_n\}$ be a corresponding Weyl sequence. We define another sequence $\{v_n\}$ by

$$v_n \equiv \frac{(A - \lambda)u_n}{\|(A - \lambda)u_n\|}.$$

Then $\|v_n\| = 1$ and $v_n \in D((A - \lambda)^{-1})$ (which, of course, may not be dense). Since $\|(A - \lambda)u_n\| \to 0$, given any $N > 0$ $\exists n_0$ such that $n > n_0$ implies

$$\|(A - \lambda)^{-1} v_n\| = \|(A - \lambda)u_n\|^{-1} > N.$$

Hence, $(A - \lambda)^{-1}$ is an unbounded operator, and so $\lambda \in \sigma(A)$, by definition. □

Unlike the self-adjoint case, $W(A)$ is not necessarily equal to $\sigma_{\text{ess}}(A)$, in general. However, these two sets are related as follows.

Theorem 10.10. *Let A be a closed operator on $\mathcal{H}$ with $\rho(A) \neq \emptyset$. Then $W(A) \subset \sigma_{\text{ess}}(A)$ and the boundary of $\sigma_{\text{ess}}(A)$ is contained in $W(A)$. If, in addition, each connected component of the complement of $W(A)$ in $\mathbb{C}$ contains a point of $\rho(A)$, then $W(A) = \sigma_{\text{ess}}(A)$. The converse of this last statement also holds.*

To explain the second statement, recall that $\sigma_{\text{ess}}(A)$ is some closed, nonempty subset of $\mathbb{C}$ which may have a nonempty interior. For example, in $L^2(\mathbb{R}^2)$, multiplication by $(x + iy)\chi_D$, where χ_D is the characteristic function on the unit disk, has spectrum equal to the closed unit disk. Hence, the boundary of σ_{ess} is S^1. The theorem says that among the points of σ_{ess} determined by Weyl sequences are exactly those on the boundary (S^1 in our example). Clearly, if Int $\sigma_{\text{ess}} \neq \emptyset$, the complement of the boundary contains Int σ_{ess}, which might not contain a point of $\rho(A)$.

Proof of Theorem 10.10.

(1) Let $\lambda \in W(A)$, and let $\{u_n\}$ be a corresponding Weyl sequence. If $\lambda \in \sigma_d(A)$, then there exists an N_0-dimensional, A-invariant subspace where $N_0 < \infty$. Let P be the corresponding Riesz projection. Since $u_n \overset{w}{\to} 0$, it must eventually leave $\operatorname{Ran} P$, that is, $(1-P)u_n \equiv w_n \overset{w}{\to} 0$ and $\|w_n\| \to 1$. Consequently, $(1-P)(A-\lambda)$ is not invertible. But this means $\lambda \in \sigma((1-P)A)$, which contradicts the fact that $\lambda \in \sigma_d(A)$.

(2) Let $\lambda \in \sigma_{\text{ess}}(A)$ and lie on the boundary. There exists a sequence $z_n \in \rho(A)$ such that $z_n \to \lambda$ (recall that $\rho(A)$ is open and that $\rho(A) \neq \emptyset$ by assumption). We can choose $f \in \mathcal{H}$ such that $u_n \equiv (A - z_n)^{-1} f$ does not vanish for all n. Define $w_n \equiv u_n \|u_n\|^{-1}$. Then $\|w_n\| = 1$ and $w_n \overset{w}{\to} 0$. For given $\varepsilon > 0$, choose N such that $|z_n - z_N| < \varepsilon$ for all $n > N$. Then by the first resolvent formula,

$$\langle h, w_n \rangle = \|u_n\|^{-1} \{ \langle h, (A - z_N)^{-1} f \rangle + (z_n - z_N) \langle (A - z_N)^{-1\star} h, u_n \rangle \},$$

for any $h \in \mathcal{H}$, so

$$|\langle h, w_n \rangle| \leq c_0 \|u_n\|^{-1} + \varepsilon c_1 < \varepsilon c.$$

Finally, we compute

$$(A - \lambda) w_n = (A - z_n) u_n \|u_n\|^{-1} + (z_n - \lambda) w_n$$

and find

$$\|(A - \lambda) w_n\| \leq \|f\| \, \|u_n\|^{-1} + |z_n - \lambda|,$$

which vanishes as $n \to \infty$. Hence there exists a Weyl sequence for λ, proving the second statement.

(3) If $W(A) = \sigma_{\text{ess}}(A)$, then we show that $\operatorname{Int} \sigma_{\text{ess}}(A) = \emptyset$. Let $\mathbb{C} \setminus W(A) = \bigcup_{i=1}^{k} C_i$, where C_i is a connected, open set. If $C_i \cap \rho(A) = \emptyset$, then $C_i \subset \sigma(A)$, which, as C_i is open, means $C_i \cap \sigma_{\text{ess}}(A) \neq \emptyset$, a contradiction. Thus $C_i \cap \rho(A) \neq \emptyset$ for all i. Conversely, if each $C_i \cap \rho(A) \neq \emptyset$, since $\mathbb{C} \setminus W(A) \subset \mathbb{C} \setminus \partial\sigma_{ess}(A)$ (where ∂K denotes the boundary of K) by part (2), the set $\{C_i\}$ covers $\operatorname{Int} \sigma_{\text{ess}}(A)$. Now let $\bigcup_i \tilde{C}_i = \mathbb{C} \setminus \partial\sigma_{\text{ess}}(A)$, so $C_j \subset \tilde{C}_i$ for some i. Hence each $\tilde{C}_i$ contains a point of $\rho(A)$. But $\tilde{C}_i$ consists entirely of $\operatorname{Int} \sigma_{\text{ess}}(A)$. Hence, $\operatorname{Int} \sigma_{\text{ess}}(A) = \emptyset$, and so $W(A) = \sigma_{\text{ess}}(A)$. □

The computation of $W(A)$ is not too easy. We would like a result, similar to Theorem 10.6 for the self-adjoint case, that tells us geometrically how to compute $W(A)$. To do this, we take $\mathcal{H} = L^2(\mathbb{R}^n)$. We introduce the notion of the *Zhislin spectrum*, $Z(A)$, which is determined by the behavior of A at infinity, in analogy with Definition 10.4.

Definition 10.11. *A Zhislin sequence $\{u_n\}$ for A and $\lambda \in \mathbb{C}$ is a sequence such that* supp $u_m \cap K = \emptyset$ *for each compact $K \subset \mathbb{R}^n$ and for all m large, and such that $\|(A - \lambda)u_n\| \to 0$. The set of all λ such that a Zhislin sequence exists for A and λ is called $Z(A)$.*

Note that $\{u_n\}$ necessarily converges weakly to zero and that $Z(A) \subset W(A)$. Our main result is the following theorem, which extends the methods of geometric spectral analysis to certain non–self-adjoint operators. Combined with Theorem 10.10, this theorem tells us when we can determine $\sigma_{\text{ess}}(A)$ by the behavior of A at infinity.

Theorem 10.12. *Let A be a closed operator on $L^2(\mathbb{R}^n)$ such that $\rho(A) \neq \emptyset$ and $C_0^\infty(\mathbb{R}^n)$ is a core. Let $\chi \in C_0^\infty(\mathbb{R}^n)$ be such that $\chi \mid B_\varepsilon(0) = 1$, for some $\varepsilon > 0$. We define $\chi_d(x) \equiv \chi(x/d)$. Suppose that for each d, $\chi_d(A - z)^{-1}$ is compact for some $z \in \rho(A)$, and that $\exists\, \varepsilon(d) \to 0$ as $d \to \infty$ such that $\forall u \in C_0^\infty(\mathbb{R}^n)$,*

$$\|[A, \chi_d]u\| \leq \varepsilon(d)(\|Au\| + \|u\|). \tag{10.15}$$

Then $W(A) = Z(A)$.

Proof. Let $\lambda \in W(A)$ and $\{u_n\}$ be a corresponding Weyl sequence. Since $C_0^\infty(\mathbb{R}^n)$ is a core, we can assume $u_n \in C_0^\infty(\mathbb{R}^n)$. For $z \in \rho(A)$,

$$\|\chi_d u_n\| = \|\chi_d(A - z)^{-1}(A - z)u_n\| \to 0, \tag{10.16}$$

since $\chi_d(A - z)^{-1}$ is compact for each d and $u_n \stackrel{w}{\to} 0$. We can also compute

$$\begin{aligned} \|(A - \lambda)(1 - \chi_d)u_n\| &\leq \|(A - \lambda)u_n\| + \|[A, \chi_d]u_n\| \\ &\leq (1 + \varepsilon(d))\|(A - \lambda)u_n\| + \varepsilon(d)(|\lambda| + 1). \end{aligned} \tag{10.17}$$

For each $d = 1, 2, \ldots$, we choose a subsequence $n(d) \to \infty$ such that

$$\|(1 - \chi_d)u_{n(d)}\| \to 1,$$

which is possible by (10.16). Consequently, by (10.17),

$$\|(A - \lambda)\xi_{n(d)}\| \leq c_0\varepsilon(d) \to 0,$$

where $\xi_{n(d)} \equiv (1 - \chi_d)u_{n(d)}$. Since supp $\xi_{n(d)}$ leaves every compact as $d \to \infty$, it follows that $\lambda \in Z(A)$. □

If we combine Theorems 10.10 and 10.12, we obtain a useful result for the computation of $\sigma_{\text{ess}}(A)$ in certain circumstances (see Theorem 16.19).

Corollary 10.13. *Let A be a locally compact, closed operator on $L^2(\mathbb{R}^n)$ such that $\rho(A) \neq \emptyset$ and $C_0^\infty(\mathbb{R}^n)$ is a core. Suppose that (10.15) holds and that each connected component of the complement of $Z(A)$ contains a point of $\rho(A)$. Then $Z(A) = W(A) = \sigma_{\text{ess}}(A)$.*

11

Semiclassical Analysis of Schrödinger Operators I: The Harmonic Approximation

11.1 Introduction

In this and the next chapter, we begin a study of the behavior of the spectrum of Schrödinger operators in the *semiclassical regime*. There is much literature on these topics, and we provide an outline in the Notes to this chapter. The material here follows a part of the work of Simon [Sim5]. In quantum mechanics, the Laplacian plays the role of the energy of a free particle. But, the Laplacian apparently has the dimensions of (length)$^{-2}$. This is because in our discussions of the Schrödinger operator, we have chosen to work in simple units in which the mass $m = 1/2$ and Planck's constant h is taken to be 2π. In these units, the coefficient of the kinetic energy $H_0 = -\Delta$ is 1. If we restore these constants, the actual differential operator in quantum mechanics representing the energy of a free particle is $H_0 = -(h^2/8\pi^2 m)\Delta$. The Planck constant h is approximately 6.624×10^{-27} ergs/sec. This is very small except on atomic scales. That is, the length scales over which quantum effects are important depend on h (for example, the Compton wavelength of an electron or the Bohr radius of an atom). This observation provides us with one way to understand the transition from classical to quantum phenomena. We consider quantum theory in which Planck's constant has been replaced by a small parameter. We then try to understand quantum mechanics in terms of the classical theory obtained as the limit of this quantum theory as the parameter is taken to zero. We call this limit $h = 0$ the *classical limit* since all quantum effects have been suppressed. The regime in which this parameter is nonzero, but taken arbitrarily small, is called the *semiclassical regime*. In this regime, the behavior of the quantum mechanical system should be dominated by the limiting $h = 0$ classical system.

Quantum effects appear as small fluctuations about classical behavior and many times can be calculated when the exact quantum mechanical solution is unknown.

We will study various aspects of this regime, beginning with the behavior of eigenvalues and eigenfunctions. In Chapter 16, we will begin a study of another aspect of quantum systems—resonances—in the semiclassical regime.

We now restore Planck's constant in the Schrödinger equation and write $\hbar$ for $h/2\pi$. We will continue to set the mass $m = 1/2$. The Schrödinger operator is $H(\hbar) \equiv -\hbar^2\Delta + V$, on $L^2(\mathbb{R}^n)$. We treat h as an adjustable parameter of the theory. We will study the semiclassical approximation to the eigenvalues and eigenfunctions of $H(h) = -\hbar^2\Delta + V$ for potentials V with $\inf_{\|x\|>R} V > 0$, in particular when $\lim_{\|x\|\to\infty} V(x) = \infty$, for some $R > 0$. By this we mean that we will identify the leading behavior of eigenvalues $e(h)$ and eigenfunctions $\psi_h(x)$ of $H(h)$, as h becomes arbitrarily small.

Because the small parameter h appears in front of the differential operator $-\Delta$, it may not be clear what is happening as h is taken to be small. It is more convenient, and perhaps more illuminating, to change the scaling. Letting $\lambda \equiv 1/\hbar$, we rewrite the Schrödinger operator as

$$H(\lambda) = -\Delta + \lambda^2 V = \hbar^{-2} H(h).$$

Looking at $H(\lambda)$, we see that the semiclassical approximation involves $\lambda \to \infty$. This limit has the physical interpretation of the potential becoming very large, at least near infinity. The tunneling analysis of Chapter 3, and the work with Zhislin sequences in Chapter 10, suggests that the conditions $\inf_{\|x\|>R} V(x) > 0$ and λ large imply (1) that the infimum of the essential spectrum is greater than zero, and (2) that there may exist low-lying eigenvalues of $H(\lambda)$ below the bottom of the essential spectrum. It is the behavior of these eigenvalues as $\lambda \to \infty$ which we now study.

11.2 Preliminary: The Harmonic Oscillator

By a harmonic oscillator Hamiltonian, we mean the following. Let $A \in M_n(\mathbb{R})$, a real $n \times n$ matrix, and let $A > 0$, that is, A is a positive, definite, and therefore symmetric, matrix. We define

$$K(\lambda) \equiv -\Delta + \lambda^2 \langle x, Ax \rangle, \tag{11.1}$$

where $\langle x, Ax\rangle \equiv \sum_{i,j=1}^n A_{ij}x_i x_j$. Note that the Euclidean quadratic form $\langle \cdot, \cdot \rangle$ is bounded from below by

$$\langle x, Ax \rangle \geq \lambda_{\min} \|x\|^2.$$

Here $\lambda_{\min}$ is the smallest eigenvalue of A and is strictly positive. The lower bound is, therefore, strictly positive for $x \neq 0$. As a consequence, we see that $K(\lambda)$ is positive with a lower bound strictly greater than zero. Since the harmonic oscillator is continuous and $V_{\text{har}}(x) \equiv \langle x, Ax \rangle \to \infty$, as $\|x\| \to \infty$, the harmonic oscillator Hamiltonian (11.1) is self-adjoint by Theorem 8.14. Moreover, the spectrum of $K(\lambda)$, $\sigma(K(\lambda))$, is purely discrete by Theorem 10.7.

We would like to find out how the eigenvalues of $K(\lambda)$ depend on λ. We do this by performing a similarity transformation on $K(\lambda)$, (11.1). First, some general comments.

Definition 11.1. *Two operators A and B, with $D(A) = D(B) = D$, are called similar if $\exists$ a bounded, invertible operator C such that $CD \subset D$ and $A = CBC^{-1}$.*

Problem 11.1. If A and B are similar, then $\sigma(A) = \sigma(B)$.

We now introduce a representation of the multiplicative group $\mathbb{R}_+$ on $L^2(\mathbb{R}^n)$. This is called the *dilation group*. For $\theta \in \mathbb{R}^+$, we define a map on any $\psi \in C_0^\infty(\mathbb{R}^n)$ by

$$U(\theta): \quad \psi(x) \to \theta^{\frac{n}{2}} \psi(\theta x). \tag{11.2}$$

In Problem 11.2, we ask the reader to verify that this map is well defined and extends to a bounded operator on $L^2(\mathbb{R}^n)$. Furthermore, for each $\theta \in \mathbb{R}_+$, $U(\theta)$ is unitary on $L^2(\mathbb{R}^n)$ and

$$U(\theta)^* = U(\theta)^{-1} = U(\theta^{-1}).$$

The map $\theta \in \mathbb{R}_+ \to U(\theta)$ enjoys another very important property. For $\theta, \theta' \in \mathbb{R}_+$, one can check from (11.2) that

$$U(\theta)U(\theta') = U(\theta\theta').$$

That is, the map from the multiplicative group $\mathbb{R}_+$ into the unitary operators $U(\theta)$ preserves all the group structure. We call such a map a *unitary representation of the group* $\mathbb{R}_+$.

Problem 11.2. Verify that the map $\theta \in \mathbb{R}_+ \to U(\theta)$ is a unitary representation of $\mathbb{R}_+$. In addition, prove that the map is strongly continuous, that is, for any $\theta, \theta' \in \mathbb{R}_+$, we have $s - \lim_{\theta' \to \theta} U(\theta') = U(\theta)$.

We now claim that $U(\lambda^{-1/2})$ implements a similarity transformation on $K(\lambda)$ by

$$U\left(\lambda^{-\frac{1}{2}}\right) K(\lambda) U\left(\lambda^{-\frac{1}{2}}\right)^{-1} = \lambda K, \tag{11.3}$$

where

$$K \equiv -\Delta + \langle x, Ax \rangle. \tag{11.4}$$

That is, the transformation $x \to \lambda^{-1/2} x$ is implemented by $U(\lambda^{-1/2})$, and its effect is to scale λ out of $K(\lambda)$.

Problem 11.3. Carefully prove (11.3).

As a consequence of Problems 11.1 and 11.3, $\sigma(K(\lambda)) = \lambda\sigma(K)$, where $\sigma(K)$ is independent of λ. Hence the eigenvalues of $K(\lambda)$ depend linearly on λ. Moreover, the multiplicities of the related eigenvalues are the same. Now we let $\{e_n\}$ denote the eigenvalues of K, so that $\sigma(K(\lambda)) = \{\lambda e_n\}$. The eigenfunctions are related through the unitary operator $U(\lambda^{-1/2})$. If $K(\lambda)\psi_n = e_n(\lambda)\psi_n$, then it follows from (11.3) that $U(\lambda^{1/2})\psi_n \equiv \tilde{\psi}_n$ is an eigenfunction of K with eigenvalue $e_n = \lambda^{-1}e_n(\lambda)$. We remark that the eigenvalues of K are

$$\left\{\sum_{i=1}^{n}(2n_i+1)\omega_i \,|\, n_i \in \mathbb{Z}_+ \cup \{0\}\right\}, \tag{11.5}$$

where $\{\omega_i^2\}_{i=1}^n$ are the eigenvalues of A.

11.3 Semiclassical Limit of Eigenvalues

We study the behavior of the eigenvalues of $H(\lambda) \equiv -\Delta + \lambda^2 V$ on $L^2(\mathbb{R}^n)$ as $\lambda \to \infty$. Here, we assume V satisfies
(V1) $V \in C^3(\mathbb{R}^n)$, $V \geq 0$, $\lim_{\|x\|\to\infty} V(x) = \infty$;
(V2) V has a single, nondegenerate zero at $x_0 = 0$: $V(0) = 0$, $V'(0) = 0$ and

$$A \equiv \frac{1}{2}\left[\left.\frac{\partial^2 V}{\partial x_i \partial x_j}\right|_{x_0=0}\right] > 0.$$

Remarks 11.2.

(1) The potential V need not be globally C^3, but C^3 only in a neighborhood of x_0 and, by modifying the following proof, we only need $\inf_{\omega\in S^{n-1}}(\lim_{\|x\|\to\infty} V(x)) > 0$, where $x = \|x\|w$, for $w \in S^{n-1}$.

(2) The potential V can have several zeros, but they must be nondegenerate.

By Theorem 10.7, $\sigma(H(\lambda))$ is discrete. All of the eigenvalues are positive, and infinity is the only possible accumulation point of eigenvalues. We label these eigenvalues by $e_n(\lambda)$, where we list the eigenvalues in increasing size, including multiplicity:

$$e_1(\lambda) < e_2(\lambda) \leq e_3(\lambda) \leq e_4(\lambda) \leq \cdots.$$

We label the distinct values by $\tilde{e}_n(\lambda)$, with multiplicity $m_n(\lambda)$, in strict increasing order. As λ becomes large, the "well region" about $x_0 = 0$ becomes more prominent and we expect the quadratic part of the potential V to dominate (see Figure 11.1). Consequently, we introduce a comparison harmonic oscillator Hamiltonian

$$K \equiv -\Delta + \langle x, Ax\rangle,$$

where A is given in (V2). Let e_n be the eigenvalues of K listed in increasing size, counting multiplicity,

$$e_1 < e_2 \leq e_3 \leq e_4 \leq \cdots,$$

and, as above, let $\{\tilde{e}_n\}$ be the set of distinct values, ordered by size, with multiplicity m_n.

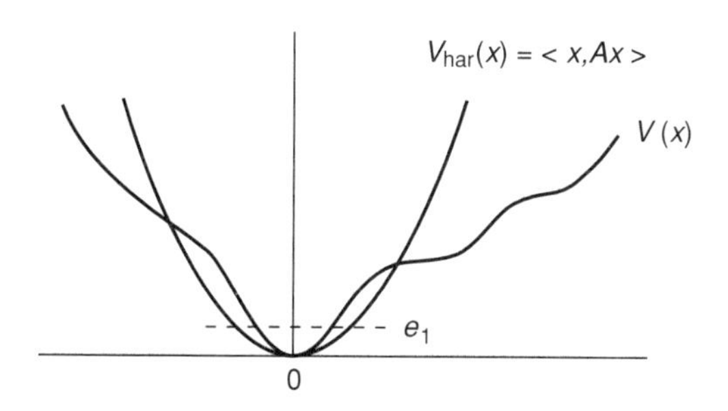

FIGURE 11.1. The harmonic approximation.

We will find it convenient to use the following notation to indicate the order of dependence of various terms on the parameter λ. We say that a real-valued function $f(\lambda)$ is $\mathcal{O}(\lambda^p)$ if $\limsup_{\lambda\to\infty} f(\lambda)\,\lambda^{-p} \leq C_0$ for some finite, positive constant C_0.

The main result of this chapter is the following theorem. It gives the leading behavior of the eigenvalues $e_n(\lambda)$ of $K(\lambda)$ in terms of $\tilde{e}_n$.

Theorem 11.3. *Assume (V1) and (V2). For each $E > 0$, there is a constant λ_E such that the following holds for all $\lambda > \lambda_E$. For each $\tilde{e}_n \in \sigma(K)$ with $\tilde{e}_n < E$, there are m_n not necessarily distinct eigenvalues $e_{k(n)}(\lambda) \in \sigma(H(\lambda))$ satisfying*

$$\lim_{\lambda\to\infty} (e_{k(n)}(\lambda)/\lambda) = \tilde{e}_n. \tag{11.6}$$

Moreover, for each such eigenvalue, we have the expansion

$$e_{k(n)}(\lambda) = \lambda\tilde{e}_n + \mathcal{O}\left(\lambda^{\frac{4}{5}}\right), \tag{11.7}$$

as $\lambda \to \infty$.

The error in the expansion (11.7) is not uniform in n. We will only prove part of the estimate (11.6) here, which simply requires an upper bound on the difference $e_{k(n)}(\lambda) - \tilde{e}_n$. We will prove under the conditions of the theorem that there exists a constant $C_n > 0$ such that, for all large λ,

$$\mid e_{k(n)}(\lambda)/\lambda - \tilde{e}_n \mid \leq C_n\lambda^{\frac{4}{5}}.$$

We refer the reader to [Sim5] for the complete proof. As one might suspect, if the potential V is smooth, one can continue the expansion (11.7) in powers of λ.

The theorem states that given any $E > 0\ \exists\ \lambda_E$ such that for $\lambda > \lambda_E$, there is a bijection between $\sigma(K) \cap [0, E]$ and $\sigma(H(\lambda)) \cap [0, \lambda E]$, including multiplicities. Our philosophy is to view K, an operator whose spectrum is known, as the $\lambda = \infty$ limit of $H(\lambda)$. The theorem makes this idea precise: The low-lying eigenvalues of $H(\lambda)$ approach those of K.

This problem is our first example of perturbation theory. It is rather singular since the known operator K corresponds to the parameter value $\lambda = \infty$. However, we can handle this type of perturbation using geometric methods, like the use of partitions of unity and the decay of eigenfunctions in classically forbidden regions. In Chapter 15, we will make a systematic study of a simpler theory, *analytic perturbation theory*. We will continue our study of singular and geometric perturbation theory in Chapters 16, 19, and 20.

As a final remark, note that in terms of the parameter h, we find the following behavior for the eigenvalues of $H(h)$ as $h \to 0$:

$$e_{k(n)}(h) = h\tilde{e}_n + \mathcal{O}\left(h^{\frac{6}{5}}\right).$$

Proof of Theorem 11.3.

(1) We scale $H(\lambda)$ using $U(\lambda^{-1/2})$ introduced in (11.2):

$$\begin{aligned} H_\lambda &\equiv \lambda^{-1} U\left(\lambda^{-\frac{1}{2}}\right) H(\lambda) U\left(\lambda^{-\frac{1}{2}}\right)^{-1} \\ &= \lambda^{-1}[\lambda p^2 + \lambda^2 V_\lambda] = p^2 + \lambda V_\lambda\,, \end{aligned} \tag{11.8}$$

where $V_\lambda(x) = V(\lambda^{-1/2}x)$, and $p \equiv -i\nabla$. The eigenvalues of H_λ are $e_n^\lambda \equiv \lambda^{-1} e_n(\lambda)$ with the same multiplicities. We show that

$$|\, e_{k(n)}^\lambda - e_n \,| \leq C_n \lambda^{-\frac{1}{5}}, \tag{11.9}$$

which implies the upper bound part of (11.7).

(2) The idea of the proof is to show that for large λ, H_λ is "near" K (defined in (11.4)) and hence their eigenvalues are close. The operator H_λ will be the most like K when acting on wave functions localized near $x_0 = 0$. Near $x_0 = 0$, we use the nondegeneracy condition (V2) and Taylor's theorem with remainder, to expand the potential, and we obtain

$$\begin{aligned} \lambda V_\lambda(x) &= \lambda V_\lambda(0) + \lambda (DV_\lambda)(0)\left(\lambda^{-\frac{1}{2}}x\right) \\ &\quad + \lambda\left\langle \lambda^{-\frac{1}{2}}x, A\lambda^{-\frac{1}{2}}x\right\rangle + \lambda\mathcal{O}\left(\left(\lambda^{-\frac{1}{2}}x\right)^3\right) \\ &= \langle x, Ax\rangle + \lambda^{-\frac{1}{2}}\mathcal{O}(x^3). \end{aligned} \tag{11.10}$$

This is the reason for introducing the scaling of part (1). Note that the $\mathcal{O}(x^3)$-term decays as $\lambda \to \infty$. We will implement this strategy by localizing H_λ near $x_0 = 0$ and estimating the remainder using the decay results of Chapter 3.

To this end, we define a cut-off function $J \in C_0^\infty(\mathbb{R}^n)$, $J \geq 0$ and

$$J(x) = \begin{cases} 1, & \|x\| < 1, \\ 0, & \|x\| > 2, \end{cases}$$

and define

$$J_\lambda(x) \equiv J\left(\lambda^{-\frac{1}{10}}x\right). \tag{11.11}$$

Then, supp (J_λ) grows like $\lambda^{1/10}$. Note that this is slower than the growth of the level sets of V_λ, which grow as $\lambda^{1/2}$.

Claim 1. $\|(H_\lambda - K)J_\lambda\| = \mathcal{O}(\lambda^{-1/5})$.
The claim follows from (11.10):

$$\begin{aligned}\|(H_\lambda - K)J_\lambda\| &= \|(\lambda V_\lambda - \langle x, Ax\rangle)J_\lambda\| \\ &= \|J_\lambda \lambda^{-\frac{1}{2}}\mathcal{O}(x^3)\| = \mathcal{O}\left(\lambda^{-\frac{1}{5}}\right),\end{aligned}$$

since $x = \mathcal{O}(\lambda^{1/10})$ on supp(J_λ).

(3) Let ϕ_n be an eigenfunction of K with $\|\phi_n\| = 1$ and eigenvalue e_n: $K\phi_n = e_n\phi_n$. We define an *approximate wave function* for H_λ by $\psi_n \equiv J_\lambda\phi_n$.

Claim 2. $\|(H_\lambda - e_n)\psi_n\| \leq c\lambda^{-1/5}$ for some constant c, $0 < c < \infty$.
To show this, we simply compute the left side:

$$\begin{aligned}&\|(H_\lambda - e_n)\phi_n\| \leq \|(H_\lambda - K)J_\lambda\phi_n\| + \|(K - e_n)J_\lambda\phi_n\| \\ &\leq \|(H_\lambda - K)J_\lambda\| + \|J_\lambda\|_\infty\|(K - e_n)\phi_n\| + \|[p^2, J_\lambda]\phi_n\|.\end{aligned} \tag{11.12}$$

By claim 1, the first term is $\mathcal{O}(\lambda^{-1/5})$. The second term vanishes because of the eigenvalue equation. Hence, we have to evaluate $\|[p^2, J_\lambda]\phi_n\|$. This is

$$\begin{aligned}\|[p^2, J_\lambda]\phi_n\| &\leq \|\Delta J_\lambda\|_\infty + 2\|\nabla J_\lambda \cdot p\phi_n\| \\ &\leq c_1\lambda^{-\frac{1}{5}} + 2\|\nabla J_\lambda \cdot p\phi_n\|,\end{aligned} \tag{11.13}$$

using (11.11) to evaluate ΔJ_λ. We will prove in Lemma 11.4 that $\|\nabla J_\lambda \cdot p\phi_n\| \leq D_n\, e^{-\delta\lambda^{1/10}}$, for some $\delta > 0$, so that $\|[p^2, J_\lambda]\phi_n\| = \mathcal{O}(\lambda^{-1/5})$. The claim now follows from (11.13) and (11.12).

(4) Given claim 2, we apply a result of Chapter 5, Theorem 5.9, on the determination of the spectrum of a self-adjoint operator. If $\exists\psi_n$ such that $\|\psi_n\| = 1$ and we have a condition like $\|(H_\lambda - e_n)\psi_n\| \leq C_n\lambda^{-1/5}$, then for all λ large, there exist eigenvalues $e^\lambda_{k(n)} \in \sigma(H_\lambda)$ such that $|e^\lambda_{k(n)} - \tilde{e}_n| \leq C_n\lambda^{-1/5}$ (recall that $\sigma(H_\lambda)$ is discrete). Hence, it remains to show that ψ_n is normalizable for all λ large. The norm can be written as

$$\begin{aligned}\|\psi_n\|^2 = \langle J_\lambda^2\phi_n, \phi_n\rangle &= \|\phi_n\|^2 - \langle(1 - J_\lambda^2)\phi_n, \phi_n\rangle \\ &= 1 - \langle(1 - J_\lambda^2)\phi_n, \phi_n\rangle.\end{aligned} \tag{11.14}$$

Now supp $(1 - J_\lambda^2) \subset \{x \mid \|x\| > \lambda^{1/10}\}$. By the exponential bound of Theorem 3.4 (or by using the known harmonic oscillator wave functions), we have

$$\langle(1 - J_\lambda^2)\phi_n, \phi_n\rangle \leq d e^{-2\lambda^{1/10}\delta},$$

for some constants $d, \delta > 0$. It follows from this estimate and (11.14) that $\lim_{\lambda \to \infty} \|\psi_n\|^2 = 1$. The result of all of this is that

$$| e^{\lambda}_{k(n)} - \tilde{e}_n | \leq C_n \left(\lambda^{-\frac{1}{5}} \right), \tag{11.15}$$

and the upper bound part of result (11.6) follows upon multiplying (11.15) by λ. □

Lemma 11.4. *In the notation of Theorem 11.3, there exists a constant D_n such that*

$$\| p \cdot \nabla J_\lambda \phi_n \| \leq D_n \, e^{-\delta \lambda^{1/10}}.$$

Proof. Let $\chi \in C_0^\infty(\mathbb{R}^n)$ be such that $\chi | \mathrm{supp}(\nabla J) = 1$ and $\chi = 0$ for $\|x\| > 5/4$ and $\|x\| < 3/4$. Then it suffices to prove the result for $p_j \chi_\lambda \phi_n$, where $\chi_\lambda(x) = \chi(\lambda^{-1/10} x)$; see Figure 11.2. We have for $j = 1, \ldots, n$,

$$\begin{aligned} p_j \chi_\lambda \phi_n &= p_j \chi_\lambda (K+1)^{-1} (K+1) \phi_n \\ &= p_j (K+1)^{-1} \chi_\lambda (e_n + 1) \phi_n + p_j (K+1)^{-1} [p^2, \chi_\lambda] \phi_n \\ &= p_j (K+1)^{-1} (e_n + 1) \chi_\lambda \phi_n + p_j (K+1)^{-1} (p^2 \chi_\lambda) \phi_n \\ &\quad + 2 p_j (K+1)^{-1} p_l (p_l \chi_\lambda) \phi_n, \end{aligned}$$

so that (using Problem 11.4 ahead),

$$\| p_j \chi_\lambda \phi_n \| \leq c_\lambda \| \phi_n | \mathrm{supp}\, \chi_\lambda \|,$$

with $c_\lambda = c_1 + c_2 \lambda^{-1/10} + c_3 \lambda^{-1/5} \to c_1$ as $\lambda \to \infty$. For all large λ, $\mathrm{supp}\, \chi_\lambda$ lies in $\mathcal{F}_E = \{x \mid \langle x, Ax \rangle - e_n > 0\}$, and so by Theorem 3.4, $\exists\, k > 0$ and $\delta > 0$ such that

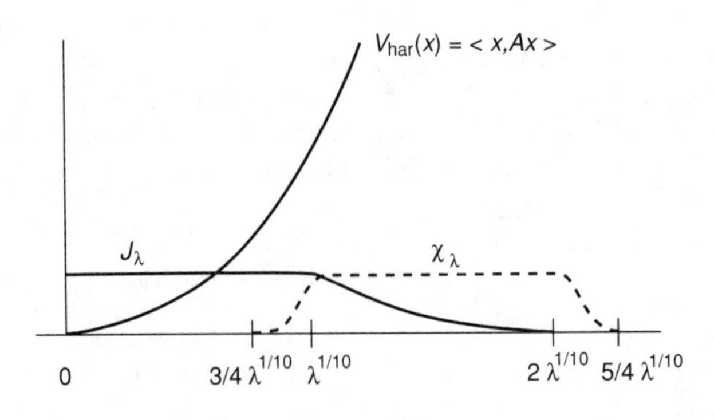

FIGURE 11.2. *Geometry for Lemma 11.4.*

$$\|\phi_n|\text{supp}\,\chi_\lambda\| \leq \text{k}e^{-\delta\lambda^{\frac{1}{10}}},$$

proving the lemma. □

Problem 11.4. Prove that $p_j(K+1)^{-1}p_l$ is bounded.

As mentioned earlier, one needs to prove a lower bound of the form

$$|\ e_{k(n)}(\lambda) - \tilde{e}_n\ | \geq F_n\lambda^{\frac{4}{5}}$$

in order to prove the expansion (11.7). If one assumes more smoothness on V and works harder, one can prove that $e_n(\lambda)$ has an asymptotic expansion in λ to all orders.

Problem 11.5. Generalize Theorem 11.3 and its proof to the case that V has finitely many nondegenerate zeros.

11.4 Notes

There is a lot of literature on the asymptotic expansion of the lowest eigenvalues of a Schrödinger operator as $h \to \infty$. The most common situation occurs when the potential has a single, nondegenerate minumun. By an *asymptotic expansion* of the lowest eigenvalue $e_1(h)$, we mean a formal series $\sum_{i=1}^{\infty} a_i h^i$ so that for any N,

$$e_1(h) - \sum_{i=1}^{N} a_i h^i = \mathcal{O}(h^{N+1}).$$

The main result of Theorem 11.3 states that the low-lying eigenvalues have asymptotic expansions with the first term $a_1 = \tilde{e}_1$. In the one-dimensional case, these expansions were computed by Combes, Duclos, and Seiler [CDS]. The discussion of this chapter follows part of Simon [Sim5]. An extensive study of the asymptotic expansions can be found in the papers of Helffer and Sjöstrand [HSj1], [HSj3]. The question of the asymptotic behavior when the minimum is not nondegenerate has been studied by Martinez and Rouleux [MR].

12

Semiclassical Analysis of Schrödinger Operators II: The Splitting of Eigenvalues

We have seen that the eigenfunction of a Schrödinger operator decays exponentially in the classically forbidden region. This is the content of Theorem 3.4. In general, an eigenfunction at energy E will decay in regions where $V(x) > E$. A striking manifestation of this is *quantum tunneling*: the capacity of a quantum mechanical particle to tunnel through a classically forbidden region of compact support. As is evident from the WKB-approximation to the eigenfunction given in (3.6), quantum tunneling is suppressed in the semiclassical regime. In this chapter, we continue our discussion of the semiclassical approximation and study Schrödinger operators with double-well potentials. It provides a nice example of how quantum tunneling influences the low-lying eigenvalues of a Schrödinger operator. The calculations in this chapter follow part of the paper of Simon [Sim6].

12.1 More Spectral Analysis: Variational Inequalities

Before continuing our discussion of the semiclassical limit of quantum mechanics, we need to develop one more tool in spectral analysis. In Chapter 6, we presented characterizations of the discrete spectrum. We now discuss the *variational inequalities* that allow one to estimate the eigenvalues of a self-adjoint operator. We begin with a theorem that expresses each eigenvalue of a self-adjoint operator, lying below the bottom of the essential spectrum, as the infimum of a quadratic form over certain subspaces of the Hilbert space. These characterizations of the eigenvalues allow one to estimate the eigenvalues from above through the use of trial functions. This is the content of the variational inequalities, which we state as Corollary 12.2.

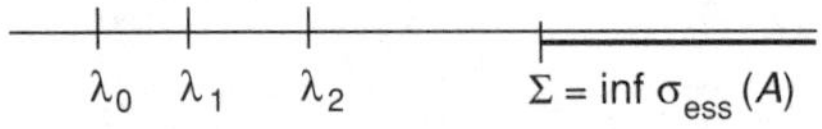

$\lambda_0 \quad \lambda_1 \quad \lambda_2 \qquad \Sigma = \inf \sigma_{ess}(A)$

FIGURE 12.1. The spectrum $\sigma(A)$ as in Proposition 12.1.

Proposition 12.1. *Let A be a self-adjoint operator that is bounded below by a finite constant $-M$. Let $\sum \equiv \inf \sigma_{\text{ess}}(A)$, and suppose that A has eigenvalues $\lambda_0 < \lambda_1 < \lambda_2 < \cdots < \sum$ (see Figure 12.1). Then, we have*

$$\lambda_0 = \inf_{\phi \in D(A)} \|\phi\|^{-2} \langle A\phi, \phi \rangle, \tag{12.1}$$

$$\lambda_1 = \inf_{\phi \in D(A) \cap K_0^{\perp}} \|\phi\|^{-2} \langle A\phi, \phi \rangle, \tag{12.2}$$

and, for $j \geq 2$,

$$\lambda_j = \inf_{\phi \in D(A) \cap K_{j-1}^{\perp}} \|\phi\|^{-2} \langle A\phi, \phi \rangle, \tag{12.3}$$

where $K_i \equiv \cup_{0 \leq j \leq i} \ker(A - \lambda_j)$, and $\ker(A - \lambda_j) \equiv$ eigenspace of A for eigenvalue λ_j. The inf is over all nonzero ϕ.

Proof.

(1) We begin with the lowest eigenvalue. Note that for any $\chi \in D(A)$:

$$\mu_0(A) \equiv \inf_{\phi \in D(A)} \|\phi\|^{-2} \langle A\phi, \phi \rangle \leq \|\chi\|^{-2} \langle A\chi, \chi \rangle. \tag{12.4}$$

Let $\phi_0 \in \ker(A - \lambda_0)$ so that $A\phi_0 = \lambda_0 \phi_0$. Then by (12.4), $\mu_0(A) \leq \lambda_0$. To show the converse, note that $\sigma(A - \lambda_0) \subset [0, \infty)$, and so by Proposition 5.12, $(A - \lambda_0) \geq 0$. Hence, we obtain a lower bound,

$$\inf_{\phi \in D(A)} \langle (A - \lambda_0)\phi, \phi \rangle \|\phi\|^{-2} \geq 0;$$

thus $\mu_0(A) \geq \lambda_0$, as is seen by writing out the left side.

(2) To treat the next eigenvalue, the idea is to remove the eigenspace corresponding to λ_0 and begin the process of part (1) again. Let P_0 be the Riesz projection for λ_0. Now, $\sigma(A(1 - P_0)) = \{\lambda_1, \lambda_2, ...\} \cup \sigma_{ess}(A)$. Applying part (1) to the self-adjoint operator $A(1 - P_0)$, we obtain

$$\mu_0(A(1 - P_0)) = \lambda_1.$$

The left side of this expression can be written as

$$\inf_{\phi \in D(A)} \langle A(1 - P_0)\phi, (1 - P_0)\phi \rangle \|\phi\|^{-2}, \tag{12.5}$$

using the fact that $[A, P_0] = 0$. Let $f = (1 - P_0)\phi$ so that $\|f\| \leq \|\phi\|$. Rewriting (12.5) using f, we get

$$\lambda_1 \leq \inf_{f \in \text{Ran}(1 - P_0)} \langle Af, f \rangle \|f\|^{-2} = \inf_{f \perp \ker(A - \lambda_0)} \langle Af, f \rangle \|f\|^{-2},$$

since Ran $P_0 = \ker(A - \lambda_0)$. On the other hand, let ϕ_1 be such that $A\phi_1 = \lambda_1\phi_1$. Then, by Theorem 5.5, $\phi_1 \perp \ker(A - \lambda_0)$, so the principle in (12.4) implies

$$\inf_{f \perp \ker(A-\lambda_0)} \langle Af, f\rangle \|f\|^{-2} \leq \langle A\phi_1, \phi_1\rangle = \lambda_1,$$

verifying the result for the second eigenvalue. The proof of the theorem is completed by induction. □

The actual variational inequalities are contained in the following corollary.

Corollary 12.2. *Let A be a self-adjoint operator satisfying the hypotheses of Theorem 12.1. The lowest eigenvalue of A satisfies*

$$\lambda_0 \leq \|\phi\|^{-2}\langle A\phi, \phi\rangle, \tag{12.6}$$

for any $\phi \neq 0$ in $D(A)$. Similarly, for $j \geq 1$, we have

$$\lambda_j \leq \|\phi\|^{-2}\langle A\phi, \phi\rangle, \tag{12.7}$$

for all $\phi \neq 0$ in $D(A) \cap K_{j-1}^{\perp}$.

Problem 12.1. Consider the Schrödinger operator for a one-dimensional harmonic oscillator $H = -d^2/dx^2 + \omega^2 x^2$. Estimate the ground state energy using a trial function of the form $\phi_\alpha \equiv e^{-\alpha x^2}$, $\alpha > 0$. Note that $\int_{\mathbb{R}} e^{-u^2}\, du = \pi^{1/2}$. After computing the right side of (12.6), minimize with respect to α.

12.2 Double-Well Potentials and Tunneling

In this section, we consider quantum phenomena of tunneling as exhibited in the splitting of the two lowest eigenvalues of the Schrödinger operator with a symmetric double-well potential (see Figure 12.2). Unlike the classical situation, the two wells are not isolated for states with energy below V_M because the particle can tunnel through the barrier.

The physical situation can be easily seen by examining the one-dimensional problem. Suppose that V is a smooth, positive, even function with exactly two nondegenerate zeros at $x = \pm 1$, respectively, with $V(0) = 1$, and $\lim_{\|x\|\to\infty} V(x) =$

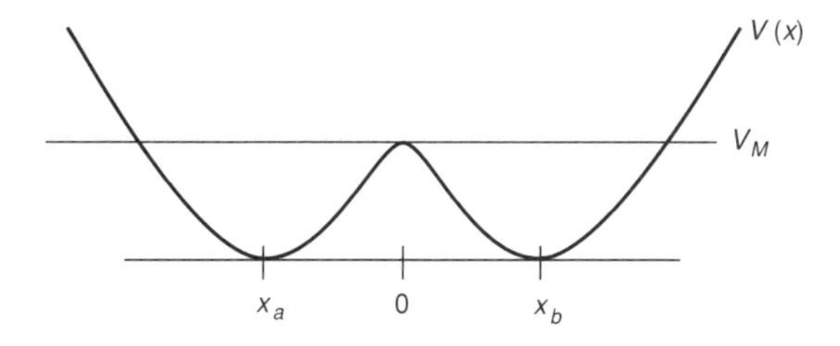

FIGURE 12.2. Symmetric double-well potential.

∞. We consider the Schrödinger operator $H(\lambda) = -d^2/dx^2 + \lambda^2 V$ acting on $L^2(\mathbb{R})$. In the limit that $\lambda \to \infty$, the two potential wells become effectively separated by an infinitely high potential barrier. From expressions (3.5) and (3.6) (with $1/h$ replaced by λ), we see that in this limit the wave functions no longer penetrate into the barrier. The two wells become decoupled and the Hilbert space decomposes into a direct sum $L^2((-\infty, 0]) \oplus L^2([0, \infty))$. We consider this uncoupled situation as the classical limit and seek to find the quantum effects that become manifest as the height of the potential barrier separating the two wells becomes finite. For the $\lambda = \infty$ or unperturbed case, the first eigenvalue for the Schrödinger operator has two independent (and even orthogonal) eigenfunctions: one centered in the well containing $x = +1$ and a reflected eigenfunction centered in the well containing $x = -1$. We say that the *ground state* is doubly degenerate. When the barrier height is decreased, the wave functions can again penetrate the potential barrier. The interaction of the two wave functions causes a splitting of the previously degenerate ground state into two nondegenerate eigenvalues. It is the size of this difference in terms of the semiclassical parameter λ that we want to estimate. One suspects that it is exponentially small in λ. This is precisely the type of situation in which geometric methods are effective.

Models in Euclidean quantum field theory of the type

$$H(\phi) = \int (\|\nabla\phi\|^2 + V(\phi))dx,$$

where

$$V(\phi) \equiv (\|\phi\|^2 - a^2)^2,$$

have such potentials, and the tunneling phenomena associated with them play an important role in the study of phase transitions.

We consider Schrödinger operators of the form $H(\lambda) = -\Delta + \lambda^2 V$, where the potential V satisfies the following two assumptions:

(A1) The potential $V \in C(\mathbb{R}^n)$, $V \geq 0$, and $\lim_{\|x\|\to\infty} V(x) = \infty$.
(A2) The potential V has two nondegenerate minima located at x_a and x_b, respectively, and we set $V(x_a) = 0 = V(x_b)$.

Note that we do not assume that V is symmetric at this stage. The Schrödinger operator $H(\lambda)$ is self-adjoint by Theorem 8.14. By Theorem 9.12 and assumption (A1), $\sigma(H(\lambda))$ is purely discrete, and so $\inf \sigma(H)$ is an eigenvalue $E_0 > 0$, called the *ground state energy*.

An eigenfunction corresponding to the eigenvalue E_0 is called a *ground state*. We first summarize some known facts about the ground state Ω_λ of $H(\lambda)$:

(1) The eigenfunction Ω_λ is a nondegenerate, positive function.

(2) If $V(x)$ is symmetric under reflection $x \to -x$, then so is Ω_λ: $\Omega_\lambda(-x) = \Omega_\lambda(x)$.

These two results are standard, and the proofs can be found in [RS4]. The nondegeneracy and positivity of Ω_λ follow from the fact that H generates a positivity-preserving semigroup, which can be seen, in $\mathbb{R}^3$, by using the explicit formula for $e^{t\Delta}$, $t \geq 0$, and the Trotter product formula.

The next property of Ω_λ that we need concerns the localization of the wave function to the wells. In the absence of any symmetry linking the two wells, it may happen that the wave function Ω_λ is strongly localized in one well with an exponentially small amplitude in the other. We wish to avoid this case here, so we assume that the ground state wave function Ω_λ is localized in both wells in the sense that

(A3) $\exists\, \delta > 0$ and N such that $\int_{\|x-x_k\|<\delta} |\Omega_\lambda(x)|^2 dx \geq c\lambda^{-N}$, for $k = a, b$ and some $c > 0$.

Under the additional assumption (A3), we will prove that the splitting between the ground state and the first excited state E_1 is exponentially small. We mention that the general situation for double wells is this: Either the ground state wave function is strongly localized in one well in the sense that the amplitude is exponentially small in the other, or there exists an eigenvalue E_1 such that $|E_0 - E_1|$ is exponentially small. If there exists a Euclidean symmetry T of order two such that $Tx_a = x_b$ and $V(Tx_a) = V(x_a)$, then the second situation always occurs.

We recall the definition of the geodesic distance in the Agmon metric (with $V \geq 0$ and $E = 0$) from Definition 3.2:

$$\rho(x, y) = \inf_{\gamma \in P_{x,y}} \left[\int_0^1 [V(\gamma(t))]^{\frac{1}{2}} |\dot{\gamma}(t)| dt \right],$$

where $\gamma \in P_{x,y} \equiv \{\gamma : [0,1] \to \mathbb{R}^n, \;\; \gamma \in AC[0,1], \;\; \gamma(0) = x, \;\; \gamma(1) = y\}$. The main theorem of this section follows.

Theorem 12.3. *Let V satisfy assumptions (A1)–(A3), and let $E_0(\lambda) < E_1(\lambda)$ be the two lowest eigenvalues of $H(\lambda) = -\Delta + \lambda^2 V$. Then*

$$\liminf_{\lambda \to \infty} \left\{ -\frac{1}{\lambda} \log\left[E_1(\lambda) - E_0(\lambda)\right] \right\} \geq \rho(x_a, x_b). \tag{12.8}$$

Remarks 12.4.

(1) By using exponentially small lower bounds on the ground state wave function, one can actually prove equality in (12.8).

(2) The geodesic in the Agmon metric giving $\rho(x_a, x_b)$ is called an *instanton*. It is possible to interpret $\rho(a_a, x_b)$ as the "action" of the instanton, which means that formula (12.8) indicates that the eigenvalue splitting is determined by the action of the instanton.

(3) We observe that (12.8) roughly states that

$$E_1(\lambda) - E_0(\lambda) \leq c e^{-\lambda \rho(x_a, x_b)}.$$

Equality in (12.8) is equivalent to an exponentially small lower bound on this eigenvalue difference.

12.3 Proof of Theorem 12.3

The proof of Theorem 12.3 requires some preliminary lemmas. A key ingredient comes from the exponential decay of eigenfunctions described in Chapter 3. Indeed, the strategy is to reduce the question of eigenvalue splitting to exponential decay of eigenfunctions. This strategy will appear again in Chapters 22 and 24 when we estimate the width of quantum resonances.

Lemma 12.5. *For any real, smooth, bounded function f with a bounded gradient,*

$$\langle H(\lambda) f\Omega_\lambda, f\Omega_\lambda\rangle = E_0(\lambda)\| f\Omega_\lambda\|^2 + \| |\nabla f|\Omega_\lambda\|^2. \tag{12.9}$$

Proof. We formally compute

$$[f, H(\lambda)] = [f, p^2] = \Delta f + 2\nabla \cdot (\nabla f),$$

and hence,

$$[f, [f, H(\lambda)]] = -2|\nabla f|^2 = f^2 H + Hf^2 - 2fHf,$$

where the last line is obtained by expanding the double commutator. Solving for fHf, we see that the left side of (12.9) is

$$\langle fH(\lambda) f\Omega_\lambda, \Omega_\lambda\rangle = E_0(\lambda)\| f\Omega_\lambda\|^2 + \| |\nabla f|\Omega_\lambda\|^2. \qquad \square$$

Problem 12.2. Verify the details of the proof of Lemma 12.5 using the fact that the classically allowed region for E_0 is compact, so that the eigenfunction decays exponentially outside this region (see Theorem 3.4 and (12.15)).

Next, we describe some geometry related to $\rho(x, y)$. Let $B \equiv \{x \mid \rho(x_a, x) = \rho(x_b, x)\}$. We call the hypersurface B the *geodesic bisector*. Let us define a function d by

$$d(x) \equiv [\rho(x, x_a) - \rho(x, x_b)]\rho(x_a, x_b)^{-1}, \tag{12.10}$$

so that

$$d(x) = \begin{cases} -1, & x = x_a, \\ 0, & x \in B, \\ +1, & x = x_b. \end{cases}$$

By Proposition 3.3, the function d is differentiable almost everywhere. We need d to be differentiable, and so we replace d by $d * I_\delta$, where I_δ is an approximate identity as in Definition 8.10. However, we will work with d for simplicity, assuming that it is everywhere differentiable.

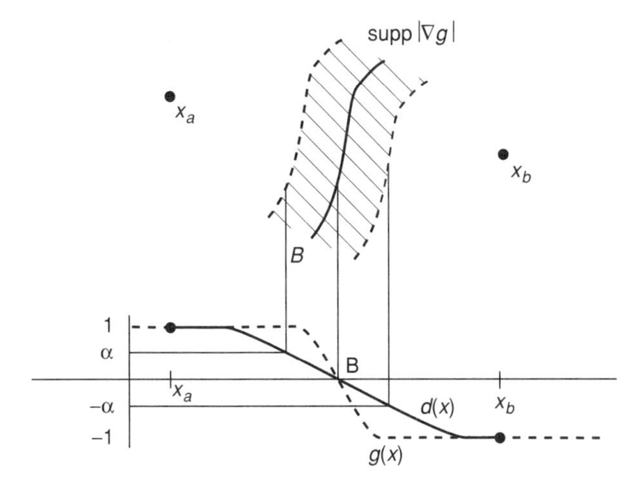

FIGURE 12.3. The functions d and g, and the set B.

Problem 12.3. Trace through the arguments below with d replaced by d_δ.

For any $\alpha > 0$ fixed, choose $h_\alpha \in C^\infty(\mathbb{R})$, with $|h_\alpha| \leq 1$, and such that

$$h_\alpha(x) = \begin{cases} +1, & x \in [\alpha, \infty), \\ -1, & x \in (-\infty, -\alpha], \end{cases}$$

and $h_\alpha(-x) = -h_\alpha(x)$. We now define

$$g(x) \equiv h_\alpha(d(x)),$$

and note that $g \in C^1(\mathbb{R}^n)$ and $\text{supp}(\nabla g)$ lies in a small neighborhood $\{x \mid |d(x)| < \alpha\}$ about B, the size of which depends on α; see Figure 12.3. We remark that if $V(-x) = -V(x)$, we can take g to be antisymmetric: $g(-x) = -g(x)$.

Finally, we define, for any operator A with $\Omega_\lambda \in D(A)$,

$$\langle A \rangle_\lambda \equiv \langle A\Omega_\lambda, \Omega_\lambda \rangle .$$

We call this the *matrix element* of A in the state Ω_λ. We need the following function,

$$f = g - \langle g \rangle_\lambda, \tag{12.11}$$

for which

$$\langle f \rangle_\lambda = 0 .$$

Lemma 12.6. *Let $E_0(\lambda)$ and $E_1(\lambda)$ be the first two eigenvalues of $H(\lambda)$. An upper bound for the difference of these two eigenvalues is given by*

$$E_1(\lambda) - E_0(\lambda) \leq \| |\nabla f| \, \Omega_\lambda \|^2 \, \| f \Omega_\lambda \|^{-2}. \tag{12.12}$$

Proof. Because of (12.11), it is easy to check that $f\Omega_\lambda \perp \Omega_\lambda$. Hence, we can apply the variational inequality, Corollary 12.2, in the form of (12.7) to compute an upper bound for $E_1(\lambda)$:

$$\begin{aligned} E_1(\lambda)\| &= \inf_{\phi\perp\Omega_\lambda} \|\phi\|^{-2}\langle H(\lambda)\phi, \phi\rangle \\ &\leq \langle H(\lambda)f\Omega_\lambda, f\Omega_\lambda\rangle \|f\Omega_\lambda\|^{-2}. \end{aligned} \tag{12.13}$$

Applying Lemma 12.5, we obtain the formula. □

Proof of Theorem 12.3.

(1) From the formula of Lemma 12.6, we obtain

$$\begin{aligned} \Delta E &\equiv \liminf_{\lambda\to\infty} \left(-\frac{1}{\lambda}\log[E_1(\lambda) - E_0(\lambda)]\right) \\ &\geq \liminf_{\lambda\to\infty} \left(-\frac{2}{\lambda}\log \||\nabla f|\,\Omega_\lambda\| + \frac{2}{\lambda}\log\,\|f\Omega_\lambda\|\right). \end{aligned} \tag{12.14}$$

Let ρ be the function

$$\rho(x) \equiv \min(\rho(x_a, x)), \rho(x_b, x)).$$

Note that for $x \notin \mathcal{F}_{E_0} = \{x | V(x) - \lambda^{-2}E_0(\lambda) > 0\}$, $\rho(x) = 0$. Using the Agmon method discussed in Chapter 3, one shows that in the L^2-sense

$$|\Omega_\lambda(x)| \leq ce^{-\lambda(\rho(x)-\epsilon)}, \tag{12.15}$$

for any $\epsilon > 0$. In the next section, we sketch the derivation of this result and its relation to the usual result. Recalling that $|\nabla f| = |\nabla g|$, and that supp$|\nabla g|$ is localized about the geodesic bisector B, we obtain from (12.15)

$$\begin{aligned} \||\nabla f|\,\Omega_\lambda\|^2 &\leq c_1\|e^{\lambda(\rho-\epsilon)}\Omega_\lambda\|^2 e^{-2\lambda(\tilde{\rho}-\frac{\epsilon}{2})} \\ &\leq ce^{-2\lambda(\tilde{\rho}-\frac{\epsilon}{2})}, \end{aligned} \tag{12.16}$$

where

$$\tilde{\rho} \equiv \min\{\rho(x)|x \in \text{supp}\,|\nabla g|\}.$$

Now for any $\epsilon > 0$ there exists an α sufficiently small such that

$$\tilde{\rho} \geq \frac{\rho(x_a, x_b) - \epsilon}{2},$$

since

$$\min\{\rho(x)|x \in B\} = \frac{1}{2}\rho(x_a, x_b).$$

Combining (12.14) with (12.15) and (12.16), we obtain for any $\epsilon > 0$,

$$\Delta E \geq \rho(x_a, x_b) - \epsilon + \liminf_{\lambda\to\infty}\left(\frac{2}{\lambda}\log\,\|f\Omega_\lambda\|\right).$$

(2) If suffices to show that

$$\liminf_{\lambda\to\infty}\left(\frac{2}{\lambda}\log\|f\Omega_\lambda\|\right)=0. \tag{12.17}$$

This follows from assumption (A3), as we now show. Note that

$$\|f\Omega_\lambda\|^2=\langle f^2\rangle_\lambda=\langle g^2\rangle_\lambda-\langle g\rangle_\lambda^2. \tag{12.18}$$

We now recall that supp $|\nabla g|$ lies in $\mathcal{F}_{E_0}$, and hence Ω_λ is exponentially small there as given in (12.15). Consequently, for some $\delta_1,\delta_2>0$, we have

$$\langle g^2\rangle_\lambda=\int_{g=1}\Omega_\lambda^2+\int_{g=-1}\Omega_\lambda^2+\mathcal{O}(e^{-\lambda\delta_1}) \tag{12.19}$$

and

$$\langle g\rangle_\lambda^2=(1-2\int_{g=1}\Omega_\lambda^2)^2+\mathcal{O}(e^{-\lambda\delta_2}). \tag{12.20}$$

Substituting estimates (12.19) and (12.20) into (12.18), we obtain

$$\langle f^2\rangle_\lambda\geq 1-(1-2\int_{g=1}\Omega_\lambda^2)^2-\mathcal{O}(e^{-\lambda\delta}),$$

for some $\delta>0$. By (A3) and the fact that $\{x|g(x)=1\}$ is a neighborhood of x_b, we have

$$\int_{g=1}\Omega_\lambda^2\geq c\lambda^{-N},$$

and so the result (12.17) follows. □

12.4 Appendix: Exponential Decay of Eigenfunctions for Double-Well Hamiltonians

We discuss the exponential decay formula (12.15) and its relation to the decay described in Chapter 3. Recall that

$$\rho(x)=\min(\rho(x_a,x),\rho(x_b,x)), \tag{12.21}$$

where

$$\rho(x,y)=\inf_\gamma\left\{\int_0^1[V(\gamma(t))]^{\frac{1}{2}}|\dot\gamma(t)|dt\right\}, \tag{12.22}$$

with $\gamma \in AC[0,1]$, $\gamma(0) = x$, and $\gamma(1) = y$. Since $V \geq 0$, this is well defined for all $x, y \in \mathbb{R}^n$. On the other hand, for an energy eigenvalue $E(\lambda)$ of $H(\lambda)$, the distance in the Agmon metric introduced in Chapter 3 is

$$\rho_E(x, y) = \inf_{\gamma} \left[\int_0^1 [(\lambda^2 V(\gamma(t)) - E(\lambda))_+]^{\frac{1}{2}} |\dot{\gamma}(t)| dt \right]. \tag{12.23}$$

Let $S_E(\lambda)$ be the set

$$S_E(\lambda) \equiv \{x \mid \lambda^2 V(x) = E(\lambda)\}, \tag{12.24}$$

that is, the boundary of the classically forbidden region $\mathcal{F}_E$ for energy E. The decay result of Chapter 3 for the eigenfunction ψ_E, $H\psi_E = E\psi_E$, is

$$\|e^{(\tilde{\rho}-\epsilon)}\psi_E\| < c, \tag{12.25}$$

where

$$\tilde{\rho}(x) \equiv \inf_{y \in S_E(\lambda)} \rho_E(x, y). \tag{12.26}$$

To see the relation between (12.15) and (12.25), we first note that by the harmonic approximation (Chapter 11), $E(\lambda) = \mathcal{O}(\lambda)$ and, consequently, (12.22) is the leading asymptotic contribution to $\lambda^{-1}\rho_E(x, y)$ as $\lambda \to \infty$, that is,

$$\lim_{\lambda \to \infty} \frac{\rho_E(x, y)}{\lambda} = \rho(x, y).$$

Next, note that if $y \in S_E(\lambda)$, then $y = \mathcal{O}(\lambda^{-1/2})$, as follows from (12.24) and the harmonic approximation. This implies that for λ sufficiently large, $S_E(\lambda) = S_E^a(\lambda) \cup S_E^b(\lambda)$, where $S_E^i(\lambda)$ are disjoint surfaces around x_i, $i = a, b$. Consequently, in the large λ limit,

$$\tilde{\rho}(x) \to \min_{i=a,b} \left(\inf_{y \in S_E^i} \rho_E(x, y) \right)$$

and

$$\lim_{\lambda \to \infty} \frac{\tilde{\rho}(x)}{\lambda} = \min_{i=a,b} \left(\lim_{\lambda \to \infty} \inf_{y \in S_E^i} \left(\frac{\rho(x, y)}{\lambda} \right) \right) = \rho(x), \tag{12.27}$$

where ρ is given in (12.21) and (12.22). As a consequence of (12.27), we see that (12.15) is indeed the leading asymptotic contribution to the decay rate of ψ_E.

We remark that result (12.15) can also be derived directly using the Agmon method described in Chapter 3.

12.5 Notes

The behavior of the eigenvalues and eigenfunctions for double- and multiple-well problems in the semiclassical regime has been studied by many authors. In the one-dimensional case, techniques of ordinary differential equations can be used quite effectively to obtain sharp estimates. We mention the work of Gérard and Grigis [GG], Harrell [H1], Kirsch and Simon [KS1], and Nakamura [N1]. A typical result [N1] in one dimension for the lower bound on the difference of the first two eigenvalues is

$$\liminf_{\lambda \to \infty} \lambda^{-1} \log[E_1(\lambda) - E_0(\lambda)] \geq -\int_{-1}^{1} \sqrt{V(x)}\, dx.$$

Combined with upper bounds of the type obtained here, one sees that this result is optimal.

In multiple dimensions, we mention, in addition to the work of Simon [Sim6] on which this discussion is based, the papers of Helffer and Robert [HR], which give the asymptotic expansions of the eigenvalues; Helffer and Sjöstrand [HSj1] [HSj3]; Kirsch and Simon [KS2]; Martinez [M1]; and, for the case of degenerate wells, Martinez and Rouleux [MR]. All of these papers deal with the semiclassical regime.

There are some related problems concerning estimates of eigenvalue differences for the Laplacian and Schrödinger operators on bounded domains with Dirichlet boundary conditions. For the Dirichlet Laplacian with dumbbell domains, which consist of two regions connected by a tube of width ϵ, the eigenvalue difference can be estimated from above and below for ϵ sufficiently small (see [BHM1] and references therein). The upper bounds are obtained in a manner similar to that discussed here. We will also examine a related problem in Chapter 23. There are some results available when there is no semiclassical parameter. Singer, Wong, Yau, and Yau [SWYY] obtained lower bounds for the eigenvalue difference for a Schrödinger operator on a convex domain with Dirichlet boundary conditions when the potential is nonnegative and convex.

13

Self-Adjointness: Part 2. The Kato–Rellich Theorem

The Kato–Rellich theorem and the Kato inequality form the basic tools for proving self-adjointness of Schrödinger operators $H = -\Delta + V$. The Kato inequality allows us to consider positive potentials that grow at infinity. These Schrödinger operators typically have a spectrum consisting only of eigenvalues. The semiclassical behavior of these eigenvalues near the bottom of the spectrum was studied in Chapters 11 and 12. The Kato–Rellich theorem will permit us to consider the other extreme: $V \to 0$ as $\|x\| \to \infty$. Families of such potentials, which might not necessarily be small in operator norm, will be shown to be small perturbations *relative* to the Laplacian $-\Delta$, in a sense to be defined ahead. We will see in the next chapter that the essential spectrum of such operators $H = -\Delta + V$ is usually $\bar{\mathbb{R}}^+ = \sigma_{\text{ess}}(-\Delta)$, that is, completely characterized by the essential spectrum of the Laplacian. We will first discuss the general theory of relatively bounded perturbations and then its application to Schrödinger operators.

13.1 Relatively Bounded Operators

We now study the following question: Suppose A is self-adjoint and B is a closed, symmetric operator such that $D(A) \cap D(B)$ is dense. What conditions must B satisfy in order that $A + B$ be self-adjoint? This question will be answered in a very satisfactory manner by the Kato–Rellich theorem. We will apply this to study perturbations of the Laplacian, which is self-adjoint, by real potential functions V.

In the following, we assume that A and B are closed operators on a Hilbert space $\mathcal{H}$.

Definition 13.1. *An operator B is called A-bounded if $D(B) \supset D(A)$.*

It is obvious, but important to note, that any bounded operator $B \in \mathcal{L}(\mathcal{H})$ is A-bounded for any linear operator A.

Proposition 13.2. *If $\rho(A) \neq \phi$ and B is A-bounded, then there exist nonnegative constants a and b such that*

$$\|Bu\| \leq a\|Au\| + b\|u\|, \tag{13.1}$$

for all $u \in D(A)$.

Proof. We equip $D(A)$ with the graph norm defined by

$$\|u\|_A \equiv (\|u\|^2 + \|Au\|^2)^{\frac{1}{2}}. \tag{13.2}$$

It is easy to see that A is a closed operator if and only if $D(A)$ is closed in the graph norm (see Appendix 3 for a discussion of this). Furthermore, this norm is induced by an inner product,

$$\langle u, v\rangle_A \equiv \langle u, v\rangle + \langle Au, Av\rangle.$$

If A is closed, the linear vector space $D(A)$ is a Hilbert space with this inner product, which we call $\mathcal{H}_A$.

Problem 13.1. Verify that $(A, D(A))$ is closed if and only if $D(A)$ is closed in the graph norm. Then show that if $z \in \rho(A)$, $R_A(z)$ is a bounded operator from $\mathcal{H}$ onto $\mathcal{H}_A$.

Since $\mathcal{H}_A = D(A)$ as sets, and B is relatively A-bounded, we have that $BR_A(z)$ is everywhere defined on $\mathcal{H}$.

Problem 13.2. Prove that $BR_A(z)$ is closed on $\mathcal{H}$.

By the closed graph theorem, Theorem A3.23, an everywhere defined closed operator is bounded. This means that there exists an $a > 0$ (depending on $z \in \rho(A)$) such that

$$\|BR_A(z)f\|_{\mathcal{H}_A} \leq a\|f\|_{\mathcal{H}}, \tag{13.3}$$

for all $f \in \mathcal{H}$. For any $f \in \mathcal{H}$, we showed that $u \equiv R_A(z)f \in D(A)$ (and each $u \in D(A)$ can be written in this way), so $f = (A - z)u$. Using this representation of f in (13.3), we get

$$\begin{aligned} \|Bu\| &= \|BR_A(z)f\| \leq a\|(A - z)u\| \\ &\leq a\|Au\| + b\,\|u\|, \end{aligned}$$

where $b \equiv a|z|$. This proves the theorem. □

Definition 13.3. *The smallest nonnegative constant a such that* (13.1) *holds for all $u \in D(A)$ is called the bound of B relative to A, or the A-bound of B.*

We remark that B may be unbounded in general. If B is relatively A-bounded, it follows from (13.1) that for all $u \in D(A)$, there exists a constant $c > 0$ such that

$$\|Bu\| \leq c\|u\|_A,$$

which is to say that $B : \mathcal{H}_A \to \mathcal{H}$ is bounded.

Problem 13.3. Consider the Laplacian on $L^2(\mathbb{R}^n)$. Prove that any partial derivative $\partial_j,\ j = 1, \ldots, n$, as a linear operator with domain $H^1(\mathbb{R}^n)$, is relatively Laplacian bounded. Compute the relative Laplacian bound of these operators.

Example 13.4. Let us consider a potential $V \in L^2(\mathbb{R}^3) + L^\infty(\mathbb{R}^3)$, as in Example 8.2. Then V is Δ-bounded with relative bound zero.

Proof. Let $V = V_1 + V_2$, with $V_1 \in L^2(\mathbb{R}^3)$ and $V_2 \in L^\infty(\mathbb{R}^3)$. The multiplication operator V_2 is a bounded operator,

$$\|V_2 u\| \ \le\ \|V_2\|_\infty \|u\|,$$

and so V_2 is Δ-bounded with relative bound zero. We consider $V = V_1 \in L^2(\mathbb{R}^3)$. Since Δ is self-adjoint (see Example 8.4), any nonzero $i\lambda,\ \lambda \in \mathbb{R}$, is in $\rho(\Delta)$. The Hölder inequality, Theorem A2.4, provides the necessary inequality,

$$\|fg\|_2 \ \le\ \|f\|_2 \|g\|_\infty,$$

so we have,

$$\|V R_\Delta(i\lambda) f\| \ \le\ \|V\|_2\ \|R_\Delta(i\lambda) f\|_\infty, \tag{13.4}$$

provided, of course, the infinity norm is finite. We will now prove this. In Chapter 4, we computed the integral representation for $R_\Delta(i\lambda)f$, (4.18), which we recall here:

$$(R_\Delta(i\lambda) f)(x) = (4\pi^2)^{-1} \int_{\mathbb{R}^3} e^{-(i\lambda)^{\frac{1}{2}} \|x-y\|} \|x - y\|^{-1} f(y) dy. \tag{13.5}$$

Applying the Young inequality, Theorem A4.7, to this convolution, we find

$$\|R_\Delta(i\lambda) f\|_\infty \ \le\ \|f\|_2\ \|G_\lambda\|_2, \tag{13.6}$$

where

$$\|G_\lambda\|_2^2 \ \equiv (2\pi)^{-4} \int e^{-2Re(i\lambda)^{\frac{1}{2}} \|x\|} \|x\|^{-2} dx.$$

Now it is easy to check that $\lim_{\lambda\to\infty} \ \|G_\lambda\|_2 \ = 0$, and so for any $\epsilon > 0$ we can find $\lambda > 0$ such that

$$\|G_\lambda\|_2 \ \le \epsilon \|V\|_2^{-1}. \tag{13.7}$$

Thus, from (13.4), (13.6) and (13.7), it follows that for all λ large enough,

$$\|V R_\Delta(i\lambda) f\|_\infty \ \le \epsilon \|f\|_2. \tag{13.8}$$

From (13.3) and (13.1), this implies that

$$\|Vu\| \ \le \epsilon \|\Delta u\| + \epsilon \lambda \|u\|,$$

for any $\epsilon > 0$ and all $u \in D(A)$. □

Notice from this last inequality that the constant in front of the Δu-term can be made small only by increasing the coefficient of the last term. This phenomenon commonly occurs: The constant a in (13.1) can be made small only at the expense of the second constant b.

The reason that relatively bounded operators are so important is due to the following theorem.

Theorem 13.5 (Kato–Rellich theorem). *Let A be self-adjoint, and let B be a closed, symmetric, and A-bounded operator with relative A-bound less than one. Then $A + B$ is self-adjoint on $D(A)$.*

The proof of the theorem relies on the following lemma.

Lemma 13.6. *Let A be self-adjoint and B be A-bounded with relative bound a. Then*

$$\|B(A - i\lambda)^{-1}\| \le a + |\lambda|^{-1}b,$$

for all real $\lambda \neq 0$ and for some $b > 0$. The relative A-bound of B is given by

$$a = \lim_{|\lambda| \to \infty} \|BR_A(i\lambda)\|.$$

Proof. Since A is self-adjoint, we have the identity

$$\|(A - i\lambda)u\|^2 = \|Au\|^2 + |\lambda|^2\|u\|^2,$$

so that

$$\|Au\| \le \|(A - i\lambda)u\|.$$

Setting $u \equiv R_A(i\lambda)v$ in this equation yields

$$\|AR_A(i\lambda)\| \le 1.$$

Also, from Corollary 5.7, we have the a priori bound,

$$\|R_A(i\lambda)\| \le |\lambda|^{-1}.$$

We use these to estimate $BR_A(i\lambda)$ as follows:

$$\begin{aligned} \|BR_A(i\lambda)u\| &\le a\|AR_A(i\lambda)u\| + b\|R_A(i\lambda)u\| \\ &\le (a + b|\lambda|^{-1})\|u\|. \end{aligned}$$

This proves the first result. It also shows that

$$a \ge \limsup_{|\lambda| \to \infty} \|BR_A(i\lambda)\|.$$

On the other hand, we have the simple estimate

$$\begin{aligned} \|Bu\| &= \|BR_A(i\lambda)(A - i\lambda)u\| \\ &\le \|BR_A(i\lambda)\|\{\, \|Au\| + |\lambda|\ \|u\| \,\}. \end{aligned}$$

This inequality, and the definition of the relative bound a, imply that

$$a \leq \liminf_{|\lambda| \to \infty} \| B R_A(i\lambda) \|.$$

These two inequalities for a prove the second statement of the lemma. $\square$

Proof of Theorem 13.5. It is left as Problem 13.4 to prove that $A + B$ is closed on $D(A)$. Since $a < 1$, it follows from Lemma 13.6 that

$$\| B R_A(i\lambda) \| < 1,$$

for λ sufficiently large. Writing

$$A + B - i\lambda = [1 + B R_A(i\lambda)](A - i\lambda),$$

it follows that $1 + B R_A(i\lambda)$ is invertible by Theorem A3.30. Since $(A - i\lambda)$ is invertible, it follows that $(A + B - i\lambda)$ is invertible. Consequently, $\text{Ran}(A + B - i\lambda) = \mathcal{H}$ and, by Theorem 8.3, $A + B$ is self-adjoint on $D(A)$. $\square$

Problem 13.4. Prove that $A + B$ is closed on $D(A)$.

Because of the fundamental importance of the Kato–Rellich theorem, we offer another proof, which does not use Lemma 13.6.

Second proof of Theorem 13.5. By the triangle inequality and the symmetry of $A + B$, we have

$$\begin{aligned} 2\|(A + B - i\lambda)u\| &\geq \|(A + B)u\| + \lambda\|u\| \\ &\geq \|Au\| - \|Bu\| + \lambda\|u\|, \end{aligned} \tag{13.9}$$

for any $u \in D(A)$ and $\lambda > 0$. By the relative boundedness inequality (13.1) with $a < 1$ and $\lambda > b$, (13.9) becomes

$$2\|(A + B - i\lambda)u\| \geq (1 - a)\|Au\| + (\lambda - b)\|u\|. \tag{13.10}$$

Hence, we obtain

$$\begin{aligned} 4\|(A + B - i\lambda)u\|^2 &\geq (1 - a)^2\|Au\|^2 + (\lambda - b)^2\|u\|^2 \\ &\geq (1 - a)^2 \left[\|Au\|^2 + \mu^2\|u\|^2 \right] \\ &\geq (1 - a)^2\|(A - i\mu)u\|^2, \end{aligned}$$

where $\mu \equiv (\lambda - b)(1 - a)^{-1} > 0$, and we used the self-adjointness of A to obtain the last line. If $(A + B - i\lambda)u = 0$, then $(A - i\mu)u = 0$ and by Theorem 8.3, the basic criteria for self-adjointness, $\ker (A \pm i\mu) = \{0\}$ and so $u = 0$. Hence, $\ker (A + B - i\lambda) = \{0\}$, and thus by Theorem 8.3, the operator $A + B$ is self-adjoint. $\square$

Problem 13.5. A self-adjoint operator A is *semibounded* if there exists a finite, nonnegative constant M_A such that for all $u \in D(A)$,

$$\langle u, Au \rangle \geq -M_A \|u\|^2.$$

Prove, under the hypotheses of the Kato–Rellich theorem, that if A is semibounded, then the self-adjoint operator $A + B$ is semibounded with constant $M_{A+B} = M_A - \max\left\{b(1-a)^{-1}, aM_A + b\right\}$. (*Hint*: For $\lambda < -M_A$, compute an estimate for $\|BR_A(\lambda)\|$, as in the proof of Theorem 13.5.)

13.2 Schrödinger Operators with Relatively Bounded Potentials

We now explore the self-adjointness and spectral properties of Schrödinger operators with relatively bounded potentials. These results parallel those in Theorems 8.14 and 10.7 for nonrelatively bounded potentials obtained with the Kato inequality.

Theorem 13.7. *Let $V \in L^2(\mathbb{R}^3) + L^\infty(\mathbb{R}^3)$ and be real. Then the operator $H \equiv -\Delta + V$, defined on $D(\Delta) = H^2(\mathbb{R}^3)$, is self-adjoint.*

Proof. This is an immediate consequence of Example 13.4 and Theorem 13.5. □

The Kato–Rellich theorem is a cornerstone in the theory of self-adjointness for Schrödinger operators $H = -\Delta + V$. The result can be used to prove the following generalization of Theorem 13.7. Suppose $V \in L^p(\mathbb{R}^n) + L^\infty(\mathbb{R}^n)$, with $p > 2$ if $n = 4$ and $p > n/2$ if $n \geq 5$. We will call such potentials *Kato–Rellich potentials*. Then, the Schrödinger operator $H = -\Delta + V$, for V a real Kato–Rellich potential, is self-adjoint on $H^2(\mathbb{R}^n)$. We refer the reader to [RS2] for the proof. In a general sense, the real Kato–Rellich potentials form the natural class of potentials associated with the Kato–Rellich theorem.

We can combine the Kato–Rellich theorem and the Kato inequality to obtain the following result.

Corollary 13.8. *Let the real potential $V = U + W$, where $U \in L^2_{\text{loc}}(\mathbb{R}^n)$, $U \geq 0$, and W is $(-\Delta)^{1/2}$-bounded. Then $H = -\Delta + V$ is self-adjoint on $D(-\Delta + U)$ and essentially self-adjoint on $C_0^\infty(\mathbb{R}^n)$.*

Proof. Set $H_0 \equiv -\Delta + U$. By Theorem 8.14, H_0 is self-adjoint on $D(H_0)$ and has $C_0^\infty(\mathbb{R}^n)$ as a core. We prove that W is relatively H_0-bounded with relative bound zero. Since W is $(-\Delta)^{\frac{1}{2}}$-bounded, we have for some $a, b \geq 0$,

$$\|Wu\| \leq a\|(-\Delta)^{\frac{1}{2}} u\| + b\|u\|, \tag{13.11}$$

for all $u \in D((-\Delta)^{1/2})$. Next, we show that $(-\Delta)^{1/2}$ is H_0-bounded with relative bound zero. For any $u \in D(H_0) \subset D(\Delta)$,

$$\|(-\Delta)^{\frac{1}{2}} u\|^2 = \langle -\Delta u, u \rangle \leq \langle H_0 u, u \rangle \leq \|H_0 u\| \, \|u\|, \tag{13.12}$$

using the positivity of U. Now for any $a,\ b \in \mathbb{R}$, and for any $\epsilon > 0$,

$$ab \leq \epsilon^2 a^2 + (4\epsilon^2)^{-1} b^2.$$

Applying this to the right side of (13.12), we get

$$\|(-\Delta)^{\frac{1}{2}} u\| \leq \epsilon \|H_0 u\| + (2\epsilon)^{-1} \|u\|. \tag{13.13}$$

Substituting (13.13) into the right side of (13.11), we obtain a relative bound for W, for all $u \in D(H_0)$,

$$\|Wu\| \leq a\epsilon \|H_0 u\| + \left(\epsilon b + \frac{1}{4}\epsilon\right) \|u\|.$$

Since $\epsilon > 0$ is arbitrary, this proves that W is H_0-bounded with relative bound zero. By the Kato–Rellich theorem, $-\Delta + U + W = H$ is self-adjoint on $D(H_0)$. □

Problem 3.6. Under the hypotheses of Corollary 13.8, prove that $C_0^\infty(\mathbb{R}^n)$ is a core for H.

Finally, we prove that $-\Delta$ determines the essential spectrum of $H = -\Delta + V$ for a certain class of relatively bounded potentials V. This result should be compared with Theorem 10.7.

Theorem 13.9. *Assume that V is real and Δ-bounded with relative Δ-bound < 1, and that $V(x) \to 0$ as $\|x\| \to \infty$. Then $H = -\Delta + V$ is self-adjoint on $D(H) = H^2(\mathbb{R}^n)$ and*

$$\sigma_{\text{ess}}(H) = \sigma(-\Delta) = [0, \infty).$$

Proof.

(1) The Schrödinger operator $H = -\Delta + V$ is self-adjoint on $H^2(\mathbb{R}^n)$, by Theorem 13.5, since the relative Δ-bound of V is less than one. We first verify the hypotheses of Theorem 10.6 in order to be able to conclude that $\sigma_{\text{ess}}(H) = Z(H)$. By the Δ-boundedness of V, the local compactness of $-\Delta$ (see Example 10.2), and the second resolvent identity, Proposition 1.9, it follows that H is locally compact. We must verify (10.7). Choose $\phi \in C_0^\infty(\mathbb{R}^n)$ as in Theorem 10.7, and set $\phi_n(x) = \phi(x/n)$. We compute the commutator,

$$[H, \phi_n](H - i)^{-1} = (\, p^2\phi_n + 2(p\phi_n) \cdot p \,)(H - i)^{-1}. \tag{13.14}$$

Since $\phi_n(x) = \phi(x/n)$, we have the estimates $\|p_i \phi_n\| = \mathcal{O}(n^{-1})$ and $\|p^2 \phi_n\| = \mathcal{O}(n^{-2})$. Moreover, by Problem 13.3 and the relative boundedness of V, one easily shows that $p_k(H - i)^{-1}$ is bounded. Hence, there is a constant $c > 0$ such that

$$\|[H, \phi_n](H - i)^{-1}\| \leq cn^{-1},$$

which verifies (10.7), and hence $Z(H) = \sigma_{\text{ess}}(H)$.

(2) We have already shown in Theorem 7.6 that

$$\sigma_{\text{ess}}(-\Delta) = \sigma(-\Delta) = [0, \infty).$$

We first show that $\sigma_{\text{ess}}(H) \subset [0, \infty)$. Let $\lambda \in \sigma_{\text{ess}}(H)$ and $\{u_n\}$ be a Zhislin sequence for H and λ. Since supp $u_k \subset \mathbb{R}^n \setminus B_k(0)$,

$$\lim_{n \to \infty} \|Vu_n\| = 0.$$

Since $D(H) = D(-\Delta)$, by the relative boundedness of V, we have

$$\|(\lambda + \Delta)u_n\| \leq \|(\lambda - H)u_n\| + \|Vu_n\|.$$

Since this upper bound vanishes as $n \to \infty$, this sequence $\{u_n\}$ is also a Weyl sequence for $-\Delta$ and λ. Hence, $\lambda \in [0, \infty)$. To prove the converse, note that $-\Delta$ is locally compact by Example 10.2 (1) and that $Z(-\Delta) = \sigma_{\text{ess}}(-\Delta)$, by the same calculation as in (13.14). Hence, if $\lambda \in \sigma_{\text{ess}}(-\Delta) = [0, \infty)$, there exists a Zhislin sequence $\{w_n\}$ for $-\Delta$ and λ. Again, the vanishing of V at infinity implies that $\|Vw_n\| \to 0$ as $n \to \infty$, so

$$\|(\lambda - H)w_n\| \leq \|(\lambda + \Delta)w_n\| + \|Vw_n\| \to 0.$$

This proves that $\{w_n\}$ is a Zhislin sequence for H and λ. Hence, $[0, \infty) \subset \sigma_{\text{ess}}(H)$. □

Problem 13.7. Let the potential V satisfy the hypotheses of Theorem 13.9. First, prove that H is locally compact, as outlined in the proof of Theorem 13.9. Second, prove that the operator $p_i(H - i)^{-1}$ is bounded.

14

Relatively Compact Operators and the Weyl Theorem

The notion of a *relatively bounded* operator is a fruitful one for establishing the self-adjointness of operators that are perturbations of self-adjoint operators. We also want to know about the effect of the perturbation on the spectrum of the original operator. This is the topic of *perturbation theory*. As with our discussion of spectrum, we will consider the effects of perturbations on both the essential and the discrete spectra. We have already seen two extreme examples of how the spectrum can change under perturbations that preserve self-adjointness, Theorem 10.7 and Theorem 13.9. In Theorem 10.7, we saw that the effect of a perturbation by a positive, increasing potential, although it preserves the self-adjointness, may drastically alter the spectrum of the unperturbed operator, the Laplacian. Such a perturbation is not relatively bounded.

The main domain of perturbation theory concerns perturbations that, unlike the example above, preserve some character of the spectrum of the unperturbed operator. A prototype of this can be found in Theorem 13.9, which proves that perturbations by real potentials V in the Kato–Rellich class *which vanish at infinity* preserve the essential spectrum of the Laplacian. In this section, we study the effect on the *essential spectrum* of perturbations by a class of relatively bounded perturbations, which include the real Kato–Rellich potentials. We establish the stability of the essential spectrum of a self-adjoint operator under perturbations called *relatively compact perturbations*. Under these perturbations, the essential spectrum of the unperturbed and perturbed operators is the same. This is our first example of what we will call *spectral stability*. We will discuss spectral stability for *discrete eigenvalues* in Chapters 15 and 19. In general, the term spectral stability means that, under a perturbation, the local aspects of the spectrum of the perturbed operator resemble that of the unperturbed operator. We give two applications of this stability result to Schrödinger operators. For the first, we show that a perturbation of the Laplacian by a real, relatively compact potential preserves the essential spectrum. The second is Persson's formula for the bottom of the essential spectrum of a Schrödinger operator. This formula, part of the geometric spectral theory discussed in Chapter 10, is derived using Zhislin sequences and Weyl's theorem.

14.1 Relatively Compact Operators

In this section, we consider self-adjoint operators on a general Hilbert space $\mathcal{H}$.

Definition 14.1. *Let A be a closed operator with $\rho(A) \neq \phi$. An operator B is called relatively A-compact if*

(i) *$D(B) \supset D(A)$;*

(ii) *$BR_A(z)$ is compact for some (and hence for all) $z \in \rho(A)$.*

If an operator B is relatively A-compact, then part (i) of Definition 14.1 implies that it is relatively A-bounded. The notion of relative compactness is related to local compactness discussed in Chapter 10. Clearly, an operator A on $L^2(\mathbb{R}^n)$ is locally compact if and only if the characteristic function χ_B, for any bounded $B \subset \mathbb{R}^n$, is relatively A-compact. Relative compactness is quite a bit stronger that relative boundedness, as the following theorem shows.

Theorem 14.2. *Suppose A is a self-adjoint operator and B is relatively A-compact. Then B is relatively A-bounded with relative bound zero.*

To prove the theorem, we need the following lemma.

Lemma 14.3. *Let B and A be as in Theorem 14.2. Then* $\lim_{|\lambda|\to\infty} \|B(A-i\lambda)^{-1}\| = 0$.

Proof.

(1) We first show that $(A-i)(A-i\lambda)^{-1} \xrightarrow{s} 0$ as $|\lambda| \to \infty, \ \lambda \in \mathbb{R}$. Note that since A is self-adjoint,

$$\begin{aligned} \|(A-i\lambda)v\|^2 &= \|Av\|^2 + \lambda^2\|v\|^2 \\ &\geq \|Av\|^2 + \|v\|^2 \\ &= \|(A-i)v\|^2. \end{aligned}$$

Letting $v \to (A-i\lambda)^{-1}v$ in this inequality, we have

$$\|v\|^2 \geq \|(A-i)(A-i\lambda)^{-1}v\|^2,$$

so $(A-i)(A-i\lambda)^{-1}$ is uniformly bounded in λ. Next, for $u \in D(A)$, we obtain from Corollary 5.7,

$$\begin{aligned} \|(A-i)(A-i\lambda)^{-1}u\| &= \|(A-i\lambda)^{-1}(A-i)u\| \\ &\leq |\lambda|^{-1}\|(A-i)u\| \to 0, \end{aligned}$$

as $|\lambda| \to \infty$. Since $(A-i)(A-i\lambda)^{-1}$ converges strongly to zero as $|\lambda| \to \infty$ on a dense set and is uniformly bounded, it converges strongly to zero.

(2) To prove the lemma, we first write

$$B(A-i\lambda)^{-1} = B(A-i)^{-1}(A-i)(A-i\lambda)^{-1}.$$

Since $B(A-i)^{-1}$ is compact and $(A-i)(A-i\lambda)^{-1} \xrightarrow{s} 0$, the operator $B(A-i\lambda)^{-1}$ converges uniformly to zero by Theorem 9.19. □

Proof of Theorem 14.2. For any $\epsilon > 0$, choose λ such that

$$\|B(A-i\lambda)^{-1}\| < \epsilon.$$

Then for any $u \in D(A)$,

$$\begin{aligned} \|Bu\| &= \|B(A-i\lambda)^{-1}(A-i\lambda)u\| \\ &\leq \epsilon\|(A-i\lambda)u\| \\ &\leq \epsilon\|Au\| + \epsilon|\lambda|\|u\|. \end{aligned}$$

As $\epsilon > 0$ is arbitrary, B has A-bound zero. □

If we combine Theorem 14.2 with the Kato–Rellich theorem, we obtain the following corollary about relatively compact perturbations of self-adjoint operators.

Corollary 14.4. *Suppose A is a self-adjoint operator and B is symmetric and relatively A-compact. Then $A + B$ is self-adjoint on $D(A)$.*

14.2 Weyl's Theorem: Stability of the Essential Spectrum

We now turn to the question of the stability of the essential spectrum under relatively compact perturbations. First we need a lemma about relatively compact perturbations.

Lemma 14.5. *Let A and B be self-adjoint operators, and let $V \equiv A - B$. If V is A-compact, then V is B-compact.*

Proof. We use the following facts:

(a) $V(i-A)^{-1}$ is compact;

(b) $-1 \notin \sigma(V(i-A)^{-1})$;

(c) $(i-B)^{-1} = (i-A)^{-1}[1 + V(i-A)^{-1}]^{-1}$.

Fact (a) is part of the hypothesis. To prove (b), note that by the Fredholm alternative, Theorem 9.12, the point $-1 \in \sigma(V(i-A)^{-1})$ if and only if there is a $\phi \in \mathcal{H}$ such that

$$V(i-A)^{-1}\phi = -\phi.$$

This is true if and only if

$$B\psi = -i\psi \quad \text{for } \psi \equiv (i-A)^{-1}\phi.$$

But $B = B^*$, so by the basic criteria for self-adjointness, Theorem 8.3, the function $\psi = 0$. This implies $\phi = 0$, and so $-1 \notin \sigma(V(i-A)^{-1})$. Finally, to prove (c), the second resolvent formula gives

$$R_B(i) - R_A(i) = R_B(i)[-VR_A(i)],$$

or

$$R_B(i)[1 + VR_A(i)] = R_A(i).$$

Since $-1 \notin \sigma(VR_A(i))$ by (b), the operator $1 + R_A(i)$ is invertible and the result follows. Now the lemma follows from (c) since

$$V(i-B)^{-1} = V(i-A)^{-1}[1 + VR_A(i)]^{-1}$$

and the right side is compact. □

We now come to the major theorem concerning relatively compact operators. This theorem, due to Weyl, states that the essential spectrum of an operator is invariant under relatively compact perturbations. We prove the self-adjoint version of the theorem, but there are extensions of it to certain closed, non–self-adjoint operators (see Theorem 18.8).

Theorem 14.6 (Weyl's theorem). *Let A and B be self-adjoint operators, and let $A - B$ be A-compact. Then,*

$$\sigma_{\text{ess}}(A) = \sigma_{\text{ess}}(B).$$

Proof. Let $\lambda \in \sigma_{\text{ess}}(A)$. Then, by Weyl's criterion, Theorem 7.2, there exists a Weyl sequence $\{u_n\}$ for A and λ, that is, $u_n \in D(A)$, $\|u_n\| = 1$, $u_n \overset{w}{\to} 0$, and $(A-\lambda)u_n \overset{s}{\to} 0$. Also note that $(i-A)u_n \overset{w}{\to} 0$, since

$$(i-A)u_n = (\lambda - A)u_n + (i-\lambda)u_n \overset{w}{\to} 0, \tag{14.1}$$

and the first term converges strongly to zero and $u_n \overset{w}{\to} 0$. We claim that $(\lambda - B)u_n \overset{s}{\to} 0$, for we can write

$$(\lambda - B)u_n = (\lambda - A)u_n + (A-B)(i-A)^{-1}(i-A)u_n. \tag{14.2}$$

The first term on the right satisfies $(\lambda - A)u_n \overset{s}{\to} 0$. The operator $(A-B)(i-A)^{-1}$ is compact, and so by (14.1), the second term on the right in (14.2) converges strongly to zero. Hence, $\{u_n\}$ is a Weyl sequence for λ and B. This proves that $\sigma_{\rm ess}(A) \subset \sigma_{\rm ess}(B)$. Now by Lemma 14.5, $(A - B)$ is B-compact. Thus, we can switch the roles of A and B in the above argument and conclude that $\sigma_{\rm ess}(B) \subset \sigma_{\rm ess}(A)$. This completes the proof that $\sigma_{\rm ess}(A) = \sigma_{\rm ess}(B)$. □

14.3 Applications to the Spectral Theory of Schrödinger Operators

We now apply the ideas of relative compactness and the Weyl theorem to study Schrödinger operators. We take the Hilbert space $\mathcal{H} = L^2(\mathbb{R}^n)$. In Chapter 13, we introduced the class of Kato–Rellich potentials. A function V is in the Kato–Rellich class if $V \in L^p(\mathbb{R}^n) + L^\infty(\mathbb{R}^n)$, with $p = 2$ if $n \leq 3$, $p > 2$ if $n = 4$, and $p \geq n/2$ if $n \geq 5$. This class is quite natural with regard to the Kato–Rellich theorem. If the real potential V is in the Kato–Rellich class, then the Schrödinger operator $H = -\Delta + V$ is self-adjoint on $D(H) = D(\Delta) = H^2(\mathbb{R}^n)$.

It is easy to see that not all Kato–Rellich class potentials are relatively compact. The potential function $V = 1$ is not relatively compact, as follows from Problem 10.2. Note that such a perturbation of the Laplacian shifts the essential spectrum. Indeed, from Theorem 13.9, one sees that an additional condition, like the vanishing of the potential at infinity, is sufficient for the invariance of the essential spectrum. We refine the Kato–Rellich class to capture this characteristic.

Definition 14.7. *A potential function $V(x)$ is called a Kato potential if V is real and $V \in L^2(\mathbb{R}^n) + L^\infty(\mathbb{R}^n)_\epsilon$, where the ϵ indicates that for any $\epsilon > 0$, we can decompose $V = V_1 + V_2$ with $V_1 \in L^2(\mathbb{R}^n)$ and $V_2 \in L^\infty(\mathbb{R}^n)$, with $\|V_2\|_\infty < \epsilon$.*

Example 14.8. The Coulomb potential: $V(x) = c\|x\|^{-1}$, c constant. For any $\epsilon > 0$, let $\chi_\epsilon(\|x\|)$ be the function that is 1 on $\{x \mid \|x\| \leq (c\epsilon)^{-1}\}$ and vanishes outside $\{x \mid \|x\| < 2(c\epsilon)^{-1}\}$. Then, we decompose the potential as $V(x) = c\chi_\epsilon(\|x\|)\|x\|^{-1} + c(1 - \chi_\epsilon(\|x\|))\|x\|^{-1} = V_1(x) + V_2(x)$. We have $V_1 \in L^2(\mathbb{R}^n)$ and

$$\sup_{x \in \mathbb{R}^n} |c(1 - \chi_\epsilon(\|x\|))\|x\|^{-1}| \leq \epsilon.$$

Problem 14.1. Let V be a real, continuous function with $\lim_{\|x\| \to \infty} V(x) = 0$. Show that V is in the Kato class.

Theorem 14.9. *If V is a real Kato potential, then V is relatively Δ-compact.*

Proof. We give the proof for the three-dimensional case. We know from Theorem 13.7 that V is Δ-bounded, that is, $D(V) \supset D(\Delta)$. We must check that $V(-\Delta + i)^{-1}$ is compact. For any $\epsilon > 0$, we decompose V as $V = V_1 + V_2$, where $V_1 \in L^2(\mathbb{R}^3)$ and $V_2 \in L^\infty(\mathbb{R}^3)_\epsilon$. Then, for any $f \in L^2(\mathbb{R}^3)$,

$$\|V_2(-\Delta + i)^{-1} f\| \leq \|V_2\|_\infty \|(-\Delta + 1)^{-1} f\| \leq \epsilon \|f\|, \tag{14.3}$$

and it follows that $\| V_2(-\Delta+i)^{-1}\| \leq \epsilon$. Next, with the help of (4.18), we observe that $V_1(-\Delta - z)^{-1}$, $\operatorname{Im} z \neq 0$, is an integral operator with kernel given by

$$K_z(x, y) \equiv V_1(x)[4\pi^2\|x - y\|]^{-1}e^{-z^{\frac{1}{2}}\|x-y\|}.$$

Indeed, for any $f \in L^2$,

$$\begin{aligned}(V_1(-\Delta - z)^{-1}f)(x) &= V_1(x)\int_{\mathbb{R}^3}[4\pi^2\|x - y\|]^{-1}e^{-z^{\frac{1}{2}}\|x-y\|}f(y)dy \\ &= \int_{\mathbb{R}^3} K_z(x, y)f(y)dy. \end{aligned} \tag{14.4}$$

The square root is defined as the principal branch: $\operatorname{Re} z^{1/2} > 0$. This kernel is a Hilbert–Schmidt kernel for

$$\int_{\mathbb{R}^3}\int_{\mathbb{R}^3}|K_z(x, y)|^2 dxdy = \left[\int_{\mathbb{R}^3}|V_1(x)|^2 dx\right]\left[\int_{\mathbb{R}^3}[4\pi^2|u|\,]^{-2}e^{-2|u|z_1}du\right] < \infty, \tag{14.5}$$

where $z_1 = \operatorname{Re} z^{1/2}$. Here, we used a change of variable $u = x - y$. Both integrals in (14.5) converge. Hence by Theorem 9.14, the operator $V_1(-\Delta - z)^{-1}$ is compact. Now, by (14.3) we have

$$\|V(-\Delta + i)^{-1} - V_1(-\Delta + i)^{-1}\| < \epsilon,$$

so $V(-\Delta + 1)^{-1}$ is approximated by compact operators and, by Theorem 9.8, is itself compact. □

By this theorem and Weyl's theorem, we immediately obtain the next result.

Corollary 14.10. *If V is a real Kato potential, then*

$$\sigma_{\text{ess}}(-\Delta + V) = \sigma_{\text{ess}}(-\Delta) = [0, \infty).$$

Note that the last equality in the corollary was obtained in Chapter 7.

Finally, we offer an alternate proof, due to M. Schechter, of Theorem 10.7.

Alternate Proof of Theorem 10.7.

For any $E > 0$, we define $V_E(x) \equiv \max(V(x) - E, 0)$ and $W_E(x) \equiv \min(V(x) - E, 0)$, so that $V(x) - E = W_E(x) + V_E(x)$. We have sketched V, W_E, and V_E in Figure 14.1. We then have

$$H - E = -\Delta + V_E + W_E.$$

By Theorem 10.3, the operator $-\Delta + V_E$ is locally compact. Since $\operatorname{supp}(W_E)$ is compact, it follows that W_E is $(-\Delta + V_E)$-compact. Hence, by Weyl's theorem,

$$\sigma_{\text{ess}}(H - E) = \sigma_{\text{ess}}(-\Delta + V_E). \tag{14.6}$$

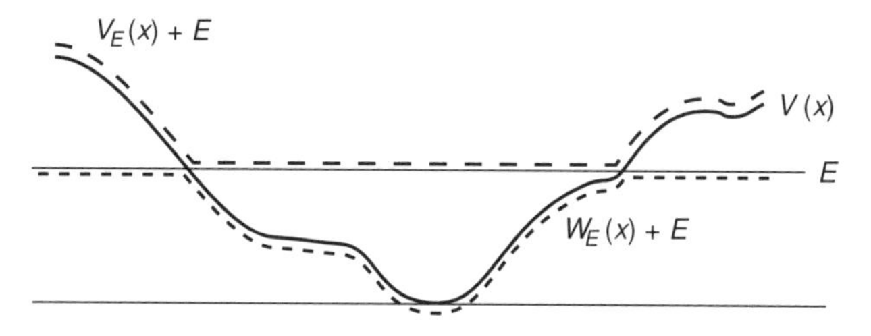

FIGURE 14.1. The potentials V_E and W_E for the potential V.

On the other hand, as $V_E \geq 0$, the operator $-\Delta + V_E \geq 0$, and so $\sigma(-\Delta + V_E) \subset [0, \infty)$. Thus

$$\sigma_{\text{ess}}(H - E) \subset [0, \infty), \quad \text{so } \sigma_{\text{ess}}(H) \subset [E, \infty).$$

Since this is true for any $E > 0$, $\sigma_{\text{ess}}(H) = \phi$, and $\sigma(H) = \sigma_d(H)$ is purely discrete. □

14.4 Persson's Theorem: The Bottom of the Essential Spectrum

Persson [Per] discovered a beautiful geometric description for the bottom of the essential spectrum of a semibounded Schrödinger operator $H = -\Delta + V$. It is based on the same principle as the modification of the Weyl criteria in the presence of local compactness: The behavior of the potential V at infinity determines $\sigma_{\text{ess}}(H)$. We state and prove a simplified version of Persson's theorem (see also [Ag1]). This theorem can be applied to other elliptic partial differential operators on unbounded domains in $\mathbb{R}^n$ or on noncompact manifolds (see, for example, [FrHi]).

Theorem 14.11. *Let V be a real-valued potential in the Kato–Rellich class, and let $H = -\Delta + V$ be the corresponding self-adjoint, semibounded Schrödinger operator with domain $H^2(\mathbb{R}^n)$. Then, the bottom of the essential spectrum is given by*

$$\inf \sigma_{\text{ess}}(H) = \sup_{\mathcal{K} \subset \mathbb{R}^n} \left[\inf_{\phi \neq 0} \left\{ \langle \phi, H\phi \rangle \, \|\phi\|^{-2} \;\middle|\; \phi \in C_0^\infty(\mathbb{R}^n \backslash \mathcal{K}) \right\} \right], \tag{14.7}$$

where the supremum is over all compact subsets $\mathcal{K} \subset \mathbb{R}^n$.

As a preliminary, we need an expression for the bottom of the spectrum. This is easily obtained from a modification of Proposition 12.1, equation (12.1), which expresses the lowest eigenvalue of a self-adjoint operator as the infimum of a quadratic form.

Proposition 14.12. *Let H be a Schrödinger operator as in Theorem 14.11. We* *have the formula,*

$$\inf \sigma(H) = \inf_{\phi \neq 0} \left\{ \langle \phi, H\phi \rangle \|\phi\|^{-2} \mid \phi \in C_0^\infty(\mathbb{R}^n) \right\}. \tag{14.8}$$

Proof. If $\inf \sigma(H)$ is in the discrete spectrum, the result follows from Proposition 12.1. Otherwise, we assume that $\lambda_0 \equiv \inf \sigma(H) \in \sigma_{\text{ess}}(H)$. Let λ denote the right side of (14.8), where we take the infimum over $\phi \in H^2(\mathbb{R}^n)$. We leave it as an exercise to pass from this result to the proposition, which is possible since $C_0^\infty(\mathbb{R}^n)$ is a core for H. Let $\{u_n\}$ be a Weyl sequence for H and λ_0. For any $\epsilon > 0$, there exists N_ϵ such that $n > N_\epsilon$ implies that

$$\langle u_n, Hu_n \rangle \geq \lambda_0 - \epsilon, \tag{14.9}$$

so that $\lambda \geq \lambda_0$. As for the reverse inequality, suppose that $\lambda < \lambda_0$. We can find a sequence of vectors v_n, with $\|v_n\| = 1$, such that

$$\lim_{n \to \infty} \langle v_n, (H - \lambda)v_n \rangle = 0. \tag{14.10}$$

Since we have

$$\|(H - \lambda)v_n\| = \inf_{\psi \neq 0} \left[| \langle \psi, (H - \lambda)v_n \rangle | \, \|\psi\|^{-1} \right], \tag{14.11}$$

we see that the right side of (14.11) can be made arbitrarily small. By the Weyl criterion for the essential spectrum, this implies that either $\lambda \in \sigma_{\text{ess}}(H)$ or λ is an eigenvalue. In either case, $\lambda \in \sigma(H)$, which contradicts the assumption that $\lambda < \lambda_0$. □

Proof of Theorem 14.11.

(1) By the hypotheses on V, $C_0^\infty(\mathbb{R}^n)$ is a core for H. Since H is semibounded, there exists a finite, nonnegative, constant M such that

$$\langle \phi, H\phi \rangle \geq -M\|\phi\|^2, \tag{14.12}$$

for all $\phi \in C_0^\infty(\mathbb{R}^n)$. By shifting from H to $H + M + 1$, we can assume that for all $\phi \in C_0^\infty(\mathbb{R}^n)$,

$$\langle \phi, H\phi \rangle \geq \|\phi\|^2. \tag{14.13}$$

For any compact subset $\mathcal{K} \subset \mathbb{R}^n$, we define

$$\Sigma(H, \mathcal{K}) \equiv \inf \left\{ \langle \phi, H\phi \rangle \|\phi\|^{-2} \mid \phi \in C_0^\infty(\mathbb{R}^n \backslash \mathcal{K}) \right\}, \tag{14.14}$$

and let

$$\Sigma(H) \equiv \sup_{\mathcal{K} \subset \mathbb{R}^n \text{ compact}} \Sigma(H, \mathcal{K}). \tag{14.15}$$

Let $\chi_{\mathcal{K}}$ be the characteristic function for $\mathcal{K}$. We leave it as Problem 14.2 to prove that the multiplication operator $\chi_{\mathcal{K}}$ is relatively H-compact. Weyl's theorem, Theorem 14.6, implies that for any compact subset $\mathcal{K} \subset \mathbb{R}^n$,

$$\sigma_{\text{ess}}(H) = \sigma_{\text{ess}}(H + \chi_{\mathcal{K}}). \tag{14.16}$$

It now follows from Proposition 14.12 and (14.13) that

$$\begin{aligned} \inf \sigma(H + \chi_{\mathcal{K}}) &= \inf_{\phi \in C_0^\infty(\mathbb{R}^n)} \langle \phi, (H + \chi_{\mathcal{K}})\phi \rangle \, \|\phi\|^{-2} \\ &\geq \inf_{\phi \in C_0^\infty(\mathbb{R}^n \backslash \mathcal{K})} \langle \phi, (H + \chi_{\mathcal{K}})\phi \rangle \, \|\phi\|^{-2} \\ &\geq \Sigma(H, \mathcal{K}). \end{aligned} \tag{14.17}$$

Results (14.16) and (14.17) together imply that

$$\inf \sigma_{\text{ess}}(H) \geq \Sigma(H), \tag{14.18}$$

which proves one part of the theorem.

(2) Conversely, let $\Sigma_0 \equiv \inf \sigma_{\text{ess}}(H)$. Since the essential spectrum of H is closed, we have that $\Sigma_0 \in \sigma_{\text{ess}}(H)$. Then by Theorem 10.6, there exists a Zhislin sequence $\{u_n\}$ for H and Σ_0. For any compact $\mathcal{K} \subset \mathbb{R}^n$, we write

$$\Sigma(H, \mathcal{K}) = \inf_{\phi \in C_0^\infty(\mathbb{R}^n \backslash \mathcal{K})} \left\{ \frac{\langle \phi, (H - \Sigma_0)\phi \rangle}{\|\phi\|^2} + \Sigma_0 \right\}. \tag{14.19}$$

We claim that for any $\epsilon > 0$, there exists a $\psi_\epsilon \in C_0^\infty(\mathbb{R}^n \backslash \mathcal{K})$ such that $\|\psi_\epsilon\| = 1$ and

$$|\langle \psi_\epsilon, (H - \Sigma_0)\psi_\epsilon \rangle| \leq \epsilon. \tag{14.20}$$

This proves that for any $\epsilon > 0$,

$$\Sigma(H, \mathcal{K}) \geq \Sigma_0 - \epsilon, \tag{14.21}$$

which, together with (14.18), proves the theorem. To prove the claim, it follows from the definition of the Zhislin sequence that there exists an index n_1 such that

$$\|(H - \Sigma_0)u_{n_1}\| \leq \epsilon, \tag{14.22}$$

and such that $\operatorname{supp} u_{n_1} \cap \mathcal{K} = \emptyset$. Because $C_0^\infty(\mathbb{R}^n)$ is a core for H, we can assume that u_{n_1} is smooth. Consequently, we have that $u_{n_1} \in C_0^\infty(\mathbb{R}^n \backslash \mathcal{K})$, and (14.20) holds. □

Problem 14.2. Let H be a Schrödinger operator as in Theorem 14.11. For any compact subset $\mathcal{K} \subset \mathbb{R}^n$, let $\chi_{\mathcal{K}}$ be the characteristic function for this subset. Prove that $\chi_{\mathcal{K}}$ is relatively H-compact. (*Hint*: Use the relative boundedness of V, the second resolvent formula, and Examples 10.2.)

15

Perturbation Theory: Relatively Bounded Perturbations

15.1 Introduction and Motivation

We have so far discussed two aspects of perturbation theory: perturbations of self-adjoint operators which preserve self-adjointness, and perturbations of self-adjoint operators which preserve the essential spectrum. This latter subject is part of what we call *spectral stability*. In this chapter, we begin a discussion of spectral stability for the discrete spectrum. The typical situation can be described as follows. Most quantum mechanical systems are described by a Hamiltonian of the form $H = -\Delta + V$. Different systems are distinguished by the potential V. For example, the Coulomb potential, $V(x) = -e^2|x|^{-1}$, where e is the unit electric charge, describes a hydrogen atom. In most situations, any given V is an idealization; there are always other effects that contribute small "corrections" to V. Hence, we may begin with a model Hamiltonian H_0, about which we have a lot of information concerning the eigenvalues, and consider the effects on H_0 caused by adding a small correction to the potential, say V_1. We are then led to consider a family of operators $H(\kappa) \equiv H_0 + \kappa V_1$, as κ varies from $\kappa = 0$ (where we have a lot of information) to $\kappa = \kappa_0$ (where we desire information). Exactly how the spectrum of $H(\kappa)$ varies with κ is one of the topics of *perturbation theory*. We will give a criteria for the "smallness" of a perturbation V_1 relative to H_0 and, in certain cases (for example, for κ sufficiently small), give very exact information about $\sigma(H(\kappa))$. We know from Weyl's theorem, Theorem 14.6, that relatively compact perturbations preserve the essential spectrum. Hence, our focus in this chapter will be on how the *discrete eigenvalues* change under a perturbation. We will return to more general questions of spectral stability in Chapter 19. Another aspect of perturbation theory, which we will not consider in this book, concerns the stability of spectral type, in particular, the absolutely continuous spectrum. This is the topic of scattering theory, and we refer the reader to Kato [K] and to Volume 3 of Reed and Simon [RS3].

15.2 Analytic Perturbation Theory for the Discrete Spectrum

Let H_0 be a simple operator in the sense that $\sigma_d(H_0)$ is computable, and consider a perturbation V. We define a family of perturbed operators $H(\kappa) = H_0 + \kappa V$, where $\kappa \in [0, \kappa_0]$. Suppose $\lambda \in \sigma_d(H_0)$. We want to know if λ survives the perturbation (i.e., if for $\kappa \neq 0$ and small, $H(\kappa)$ has an eigenvalue $\lambda(\kappa)$ near λ). Although in physical situations κ is real, we allow κ to become complex. Moreover, we want to know if $\lambda(\kappa)$ depends continuously on κ. (*Note*: As κ need not be real, the operator $H(\kappa)$ may not be self-adjoint.)

First we give a general formulation of these ideas. Consider a family of closed operators T_κ, depending on a parameter $\kappa \in B_\epsilon(0)$ for some $\epsilon > 0$, with a common domain D in a Hilbert space $\mathcal{H}$. We assume that each T_κ has a nonempty resolvent set. We consider T_κ, $\kappa \neq 0$, as an operator obtained by perturbing T_0 in the sense that

$$T_\kappa = T_0 + (T_\kappa - T_0) = T_0 + V_{\text{eff}}(\kappa), \text{ where } V_{\text{eff}}(\kappa) \equiv T_\kappa - T_0.$$

For the example in the last section, the perturbation $V_{\text{eff}}(\kappa)$ depends linearly on the parameter κ. We will study this special case in the next section.

Because we are dealing with non–self-adjoint operators, in general, we comment on the term *multiplicity* (see also Section 6.2). As in the theory of matrices, we distinguish between algebraic and geometric multiplicity.

Definition 15.1. *Let T be a closed operator, and suppose that $\Gamma \subset \rho(T)$ is a simple closed curve about a discrete eigenvalue λ of T. Let P_λ be the Riesz projection for T and Γ. Then dim(Ran P_λ) is the algebraic multiplicity of λ. The geometric multiplicity of λ is the number of linearly independent eigenvectors of T for λ. This is equal to the dimension of ker($A - \lambda$).*

We note that the algebraic multiplicity is greater than or equal to the geometric multiplicity in general. The vectors in Ran P_λ (with dim Ran $P_\lambda < \infty$) are *generalized eigenvectors* of T in the sense that there exists an $n > 0$, the algebraic multiplicity, such that for all $\psi \in \text{Ran } P_\lambda$, $(T - \lambda)^n \psi = 0$.

Problem 15.1. Prove the characterization of Ran P_λ as generalized eigenvectors stated above.

In what follows, multiplicity will always mean *algebraic multiplicity*. Of course, when T is self-adjoint, algebraic multiplicity equals the geometric multiplicity. We now introduce the notion of *stability*.

Definition 15.2. *A discrete eigenvalue λ of T_0 is said to be stable with respect to the family T_κ if*

(i) $\exists r > 0$ *such that* $\Gamma_r \equiv \{z | \, |z - \lambda| = r\} \subset \rho(T_\kappa)$ *for all small* $|\kappa|$;

(ii) *we let P_κ be the Riesz projection for T_κ corresponding to the contour Γ_r, then $P_\kappa \to P_0$ as $\kappa \to 0$ in norm.*

Proposition 15.3. *Suppose λ is a stable eigenvalue of T_0. Then for all $|\kappa|$ sufficiently small, any operator T_κ has discrete eigenvalues $\lambda_i(\kappa)$ near λ of total multiplicity equal to the multiplicity of λ.*

Proof. By part (ii) of Definition 15.2, $P_\kappa \neq 0$, for κ small, so T_κ has spectrum inside the contour Γ_r. The discreteness of the spectrum follows from the fact that the multiplicity of λ is finite and from Lemma 15.4. Hence, since P_κ is a continuous, projection-valued function of κ, dim(Ran P_κ) is a finite constant and equal to dim(Ran P_0), for all small $|\kappa|$. □

Lemma 15.4. *Let P and Q be projections. If dim(Ran P) $\neq$ dim(Ran Q), then $\|P - Q\| \geq 1$.*

Problem 15.2. Let M and N be two subspaces of $\mathcal{H}$ with $\dim M > \dim N$. Then $\exists x \in M$ such that x is orthogonal to N. (*Hint*: By considering a subspace of M, if necessary, we can assume that $\dim M$ and $\dim N$ are finite. Take orthonormal bases $\{x_i\}_{i=1}^m$ and $\{y_j\}_{j=1}^n$, $m > n$, for M and N, respectively. Let $x = \sum_{i=1}^m a_i x_i$, and solve $\langle x, y_i \rangle = \sum_{j=1}^m a_j \langle x_j, y_i \rangle = 0$ for $i = 1, \ldots, m$. The matrix $A \equiv [\langle y_j, x_i \rangle]$ is an $n \times m$ matrix. Then $\exists x \in M$, $x \neq 0$, and $\langle x, y_i \rangle = 0$, $i = 1, \ldots, n$ if and only if $Aa = 0$ has a nontrivial solution. But ker $A \neq \{0\}$, and so we can solve the equations.)

Proof of Lemma 15.4. Let dim(Ran P) $<$ dim(Ran Q). Let $u \in \text{Ran } Q \cap [\text{Ran } P]^\perp$, $\|u\| = 1$. Such a u exists by Problem 15.2. Then

$$(P - Q)u = -Qu = -u,$$

which implies that $\|(P - Q)u\| = 1$, so $\|P - Q\| \geq 1$. □

15.3 Criteria for Eigenvalue Stability: A Simple Case

We begin by considering a special case, which we will generalize in the next section. Consider a self-adjoint operator H on a Hilbert space $\mathcal{H}$ and a family of operators H_κ defined by

$$H_\kappa \equiv H + \kappa W, \tag{15.1}$$

where we assume

Condition A. *The perturbation W is H-bounded.*
Note that we do not assume that W is symmetric. By employing the techniques used in the roof of the Kato–Rellich theorem, we have

(a) for any $\kappa \in \mathbb{C}$, the operator H_κ is closed and $D(H_\kappa) = D(H)$, independent of κ;

(b) for any $u \in D(H)$, $H_\kappa u$ is an entire analytic function of κ.

Problem 15.3. Prove properties (a) and (b) of the family H_κ. Suppose, in addition, that W is symmetric. Obtain an explicit bound on the size of $|\kappa|$ in (a) so that H_κ is self-adjoint for $\kappa \in \mathbb{R}$.

We have had occasion to discuss analytic families of operators in Chapter 1, Theorem 1.2. We now study them more carefully. We begin with a simple case.

Definition 15.5. *A family of bounded operators B_κ is called analytic at κ_0 if the map $\kappa \to B_\kappa$ is analytic as an operator-valued function on a small disk $B_\epsilon(\kappa_0)$. That is, for κ near κ_0, B_κ has a uniformly convergent power series expansion: $B_\kappa = \sum_{n=0}^\infty (\kappa - \kappa_0)^n b_n$, where, for each n, b_n is a bounded operator.*

Problem 15.4.

(1) Show that B_κ is an analytic family about $\kappa = \kappa_0$ if and only if $\kappa \to B_\kappa u$ is analytic for any $u \in \mathcal{H}$. This is the notion of *strong analyticity*. (*Hint*: Use the principle of uniform boundedness (see [RS1]) to show that $\| B_\kappa \| < C_0$, for all $\kappa \in B_\epsilon(\kappa_0)$. Then use the Cauchy formula for the nth derivative of an analytic function.)

(2) Reformulate the notion of analyticity in terms of Morera's theorem [R].

(3) Show that B_κ is an analytic family about $\kappa = \kappa_0$ if and only if $\kappa \to \langle B_\kappa u, v\rangle$ is analytic for any $u, v \in \mathcal{H}$. This is the notion of *weak analyticity*.

We first study families of operators H_κ as in (15.1) which satisfy condition A.

Lemma 15.6. *Assume condition A and then let H_κ be as in* (15.1).

(i) *Let G be any bounded, open set such that $\bar{G} \subset \rho(H)$. Then $G \subset \rho(H_\kappa)$ for $|\kappa|$ sufficiently small.*

(ii) *For any $\lambda \in G$ as in (i), the map $\kappa \to (\lambda - H_\kappa)^{-1}$ is analytic in κ for $|\kappa|$ sufficiently small.*

Proof.

(1) Let $\lambda \in G$. Then $\lambda \in \rho(H)$ and

$$\lambda - H_\kappa = \lambda - H - \kappa W = (1 - \kappa W(\lambda - H)^{-1})(\lambda - H).$$

By condition A, $W(\lambda - H)^{-1}$ is bounded. Hence for $|\kappa|$ small enough, $\|\kappa W(\lambda - H)^{-1}\| < 1$, so $1 - \kappa W(\lambda - H)^{-1}$ is invertible, and hence $(\lambda - H_\kappa)$ is invertible. Thus, $\lambda \in \rho(H_\kappa)$ for $|\kappa|$ sufficiently small, that is, $G \subset \rho(H_\kappa)$ provided $|\kappa| < (\max_{\lambda \in \bar{G}} \|W(\lambda - H)^{-1}\|)^{-1}$.

(2) To show that $\kappa \to (\lambda - H_\kappa)^{-1}$ is analytic, we use the second resolvent equation:

$$(\lambda - H_\kappa)^{-1} = (\lambda - H)^{-1} + \kappa(\lambda - H)^{-1} W(\lambda - H_\kappa)^{-1}.$$

Upon iterating this equation, we obtain

$$(\lambda - H_\kappa)^{-1} = (\lambda - H)^{-1} \left[\sum_{l=0}^{\infty} (\kappa W(\lambda - H)^{-1})^l \right]. \tag{15.2}$$

Since $\|(W(\lambda - H)^{-1})^l\| \leq \|W(\lambda - H)^{-1}\|^l \leq M^l$, for all $\lambda \in \bar{G}$, the series converges absolutely for $|\kappa| < M^{-1}$. Hence, $\kappa \to (\lambda - H_\kappa)^{-1}$ is an analytic family of bounded operators for $|\kappa| < M$. □

Theorem 15.7. *Let H_κ be as in (15.1), and assume condition A. Then all discrete eigenvalues of $H_0 = H$ are stable. Moreover, if $\lambda_i(\kappa)$ are the eigenvalues of H_κ near the eigenvalue λ of H_0, then the total multiplicity of the $\lambda_i(\kappa)$'s equals the multiplicity of λ.*

Proof. Let λ be a discrete eigenvalue of H_0. Due to part (i) of Lemma 15.6, if $r > 0$ is such that $\Gamma_r \equiv \{z \mid |z - \lambda| = r\} \subset \rho(H_0)$ (such an r exists as $\lambda \in \sigma_d(H_0)$), then $\Gamma_r \subset \bigcap_\kappa \rho(H_\kappa)$, for $|\kappa|$ sufficiently small. Let $P_\kappa \equiv (2\pi i)^{-1} \int_{\Gamma_r} R(\kappa, z) dz$, where $R(\kappa, z) \equiv (H_\kappa - z)^{-1}$. Then due to part (ii) of Lemma 15.6, $\kappa \to P_\kappa$ is analytic about $\kappa_0 = 0$ (see Problem 15.5). Hence λ is stable. The remaining part of the theorem follows from Proposition 15.3. □

Problem 15.5. Prove that $\kappa \to P_\kappa$ is analytic about $\kappa = 0$. (*Hint*: Use the expansion (15.2).)

15.4 Type-A Families of Operators and Eigenvalue Stability: General Results

In this section, we discuss general analytic families of type A. Not only can we establish a stability result, but we will be able to describe the manner in which the eigenvalues $\lambda_i(\kappa)$ depend upon κ. It follows from the proof of Theorem 15.7 that $\lambda_i(\kappa)$ depends continuously on κ as $\kappa \to 0$. In many cases, we will be able to show that this dependence is, in fact, analytic.

We now introduce a family of operators generalizing (15.2) and condition A.

Definition 15.8. *Let $\mathcal{R}$ be a nonempty, open subset of $\mathbb{C}$ and suppose that for each $\kappa \in \mathcal{R}$, T_κ is a closed operator with $\rho(T_\kappa) \neq \phi$. Then the family of closed operators T_κ, $\kappa \in \mathcal{R}$, is said to be type-A if*

(i) *$D(T_\kappa)$ is independent of κ (which we call D);*

(ii) *for each $u \in D$, $T_\kappa u$ is strongly analytic in κ on $\mathcal{R}$.*

Problem 15.6. Prove that a vector-valued function u_α on an open, connected subset $\mathcal{R}$ of $\mathbb{C}$ is weakly analytic (i.e., $\alpha \in \mathcal{R} \to \langle u_\alpha, v\rangle$ analytic) if and only if it is strongly analytic (i.e., $\alpha \in \mathcal{R} \to u_\alpha$ analytic in the usual topology on $\mathcal{H}$).

In analogy with Lemma 15.6, we have the following lemma, which describes the main technical aspects of type-A analytic families.

Lemma 15.9. *Let $T_\kappa, \kappa \in \mathcal{R}$, be a type-A analytic family of operators. Then for any $\kappa_0 \in \mathcal{R}$, we have the following two statements.*

(i) *Let G be any bounded, open set such that $\bar{G} \subset \rho(T_{\kappa_0})$. Then $G \subset \rho(T_\kappa)$ for $|\kappa - \kappa_0|$ sufficiently small.*

(ii) *For any $\lambda \in G$ as in (i), $\kappa \to (\lambda - T_\kappa)^{-1}$ is analytic in κ for $|\kappa - \kappa_0|$ sufficiently small.*

Proof.

(i) Let $\lambda \in G$ and write $T_\kappa = T_{\kappa_0} + V_\kappa$, where $V_\kappa \equiv T_\kappa - T_{\kappa_0}$. Then by the second resolvent formula,

$$(\lambda - T_\kappa)^{-1} = (1 - V_\kappa(\lambda - T_{\kappa_0})^{-1})(\lambda - T_{\kappa_0})^{-1}. \tag{15.3}$$

Now $D = (\lambda - T_{\kappa_0}^{-1})\mathcal{H}$, and so $V_\kappa(\lambda - T_{\kappa_0})^{-1}$ is a closed, everywhere defined operator. Hence, by the closed graph theorem, Theorem A3.23, it is bounded. Furthermore, it is analytic in κ for $|\kappa - \kappa_0|$ sufficiently small (see Problem 15.4). Consequently, because $V_\kappa(\lambda - T_{\kappa_0})^{-1} \to 0$ as $|\kappa - \kappa_0| \to 0$, it follows that $\| V_\kappa(\lambda - T_{\kappa_0})^{-1}\| < 1$, for all κ sufficiently close to κ_0. Then $1 - V_\kappa(\lambda - T_{\kappa_0})^{-1}$ is invertible, and so $\lambda \in \rho(T_\kappa)$ by (15.3).

(ii) For any $\lambda \in G$ as in (i) and for $|\kappa - \kappa_0|$ small, we can iterate equation (15.3) to obtain

$$(\lambda - T_\kappa)^{-1} = (\lambda - T_{\kappa_0}) \sum_{n=0}^{\infty} [V_\kappa(\lambda - T_{\kappa_0})^{-1}]^n. \tag{15.4}$$

Now, for any $\epsilon > 0 \; \exists \; \delta > 0$ such that $|\kappa - \kappa_0| < \delta$ implies that $\| V_\kappa(\lambda - T_{\kappa_0})^{-1}\| < \epsilon$. Choose $\epsilon < 1$. Then as each term in the series on the right side of (15.4) is analytic on $\{\kappa||\kappa - \kappa_0| < \delta\}$, bounded by ϵ^n, and as $\sum_n \epsilon^n = (1-\epsilon)^{-1}$, it follows by the Weierstrass M-test that the series is analytic in κ on $\{\kappa \mid |\kappa - \kappa_0| < \delta\}$. □

The stability of isolated eigenvalues with respect to perturbations given by type-A families is easy to prove. If, in addition, the eigenvalue is non-degenerate, then we can also prove that the corresponding eigenvalue of T_κ is analytic in κ.

Theorem 15.10. *Let T_κ be an analytic family of type A about $\kappa_0 = 0$. Let λ be a discrete, nondegenerate eigenvalue of T_0. Then there exists an analytic family $\lambda(\kappa)$ of discrete, nondegenerate eigenvalues of T_κ such that $\lambda(0) = \lambda$, for $|\kappa|$ sufficiently small.*

Proof.

(1) By part (i) of Lemma 15.9, $\exists r > 0$ such that $\Gamma_r \subset \bigcap_\kappa \rho(T_\kappa)$ for $|\kappa|$ sufficiently small. Moreover, P_κ (the Riesz projection for T_κ and Γ_r) is analytic in κ. Hence, by the stability Theorem 15.2, T_κ has a nondegenerate, discrete eigenvalue for $|\kappa|$ small.

Problem 15.7. Prove that P_κ is analytic in κ for small $|\kappa|$.

(2) To prove the analyticity of $\lambda(\kappa)$, let ψ, $\|\psi\| = 1$, be an eigenfunction of T_0 for λ so that $T_0\psi = \lambda\psi$. Then, $P_\kappa\psi \to \psi$ as $\kappa \to 0$, so $\|P_\kappa\psi\| \to 1$. Moreover, $T_\kappa P_\kappa\psi = \lambda(\kappa)P_\kappa\psi$, so $(z - T_\kappa)^{-1}P_\kappa\psi = (z - \lambda(\kappa))^{-1}P_\kappa\psi$. Taking the inner product of this equation with ψ, we get

$$(z - \lambda(\kappa))^{-1} = \langle\psi, P_\kappa\psi\rangle^{-1}\langle\psi, (z - T_\kappa)^{-1}P_\kappa\psi\rangle. \tag{15.5}$$

Now $\langle\psi, P_\kappa\psi\rangle$ is analytic for $|\kappa|$ small and $\langle\psi, P_0\psi\rangle = 1$, so $\langle\psi, P_\kappa\psi\rangle^{-1}$ is analytic for $|\kappa|$ small. Moreover, the second factor on the right side of (15.5) is clearly analytic in κ. Hence, $(z - \lambda(\kappa))^{-1}$ is analytic in κ, and as $z \in \rho(T_0)$, this implies that $\lambda(\kappa)$ is analytic in κ for $|\kappa|$ small enough. □

We now turn to the general theorem on stability for a discrete, degenerate eigenvalue. In general, a degenerate, discrete eigenvalue λ_0 splits into several eigenvalue branches $\lambda_l(\kappa)$ under a perturbation. These are the eigenvalues of T_κ which converge to λ_0 as $\kappa \to 0$. In the case for which T_κ, $\kappa \in \mathbb{R}$ is self-adjoint, each branch is an analytic function of κ. In the non–self-adjoint case, the eigenvalues are branches of a multivalued function, that is, a function defined on a Riemannian surface.

Theorem 15.11. *Let T_κ be an analytic family of type-A about $\kappa_0 = 0$. Let λ_0 be a discrete eigenvalue of T_0. Then there exist families $\lambda_l(\kappa)$, $l = 1, \ldots, r$, of discrete eigenvalues of T_κ such that*

(i) *$\lambda_l(0) = \lambda_0$ and the total multiplicity of the eigenvalues $\lambda_l(\kappa)$, $l = 1, \ldots, r$, is equal to the multiplicity of λ_0;*

(ii) *each family $\lambda_l(\kappa)$ is analytic in $\kappa^{1/p}$ for some integer p (i.e., $\lambda_l(\kappa)$ has a Puiseux expansion); if T_κ is self-adjoint for κ real, then the eigenvalues are analytic in κ.*

Proof.

(i) Suppose that $\mathcal{H}$ is finite-dimensional and dim $\mathcal{H} = n < \infty$. Then T_κ is an analytic type-A family of matrix-valued functions, and the eigenvalues $\lambda_l(\kappa)$ are solutions of

$$\det(\lambda - T_\kappa) = 0.$$

Let $F(\lambda, \kappa) \equiv \det(\lambda - T_\kappa)$. Then F is a polynomial of degree n in λ with leading coefficient 1:

$$F(\lambda, \kappa) = \lambda^n + a_n(\kappa)\lambda^{n-1} + \ldots + a_2(\kappa)\lambda + a_1(\kappa),$$

where $a_i(\kappa)$ are analytic in κ. We know that $F(\lambda_0, 0) = 0$. It follows from a classical result (see, for example, [Kn]) that for $|\kappa|$ small, $F(\lambda, \kappa)$ has roots $\lambda_l(\kappa)$ near λ_0, and these roots are given by the branches of one or more multivalued analytic functions having at worst algebraic singularities at $\kappa = 0$. Consequently, these roots have a Puiseux expansion in $\kappa^{1/p}$, $p \in \mathbb{Z}^+$, and satisfy $\lambda_l(0) = \lambda_0$. If T_κ is self-adjoint for κ real, then $\lambda_l(\kappa)$ is real and p must be 1.

(ii) Suppose T_0 has a discrete, degenerate eigenvalue λ_0. Then dim Ran P_0 is finite. Since P_κ is analytic in κ for $|\kappa|$ small, and $P_\kappa \to P_0$, we know that dim Ran P_κ = dim Ran P_0, for all $|\kappa|$ small. Let $M_\kappa \equiv$ Ran P_κ. Then $T_\kappa P_\kappa$ restricted to M_κ is a linear transformation on a finite-dimensional space. We want to apply the method of part (i), but the subspaces M_κ depend on κ. Hence we use a simple device to map all the subspaces M_κ to M_0. We prove in Lemma 15.12 in the appendix to this chapter that for small $|\kappa|$ there exists an invertible map $S_\kappa : M_\kappa \to M_0$ that, together with its inverse, is analytic in κ. Define $E_\kappa \equiv S_\kappa P_\kappa T_\kappa S_\kappa^{-1}$; then $E_\kappa : M_0 \to M_0$ is analytic in κ by construction. Moreover, as $\sigma(P_\kappa T_\kappa | M_\kappa) = \sigma(E_\kappa)$, the results of part (1) applied to E_κ establish the theorem for $P_\kappa T_\kappa$, and hence for T_κ. □

Let us note that the idea behind the proof of this theorem is simple. The Riesz projection P_κ allows us to consider the problem for the finite-rank operator $P_\kappa T_\kappa$. The only technical problem concerns the construction of an analytic, invertible map $S_\kappa : M_\kappa \to M_0$ with analytic inverse. Once such a map is constructed, the spectral problem reduces to the study of the roots of the equation $F(\lambda, \kappa) = 0$ for κ near $\kappa_0 = 0$. This is done using the implicit function theorem.

15.5 Remarks on Perturbation Expansions

We briefly consider the Taylor expansion of the eigenvalues $\lambda(\kappa)$ about $\kappa = 0$ for the case when $\lambda_0 \equiv \lambda(0)$ is a nondegenerate, discrete eigenvalue of a type-A analytic family $H_\kappa = H + \kappa W$, where W is relatively H-bounded. The Taylor series for $\lambda(\kappa)$ is called the *Rayleigh–Schrödinger series.* This case was studied in Section 3 of this chapter. For κ near 0, the operator H_κ has a nondegenerate eigenvalue that is analytic in κ. The projection P_κ onto the eigenspace of H_κ with eigenvalue $\lambda(\kappa)$ was shown to be analytic in κ (Problem 15.5), and $P_\kappa \to P_0$.

To obtain a perturbation expansion, note that if Ω_0 is an eigenfunction of H_0 : $H_0\Omega_0 = \lambda_0\Omega_0$, then $\|P_\kappa\Omega_0\| \to \|\Omega_0\|$ as $\kappa \to 0$, whence it follows that for $|\kappa|$ small:

$$\lambda(\kappa) = \frac{\langle\Omega_0, H_\kappa P_\kappa\Omega_0\rangle}{\|P_\kappa\Omega_0\|^2} = \lambda_0 + \kappa\frac{\langle\Omega_0, WP_\kappa\Omega_0\rangle}{\|P_\kappa\Omega_0\|^2}. \tag{15.6}$$

Now P_κ is expressible in terms of the resolvent:

$$P_\kappa = \frac{1}{2\pi i}\oint_{\Gamma_r}(z - H_\kappa)^{-1}dz,$$

for a contour $\Gamma_r \subset \bigcap_\kappa \rho(H_\kappa)$, and it follows from the Neumann expansion (15.2) that

$$P_\kappa = \frac{1}{2\pi i}\sum_{l=0}^{\infty}\kappa^l\oint_{\Gamma_r}dz(z - H)^{-1}[W(z - H)^{-1}]^l. \tag{15.7}$$

Substituting (15.7) into (15.6), we find the expansion

$$\lambda(\kappa) = \lambda_0 + \kappa\left(\frac{\sum_{n=0}^{\infty}a_n\kappa^n}{\sum_{m=0}^{\infty}b_m\kappa^m}\right) = \lambda_0 + \sum_{l=1}^{\infty}\alpha_l\kappa^l, \tag{15.8}$$

where

$$a_n \equiv \frac{1}{2\pi i}\oint_{\Gamma_r}dz\langle\Omega_0, [W(z - H)^{-1}]^{n+1}\Omega_0\rangle$$

and

$$b_m \equiv \frac{1}{2\pi i}\oint_{\Gamma_r}dz\langle\Omega_0, (z - H)^{-1}[W(z - H)^{-1}]^m\Omega_0\rangle.$$

Although the terms are complicated at higher orders, in principle they can all be written out. The lowest-order term is $\alpha_1 = \langle\Omega_0, W\Omega_0\rangle$, which is a well-known expression for the eigenvalue shift to first order. By Theorem 15.10, series (15.8) has a nonzero radius of convergence.

15.6 Appendix: A Technical Lemma

In order to treat the case of degenerate eigenvalues, we use the following result of Kato (see [K]).

Lemma 15.12. *Let $M_\kappa \equiv$ Ran P_κ be as in the proof of Theorem 15.11. Then there exists, for $|\kappa|$ sufficiently small, an invertible map $S_\kappa : M_\kappa \to M_0$ such that S_κ and its inverse are analytic in κ about $\kappa = 0$.*

Proof.

(1) We first construct two matrix-valued functions U and V that are holomorphic in a neighborhood of $\kappa = 0$. Since $P_\kappa{}^2 = P_\kappa$, we have

$$P'_\kappa P_\kappa + P_\kappa P'_\kappa = P'_\kappa, \tag{15.9}$$

and upon multiplying the left and right sides of this equation by P_κ, we obtain

$$P_\kappa P'_\kappa P_\kappa = 0. \tag{15.10}$$

Let $Q_\kappa \equiv [P'_\kappa, P_\kappa]$. This is holomorphic in κ about $\kappa_0 = 0$. By (15.9), we have (suppressing the κ)

$$\begin{cases} PQ = PP'P - PP' = -PP', \\ QP = P'P, \end{cases} \tag{15.11a}$$

so that

$$[Q, P] = P', \tag{15.11b}$$

by (15.9)–(15.11). We consider the first-order linear system

$$X'(\kappa) = Q_\kappa X(\kappa). \tag{15.12}$$

By the standard theory of ordinary differential equations (e.g., the method of successive approximation; see [HiSm]), the initial-value problem for (15.12) has a unique solution given $X(0)$. Let $U(\kappa)$ be the solution corresponding to $X(0) = 1$. Similarly, let $V(\kappa)$ be the unique solution of

$$Y'(\kappa) = -Y(\kappa)Q_\kappa, \tag{15.13}$$

corresponding to initial condition $Y(0) = 1$. Both $U(\kappa)$ and $V(\kappa)$ are analytic in κ about $\kappa = 0$.

Problem 15.8. Use the method of successive approximations to prove that the initial-value problems for (15.12) and (15.13) have unique holomorphic solutions about $\kappa = 0$.

(2) The matrix-valued functions $U(\kappa)$ and $V(\kappa)$ are inverses. To see this, we write the derivative

$$(VU)' = -VQU + VQU = 0,$$

by (15.12) and (15.13). Thus, the function $(VU)(\kappa)$ is a constant, and by the initial conditions, $(VU)(\kappa) = (VU)(0) = 1$. Since the spaces are finite-dimensional, this implies that $V(\kappa) = U(\kappa)^{-1}$. Finally, we have to show that $U(\kappa) : M_0 \to M_\kappa$. This is equivalent to showing that

$$P_\kappa = U(\kappa)P_0U(\kappa)^{-1},$$

or

$$U(\kappa)P_0 = P_\kappa U(\kappa). \tag{15.14}$$

Consider the derivative of $P_\kappa U$:

$$\begin{aligned} (PU)' &= P'U + PU' = P'U + PQU \\ &= [Q, P]U + PQU = QPU, \end{aligned} \tag{15.15}$$

where we used (15.11b) and (15.12). Equation (15.15) says that PU is a solution of (15.12) with initial condition P_0. By uniqueness, this means that $P_\kappa U = UP_0$, which is (15.14). Hence, we take $S_\kappa \equiv U(\kappa)^{-1} = V(\kappa)$ in Theorem 15.11. □

16

Theory of Quantum Resonances I: The Aguilar–Balslev–Combes–Simon Theorem

16.1 Introduction to Quantum Resonance Theory

We have studied the discrete and essential spectrum of a self-adjoint operator H. In particular, if $H = -\Delta + V$ is a Schrödinger operator with V in the Kato–Rellich class, then H is self-adjoint on $D(-\Delta) = H^2(\mathbb{R}^n)$. If, in addition, $\lim_{|x|\to\infty} V(x) = 0$, then we know that $\sigma_{\text{ess}}(H) = [0, \infty)$. Under these conditions on V, it is physically reasonable that there are no bound states, that is to say, eigenvalues, of H with positive energy. This is because, as follows from the principles of quantum tunneling discussed in Chapter 3, the wave function for such an eigenvalue will tunnel through the potential barrier and eventually reach the region where $E > V$. In this unbounded region, the particle will behave like a free particle and escape to infinity. To be more specific, suppose V is such that a classical particle with positive energy $E > 0$ is trapped by potential barriers (see Figure 16.1). In contrast to the situation of Chapter 3, we now assume that the barriers decay to zero at infinity. That is, we assume that the set, which we call the *classically forbidden region at energy* E as in Chapter 3, defined by $\{x \mid V(x) > E\} \equiv \text{CFR}(E)$, separates $\mathbb{R}^n$ into two disjoint connected sets: the potential well $W(E)$ at energy E, and the exterior region $\mathcal{E}(E)$. We assume that the potential well $W(E)$ is bounded and that the complement of $W(E) \cup \text{CFR}(E)$ in $\mathbb{R}^n$, which is $\mathcal{E}(E)$, is unbounded.

Any *classical* trajectory with energy E beginning in $W(E)$ will remain in $W(E)$ for all times. We now consider a quantum state ψ_t with initial condition ψ_0 localized in $W(E)$ and such that the energy of ψ_0 is approximately E. If the potential barrier $\text{CFR}(E)$ is very large, we expect that ψ_t remains localized in $W(E)$ for a long time. However, the quantum tunneling effect will eventually cause the wave packet to

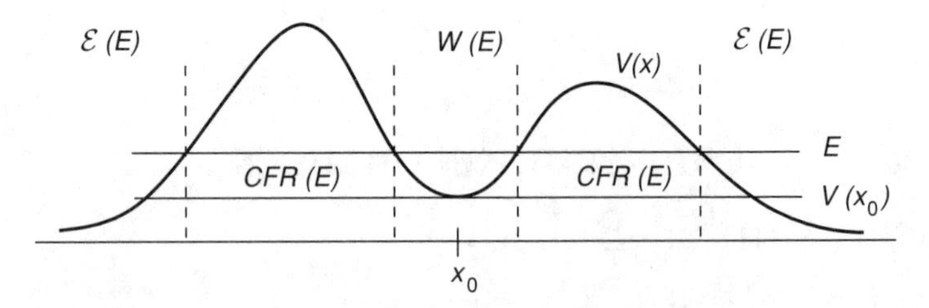

FIGURE 16.1. A typical resonance situation: the potential well $W(E)$ and the classically forbidden region CFR(E).

decay away from $W(E)$. In fact, if $B_R(0)$ is a ball of radius R centered at 0 and χ_R is the characteristic function on $B_R(0)$, one can show that

$$\lim_{T\to\infty} \frac{1}{T} \int_0^T \|\chi_R \psi_t\|^2 dt = 0. \tag{16.1}$$

This is part of the RAGE theorem; see [RS3]. Note that if H had an eigenvalue at $E > 0$ and if we take ψ_0 to be the corresponding eigenvector, then $\|\chi_R\psi_t\| = \|\chi_R\psi_0\|$, and the integral in (16.1) is a nonzero constant.

Thus, although ψ_t does not behave like a bound state, if $W(E)$ is deep and the potential barriers are high, it is suggested by the WKB-form of the wave function in the CFR(E) given in (3.6), that ψ_t remains concentrated for a long time in $W(E)$. Such an almost-bound state, characterized by the fact that it has a finite lifetime, is called a *quantum resonance*. We will give a precise definition in Section 16.2. In theoretical physics, resonances are used to describe states with finite lifetimes, such as unstable particles. They are usually associated with poles of the meromorphic continuation of the S-matrix. The famous Briet–Wigner formula associates a resonance at energy E with a large increase or bump in the scattering cross-section at energy E. However, it is quite difficult to give a mathematical description of resonances in this way. Rather, we will work with the resolvent of H and its *meromorphic continuation* to define quantum resonances.

In the remainder of this chapter, we will define quantum resonances as poles of the meromorphic continuations of certain matrix elements of the resolvent. These poles will be identified as the eigenvalues of non–self-adjoint operators constructed from H. We give a complete exposition of the theory developed by Aguilar and Combes [AC], Balslev and Combes [BC], and Simon [Sim9]. In the first two papers, dilation analytic techniques were used to prove the absence of singular continuous spectra for 2- and N-body Schrödinger operators, respectively. Simon [Sim9] applied these methods to define quantum resonances. In the following chapters, we will study the existence of resonances in the generalized semiclassical regime for various two-body Schrödinger operators. The notion of "semiclassical regime" makes precise the earlier statement "for $W(E)$ large enough." The term "generalized" refers to the fact that the parameter that we vary is not necessarily $\hbar$ (as in Chapters 11–12) but may be the electric or magnetic field strength, or so on, depending on the problem. The main idea, however, is that in varying a parameter the quantum system becomes close, in a prescribed way, to a system in which quantum tunneling is suppressed. Such a system is similar to the classical system. For such a "quasiclassical system," the resonances appear as actual bound states of an approximate Hamiltonian.

We want to emphasize that the resonances of H do not correspond directly to any spectral data for the self-adjoint operator H. Resonances can be detected by examining the spectral concentration of the continuous spectrum of H (see [RS4]), but this does not provide a convenient criterion. Since resonance energies are complex (with negative imaginary part), there are no L^2-eigenfunctions of H at these energies. (We will not discuss zero energy resonances.) In the cases for which the S-matrix for H and $H_0 = -\Delta$ exists and has a meromorphic continuation, the quantum resonances of H appear as the poles of this continuation. Therefore, they are intrinsic for this class of Schrödinger operators.

The study of resonances is closely connected with the absence of positive eigenvalues for certain families of Schrödinger operators, as indicated previously. We will, in fact, use the absence of positive eigenvalues to establish that resonances have nonzero imaginary parts. Although we will not prove the absence of positive eigenvalues in this book, we give a theorem that will suffice for the examples we study in the following chapters. This theorem can be proved using the method of Froese and Herbst [FrHe]. A textbook discussion can be found in [CFKS].

Theorem 16.1. *Suppose V is in the Kato–Rellich class and satisfies the following two properties*:

(1) *The potential vanishes at infinity,* $\lim_{\|x\|\to\infty} V(x) = 0$.

(2) *The operator $x \cdot \nabla V(x)$ is relatively Laplacian compact.*

Then $H = -\Delta + V$ has no strictly positive eigenvalues.

Before turning to the technical machinery necessary to make these ideas concerning resonances precise, let us sketch the general scheme further. Suppose we are given a Schrödinger operator $H(\lambda)$ depending on a parameter λ. We suppose that for λ large, quantum tunneling effects are strongly suppressed ($\lambda = 1/\hbar$ is the typical case). This will lead us to an approximate operator $H_0(\lambda)$, which will have only eigenvalues in some positive energy interval I that interests us. By contrast, $H(\lambda)$ will typically have essential spectrum in I. We want to compare $H(\lambda)$ and $H_0(\lambda)$ in the interval I. This seems difficult at first because of the different spectral properties of $H_0(\lambda)$ and $H(\lambda)$ in I. The first step is to replace $H(\lambda)$ by a family $H(\lambda, \theta)$, $\theta \in D \subset \mathbb{C}$, of non–self-adjoint operators. This family has the property that the essential spectrum near I has been moved off the real axis. As the essential spectrum has been deformed, we call $H(\lambda, \theta)$ a *spectral deformation family* for H. We develop the general theory of spectral deformation in Chapter 17 and study its application to Schrödinger operators in Chapter 18.

We can now apply perturbation theory to compare $H(\lambda, \theta)$ with $H_0(\lambda)$. These two operators will not be close in any usual sense of perturbation theory discussed so far. However, the theory we now develop requires only that the difference

$$V(\lambda, \theta) \equiv H(\lambda, \theta) - H_0(\lambda)$$

be relatively small when localized to the region (corresponding to CFR(E), $E \in I$) where $V(\lambda)$ is large. The perturbation theory works because, as we have seen in Chapter 3, quantum mechanical quantities, for example, wave functions, are small in such a region. It is reasonable to expect that the difference of the resolvents localized to such a region will also be small. This type of perturbation theory, which differs from the analytic perturbation theory of Chapter 15, is called *geometric perturbation theory*. We discuss this, along with other general aspects of eigenvalue stability, in Chapter 19. This will allow us to conclude the existence of complex eigenvalues of $H(\lambda, \theta)$ near the (real) eigenvalues of $H_0(\lambda)$. Finally, these complex eigenvalues are related to the poles of the meromorphic continuation of the matrix elements of the resolvent, and hence to the resonances of H, by the Aguilar–Balslev–Combes–Simon argument. This establishes the existence of resonances of H near the eigenvalues of $H_0(\lambda)$ in I for large λ. We now turn to the definition of resonances and the Aguilar–Balslev–Combes–Simon argument.

16.2 Aguilar–Balslev–Combes–Simon Theory of Resonances

The Aguilar–Balslev–Combes–Simon theory gives a precise meaning to the notion of quantum resonance. Let us consider a Schrödinger operator H with essential spectrum $[0, \infty)$ and bound states below zero. The resolvent of H, $R_H(z)$, is analytic in $\mathbb{C} \setminus \sigma(H)$. In Corollary 5.7, we proved that the operator norm of the resolvent $R_H(z)$ is bounded above by $1/|\operatorname{Im} z|$. This is a reflection of the fact that for $z \in \sigma(H)$, the resolvent is an unbounded operator on $L^2(\mathbb{R}^n)$. If we relax the condition that the resolvent be considered on the space $L^2(\mathbb{R}^n)$, it is a classical result that the resolvent of a Schrödinger operator (for suitable potentials V) remains a bounded operator, as $\operatorname{Im} z \to 0$, between other, weighted, Hilbert spaces. This result, called the *limiting absorption principle*, is discussed in Reed and Simon, Volume IV [RS4] and in Cycon, Froese, Kirsch, and Simon [CFKS]. Another way in which the boundary values of the resolvent can be controlled is to consider matrix elements of the resolvent between vectors in $L^2(\mathbb{R}^n)$ with certain nice properties. In general, there is a discontinuity in the matrix elements $\langle f, R_H(z)g\rangle$, for suitable $f, g \in L^2(\mathbb{R}^n)$, as we approach $\mathbb{R}^+$ from above and below. This can be seen for the free Schrödinger operator $H_0 = -\Delta$ by studying the explicit kernel of $(-\Delta - z)^{-1}$.

Problem 16.1. By methods of Fourier transform (see Appendix 4 and Section 4.3), construct Green's function $G_0(x, y; z) \equiv (-\Delta - z)^{-1}(x, y)$ in even and odd dimensions. These can be expressed in terms of the Bessel function $H^{(1)}$. Using these representations, study the discontinuity along $\mathbb{R}^+$ and study the meromorphic continuation of $R_0(z)$. Your computations should lead to the following formula for the kernel:

$$R_0(z^2)(x,y) = \left(\frac{i}{4}\right)\left(\frac{z}{2\pi|x-y|}\right)^{\frac{n-2}{2}} H^{(1)}_{\frac{n-2}{2}}(z|x-y|).$$

Given that there is a discontinuity across the essential spectrum, one is tempted to construct a meromorphic continuation. This is one of the accomplishments of the Aguilar–Balslev–Combes theorem. Hence, we are interested in studying the meromorphic continuation of matrix elements of $R_H(z)$ through the discontinuity along $\mathbb{R}^+$.

Definition 16.2. *The quantum resonances of a Schrödinger operator H associated with a dense set of vectors $\mathcal{A}$ in the Hilbert space $\mathcal{H}$ are the poles of the meromorphic continuations of all matrix elements $\langle f, R_H(z)g\rangle$, $f, g \in \mathcal{A}$, from $\{z \in \mathbb{C} \mid \text{Im } z > 0\}$ to $\{z \in \mathbb{C} \mid \text{Im } z \leq 0\}$.*

Of course, we have yet to demonstrate that the meromorphic continuations of such matrix elements exist. The existence of such continuations (obtained by explicit construction), the association of the poles of these continuations with the eigenvalues of certain non–self-adjoint operators associated with H, and the identification of these eigenvalues as resonances, are the main and general results of the Aguilar–Balslev–Combes–Simon theory. The dependence of the resonances on $\mathcal{A}$ is, at first, disturbing. However, as mentioned in Section 16.1, these poles can often be shown to be identical with the poles of the meromorphic continuation of Green's function or of the S-matrix for H and H_0 in scattering situations. In these cases, the resonances are independent of $\mathcal{A}$ and intrinsic to the pair (H, H_0).

To present the Aguilar–Balslev–Combes–Simon theory, we need some definitions and assumptions. We will verify these assumptions for a large class of Schrödinger operators in Chapters 18 and 20.

(A0) $H = -\Delta + V$ is a self-adjoint Schrödinger operator with domain $D(H)$ and $\sigma_{\text{ess}}(H) = [0, \infty)$ with discrete spectrum $\sigma_d(H) \subset (-\infty, 0]$.

(A1) There exists a family $\mathcal{U}$ of linear operators U_θ, $\theta \in D \equiv \{z \in \mathbb{C} \mid |z| < 1\}$, such that for $\theta \in D \cap \mathbb{R}$, U_θ is unitary, $U_\theta D(H) = D(H)$ $\forall \theta \in D$, and $U_0 = 1$. Furthermore, there exists a dense set of vectors $\mathcal{A} \subset \mathcal{H}$ such that

  (i) the map $(\psi, \theta) \in \mathcal{A} \times D \to U_\theta \psi$ is analytic on D with values in $\mathcal{H}$;
  (ii) for $\theta \in D$, $U(\theta)\mathcal{A}$ is dense in $\mathcal{A}$.

(A2) We define, for $\theta \in D \cap \mathbb{R}$, a family of unitary equivalent operators $H(\theta) \equiv U_\theta H U_\theta^{-1}$. We assume that the map

$$\theta \in D \to H(\theta),$$

is analytic of type-A as in Definition 15.8.

The family $\mathcal{U}$ satisfying (A1) and (A2) is called a *spectral deformation family* for H. We call the dense set of vectors $\mathcal{A}$ the *analytic vectors* for U_θ. We have normalized the domain D to be the unit disk for convenience. We note that condition (A2) can

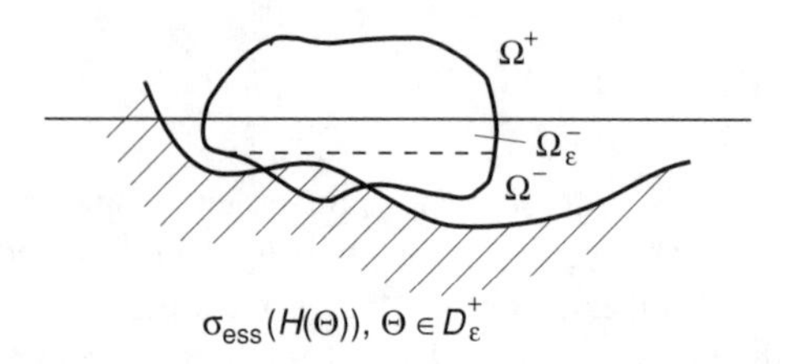

FIGURE 16.2. Typical spectrum of the spectrally deformed family $H(\theta)$.

be relaxed. We will see later that we only need analyticity of the resolvent of $H(\theta)$ in θ. This follows from weaker analyticity conditions on $H(\theta)$; see, for example, [K] or [RS4].

Let us now assume that (A0)–(A2) are satisfied by H, and we consider $\sigma(H(\theta))$. When $\theta \in D \cap \mathbb{R}$, the operators $H(\theta)$ are unitarily equivalent to H by (A2), and so $\sigma(H) = \sigma(H(\theta))$. For general $\theta \in D, \sigma(H(\theta))$ is a closed subset of $\mathbb{C}$ that may even have a nonempty interior. Furthermore, $H(\theta)$ may have complex eigenvalues that have no counterpart in $\sigma(H)$. To say more about $\sigma(H(\theta))$, we make the following assumption:

(A3) There exists an open, connected set $\Omega \subset \{z \in \mathbb{C} \mid \operatorname{Re} z > 0\}$ such that $\Omega^+ \equiv \Omega \cap \mathbb{C}^+ \neq \emptyset$, and $\Omega^- \equiv \Omega \cap \mathbb{C}^- \neq \emptyset$, and for all $\theta \in D^+ \equiv \overline{D \cap \mathbb{C}^+}$, $\sigma_{\text{ess}}(H(\theta)) \cap \Omega^+ = \emptyset$. For each $\varepsilon > 0$, there exists a subset $\Omega_\varepsilon^- \subset \overline{\Omega^-}$ such that for some $\theta \in D_\varepsilon^+ \equiv \{z \in D \mid \operatorname{Im} z > \varepsilon\}$, we have $\sigma_{\text{ess}}(H(\theta)) \cap \Omega_\varepsilon^- = \emptyset$.

In the specific cases discussed ahead, the region in the lower half-plane, Ω^-, which is free from the spectrum of $H(\theta)$ for $\theta \in D_\varepsilon^+$, can be taken larger as ε is increased. The regions defined in (A3) are shown in Figure 16.2.

Example 16.3. To illustrate assumptions (A0)–(A3), we consider the classic case of *dilation analyticity* of the Laplacian $H_0 = -\Delta$ on $L^2(\mathbb{R}^n)$. Recall that $\sigma(H_0) = \sigma_{\text{ess}}(H_0) = [0, \infty)$. Let $U_\theta, \theta \in \mathbb{R}$, be the implementation on $L^2(\mathbb{R}^n)$ of the group of dilations on $\mathbb{R}^n$, $x \to e^\theta x$. For $f \in \mathcal{S}(\mathbb{R}^n)$, U_θ is defined by

$$(U_\theta f)(x) \equiv e^{\frac{n\theta}{2}} f(e^\theta x).$$

It is easy to check that $\mathcal{U} \equiv \{U_\theta \mid \theta \in R\}$ forms a one-parameter unitary group and that $U_\theta D(H_0) = D(H_0)$, $\theta \in \mathbb{R}$. As for the set of analytic vectors $\mathcal{A}$, consider all functions of the form

$$\psi(z) = p(z)e^{-\alpha z^2},$$

for $\alpha > 0$, $z \in \mathbb{C}^n$, and any polynomial p. The set of all such functions restricted to $\mathbb{R}^n$ forms a dense set in $L^2(\mathbb{R}^n)$. Now consider the action of U_θ on such functions

$$(U_\theta \psi)(x) = e^{\frac{n\theta}{2}} p(e^\theta x)e^{-\alpha e^{2\theta} x^2}.$$

Writing $\theta = \theta_1 + i\theta_2$, we see that as long as

$$\operatorname{Re} e^{2i\theta_2} x^2 = x^2 \cos 2\theta_2 > \varepsilon x^2,$$

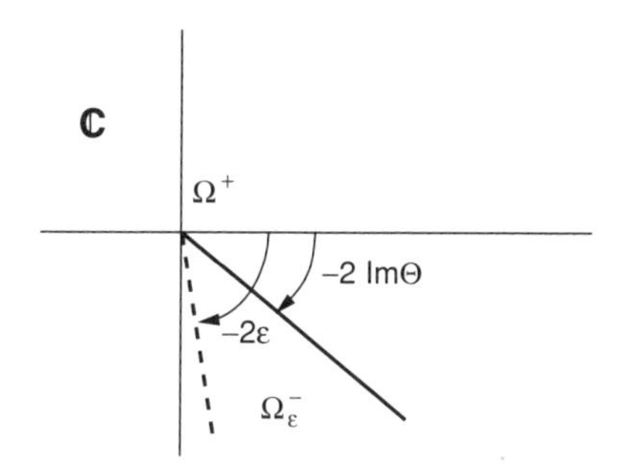

FIGURE 16.3. Spectrum of the dilated Laplacian for $0 < \varepsilon < \pi/4$.

for some $\varepsilon > 0$, the function $U_\theta \psi \in L^2(\mathbb{R}^n)$. It is easy to verify, then, that for $D \equiv \{z \mid |\mathrm{Im}\, z| < \pi/4\}$, the map

$$(\theta, \psi) \in D \times \mathcal{A} \to U_\theta \psi$$

is an L^2-analytic map. This verifies (A0) and (A1) (except for the density of $U_\theta \mathcal{A}$, which is proven in Proposition 17.10). As for (A2), for $\theta \in \mathbb{R}$ a simple computation shows that

$$H_0(\theta) = -e^{-2\theta} \Delta = e^{-2\theta} H_0.$$

This formula defines $H_0(\theta)$ for $\theta \in D$. It is clear that (A2) is satisfied. Furthermore, the formula shows that

$$\sigma_{\mathrm{ess}}(H_0(\theta)) = e^{-2i \,\mathrm{Im}\, \theta}\, \overline{\mathbb{R}^+}.$$

The effect of the spectral deformation is to rotate the essential spectrum of H_0 about the origin through an angle of $-2\,\mathrm{Im}\,\theta$. We define $\Omega \equiv \mathbb{C}^+ \cup \{z \mid \arg z > -\pi/2\}$. Hence, if we define D_ε^+ as in (A3), we can take $\Omega_\varepsilon^- = \{z \mid \arg z > -2\varepsilon\}$. Of course, we can take ε as close to $\pi/4$ as we like. This increases the size of the sector Ω_ε^-. The spectrum of this family of operators is sketched in Figure 16.3.

We will discuss in detail the technical aspects of Example 16.3 in Chapters 17 and 18, in which we present the theory of spectral deformation and its application to Schrödinger operators. However, this example contains all of the basic ideas concerning spectral deformation and resonances, and should always be kept in mind.

Given assumptions (A0)–(A3), we can now formulate the Aguilar–Balslev–Combes theorem concerning meromorphic continuations.

Theorem 16.4. *Let H be a self-adjoint Schrödinger operator with spectral deformation family $\mathcal{U}$ and analytic vectors $\mathcal{A}$ such that* (A0)–(A3) *are satisfied.*

(1) *For $f, g \in \mathcal{A}$, the function*

$$F_{fg}(z) \equiv \langle f, R_H(z) g \rangle, \tag{16.2}$$

defined for $\mathrm{Im}\, z > 0$, *has a meromorphic continuation across $\sigma_{\mathrm{ess}}(H) = \overline{\mathbb{R}^+}$ into Ω_ε^-, for any $\varepsilon > 0$.*

(2) *The poles of the continuation of* $F_{fg}(z)$ *into* Ω_ε^- *are eigenvalues of all the operators* $H(\theta)$, $\theta \in D_\varepsilon^+$, *such that* $\sigma_{\mathrm{ess}}(H(\theta)) \cap \Omega_\varepsilon^- = \emptyset$.

(3) *These poles are independent of* $\mathcal{U}$ *in the following sense. If* $\mathcal{V}$ *is another spectral deformation family for* H *with a set of analytic vectors* $\mathcal{A}_{\mathcal{V}}$ *such that* (A1)–(A3) *are satisfied and* $\overline{\mathcal{A} \cap \mathcal{A}_{\mathcal{V}}}$ *is dense, then the eigenvalues of* $\tilde{H}(\theta) \equiv V_\theta H V_\theta^{-1}$, $\theta \in D_\varepsilon^+$, *in* Ω_ε^- *are the same as those of* $H(\theta)$ *in this region.*

The Aguilar–Balslev–Combes–Simon theory identifies quantum resonances, as defined in Definition 16.2, as the eigenvalues of the spectrally deformed Hamiltonians $H(\theta)$ in the lower half-plane. In the next two chapters, we will discuss how to construct such families of operators and identify their spectrum. The price we pay for such a description is the fact that we must work with non–self-adjoint operators. Despite this, the theory is a powerful tool for investigation of resonances in quantum mechanical systems (as we will study in Chapters 20 and 23) and for the numerical computation of resonances in atomic and molecular systems [Sim10]. We also mention that the technique of spectral deformation allows one to study the perturbation of embedded eigenvalues. In particular, if H has an embedded eigenvalue λ, then λ remains in the spectrum of $H(\theta)$ and will be an isolated eigenvalue. We can then apply the methods of analytic perturbation theory to $H(\theta)$ and λ.

16.3 Proof of the Aguilar–Balslev–Combes Theorem

We now give the proof of the Aguilar–Balslev–Combes theorem. We invite the reader to follow through the proof with the simple example of the Laplacian using the representation of the resolvent given in Problem 16.1.

Proof.

(1) With $F_{fg}(z)$ defined in (16.2), assumption (A0) implies that F_{fg} is analytic on $\mathbb{C} \setminus \mathbb{R}$. Fix $z \in \mathbb{C}^+$. For $\theta \in \mathbb{R} \cap D$, U_θ is invertible with $U_\theta^{-1} = U_\theta^*$, and so we can write

$$F_{fg}(z) = \langle U_\theta f, (U_\theta R_H(z) U_\theta^{-1}) U_\theta g \rangle. \tag{16.3}$$

By (A1), $\theta \in D \to U_\theta f, U_\theta g$ are analytic maps. Furthermore, we have

$$U_\theta R_H(z) U_\theta^{-1} = R_{H(\theta)}(z). \tag{16.4}$$

Condition (A2) guarantees that $\theta \in D \to R_{H(\theta)}(z)$ is an analytic map provided $z \notin \sigma(H(\theta))$. Since we can write $U_{\bar{\theta}} f$ on the left in (16.3), we see that

$$\theta \in D \to F_{fg}(z;\theta) \equiv \langle U_{\bar{\theta}} f, R_{H(\theta)}(z) U_\theta g \rangle \tag{16.5}$$

is an analytic map provided z stays away from $\sigma(H(\theta))$. We now choose $\varepsilon > 0$ and fix $z \in \Omega_\varepsilon^+$. The function $F_{fg}(z,\theta)$, defined for $\theta \in D \cap \mathbb{R}$, can now be extended in θ into D_ε^+ by (A2) and (A3). We fix $\theta \in D_\varepsilon^+$ according to (A3) so that $\sigma_{\text{ess}}(H(\theta)) \cap \Omega_\varepsilon = \emptyset$. It follows that $F_{fg}(z,\theta)$ can be meromorphically continued in z from Ω_ε^+ into $\overline{\Omega_\varepsilon^-} \equiv \overline{\Omega_\varepsilon \cap \mathbb{C}^-}$. Now recalling (16.3), which says that $F_{fg}(z,\theta) = F_{fg}(z)$, $z \in \Omega_\varepsilon^+$, the identity principle for meromorphic functions [T1] says that there exists a function meromorphic on Ω_ε which equals $F_{fg}(z)$ on Ω_ε^+. This function provides the meromorphic continuation into Ω_ε^-.

(2) The meromorphic continuation of $F_{fg}(z)$ into Ω_ε^- is given by the matrix elements of $R_{H(\theta)}(z)$ in the states f_θ and g_θ, which denote the continuation of $U_\theta f$ and $U_\theta g$, respectively. Condition (A1) states that such vectors in $U(\theta)\mathcal{A}$, $\theta \in D$, are dense. Consequently, if $H(\theta)$ has an eigenvalue at $\lambda(\theta) \in \overline{\Omega_\varepsilon^-}$, $F_{fg}(z)$ will have a pole there. Conversely, if the continuation of $F_{fg}(z)$ has a pole at $\lambda(\theta) \in \overline{\Omega_\varepsilon^-}$, then it must be an eigenvalue of $H(\theta)$.

Problem 16.2. Verify this last statement by showing that the corresponding Riesz projection for $H(\theta)$ is nonzero.

Problem 16.3. By the above analysis, conclude that the poles of the continuation of $F_{fg}(z)$ into $\overline{\Omega_\varepsilon^-}$ are independent of θ. (*Hint*: Use uniqueness.) Conclude that if $z_i \in \overline{\Omega_\varepsilon^-}$ is an eigenvalue of $H(\theta)$, $\theta \in D_\varepsilon^+$, then it is also an eigenvalue of $H(\theta')$, $\theta' \in D_\varepsilon^+$, provided it remains away from $\sigma_{\text{ess}}(H(\theta'))$.

Problem 16.4. Prove the third part of Theorem 16.4. (*Hint*: Again use uniqueness.) With these problems, the proof of Theorem 16.4 is completed. □

We would like to call attention to certain points of Theorem 16.4.

Corollary 16.5. *Under the hypothesis of Theorem* 16.4,

(i) $\sigma_d(H(\theta)) \cap \Omega_\varepsilon^+ = \emptyset$, $\theta \in D^+$;

(ii) *any* $\lambda \in \sigma_d(H(\theta))$, $\theta \in D^+$, *is independent of* θ *provided* $\lambda \notin \sigma_{\text{ess}}(H(\theta))$;

(iii) *if* λ *is an eigenvalue of* H *in* $\Omega_\varepsilon \cap \mathbb{R}$, *then* $\lambda \in \sigma_d(H(\theta))$ *provided* $\lambda \notin \sigma_{\text{ess}}(H(\theta))$ *for* $\theta \in D^+$.

Proof.

(i) Since $F_{fg}(z;\theta) = F_{fg}(z)$ for $\theta \in D^+$ and $z \in \Omega_\varepsilon^+$, and H is self-adjoint, a pole of the left side would imply that H had a complex eigenvalue in Ω_ε^+.

(ii) The continuation of $F_{fg}(z)$ is unique and independent of θ. Hence as $\lambda \in \sigma_d(H(\theta))$ is a pole of this continuation (for some $f, g \in \mathcal{A}$), it is independent of θ.

(iii) If $\lambda \in \sigma_d(H)$, the corresponding projection is

$$P = (2\pi i)^{-1} \oint_\Gamma R_H(z) dz, \tag{16.6}$$

where Γ is a simple, closed contour about λ. If λ remains isolated from $\sigma_{\text{ess}}(H(\theta)), \theta \in D^+$, then we proceed as follows. Let $f, g \in \mathcal{A}$, and consider the matrix element

$$\langle f, Pg \rangle = \langle U_{\bar{\theta}} f, P_\theta U_\theta g \rangle, \tag{16.7}$$

for $\theta \in \mathbb{R} \cap D$, where

$$P_\theta \equiv U_\theta P U_\theta^{-1} = (2\pi i)^{-1} \oint_\Gamma R_{H(\theta)}(z) dz. \tag{16.8}$$

The right side of (16.7) has a continuation in θ into D^+. By uniqueness of the continuation and the identity (16.7), the Riesz projection $P_\theta, \theta \in D^+$, exists and is nonzero. Hence, $\lambda \in \sigma_d(H(\theta))$. Now suppose that λ is an embedded eigenvalue of H in $\Omega_\varepsilon \cap \mathbb{R}$. In Section 6.4, we proved that the projection for λ is given by

$$P = s - \lim_{\varepsilon \to 0} (-i\varepsilon)(H - \lambda - i\varepsilon)^{-1}.$$

Again, we conjugate P with $U_\theta, \theta \in \mathbb{R} \cap D$, and take matrix elements for $f, g \in \mathcal{A}$ to obtain

$$\langle f, Pg \rangle = \lim_{\varepsilon \to 0} (-i\varepsilon) \langle U_{\bar{\theta}} f, R_{H(\theta)}(\lambda + i\varepsilon) U_\theta g \rangle. \tag{16.9}$$

The right side can be continued into D^+. Hence, the weak limit exists and is nonzero:

$$P_\theta = w - \lim_{\varepsilon \to 0} (-i\varepsilon) R_{H(\theta)}(\lambda + i\varepsilon).$$

However, as $P^2 = P$, an application of the first resolvent formula shows that $P_\theta^2 = P_\theta$. Hence, P_θ is a Riesz projection for $H(\theta)$ and, as $P_\theta \neq 0$, this implies that $\lambda \in \sigma(H(\theta))$. (Note that this argument applies as long as λ remains isolated from $\sigma_{\text{ess}}(H(\theta))$ or, at most, λ remains on the boundary of $\sigma_{\text{ess}}(H(\theta))$.) □

Problem 16.5. Prove that $P_\theta^2 = P_\theta$ as in the proof of Corollary 16.5.

Problem 16.6. Suppose λ is an isolated eigenvalue of H of finite multiplicity. Assuming (A0)–(A3), use type-A analyticity to give another proof that $\lambda \in \sigma(H(\theta))$, provided λ is away from $\sigma_{\text{ess}}(H(\theta))$, $\theta \in D^+$. (*Hint*: Use analyticity and the fact that U_θ is unitary for $\theta \in \mathbb{R} \cap D$.)

It follows from Theorem 16.4 and its corollary, that the resonances of H, $\mathcal{R}(H)$, in the sector $\Omega_\varepsilon^- \subset \mathbb{C}^-$ can be given as

$$\mathcal{R}(H) \cap \Omega_\varepsilon^- = \bigcup_{\theta \in D_\varepsilon^+} \sigma_d(H(\theta)). \tag{16.10}$$

Of course, we take ε as large as possible, but it may be that by the spectral deformation method we do not get all resonances of H.

16.4 Examples of the Generalized Semiclassical Regime

We want to mention some physical situations where resonances have been shown to exist. We will only consider two-body, time-independent Hamiltonians. There are two basic types of resonances whose existence and properties can be established using the techniques described later in this section: (1) shape resonances and (2) Stark effect resonances. In addition, resonances play an important role in the study of the Zeeman effect. We will treat shape resonances in detail. The Hamiltonian $H(h) = -h^2\Delta + V$ is considered in the semiclassical regime of small h. The potential V vanishes at infinity, $V \geq 0$ (for simplicity), and forms a positive barrier that traps classical particles with energies in an interval $(0, E_0)$. By rescaling $H(h)$, we consider $H(\lambda) = -\Delta + \lambda^2 V$ as $\lambda \to \infty$. In the large λ regime, quantum tunneling through the barrier is suppressed. This is seen by noting that the distance across the barrier in the Agmon metric diverges as $\lambda \to \infty$.

We can see the same phenomena in many Stark Hamiltonians. A Stark Hamiltonian describes a particle moving in a static electric field F and potential V. It has the form $H(F) = -\Delta + V + F \cdot x$. Suppose V is an attractive Coulomb potential so $H(0)$ has bound states. When $|F| \neq 0$, the spectrum of $H(F)$ is $\mathbb{R}$ and there are no longer any eigenvalues. We think of the eigenvalues of $H(0)$ as becoming resonances since, due to F, the quantum particle can tunnel through the Coulomb barrier in the direction of F. Again, when $|F| \to 0$, the Agmon distance across this barrier diverges. Hence, the electric field F for which $|F|$ is very small is the semiclassical regime for this problem. A similar situation occurs in one dimension when V is a periodic potential. This is known as the Stark ladder problem. Each potential minimum of V contributes a resonance when $F \neq 0$, and due to the periodicity of V, we get a sequence or ladder of resonances.

Finally, we can consider a hydrogen atom in $\mathbb{R}^3$ in an external, constant magnetic field in the z-direction. This is the Zeeman effect. The Hamiltonian for a particle in the magnetic field $(0, 0, B)$ with vector potential $A = \frac{1}{2}B(-y, x, 0)$ is

$$H(B) = -\Delta + \frac{\alpha^2 B^2}{4}(x^2 + y^2) - \alpha B L_z + V,$$

where α is a constant involving the electric charge and $L_z \equiv -i\left(x\frac{\partial}{\partial y} - y\frac{\partial}{\partial x}\right)$ is the third component of angular momentum. Neglecting the L_z-term, let us analyze $H_0(B) + V$, where

$$H_0(B) = -\Delta_{x,y} + \frac{\alpha^2 B^2}{4}(x^2 + y^2) + \left(-\frac{d^2}{dz^2}\right),$$

on $L^2(\mathbb{R}^3) = L^2(\mathbb{R}^2) \otimes L^2(\mathbb{R})$. The two-dimensional operator $H_0^{(2)}(B) \equiv -\Delta_{x,y} + \frac{\alpha^2 B^2}{4}(x^2 + y^2)$ is simply a two-dimensional harmonic oscillator Hamiltonian. The spectrum is discrete (Theorem 10.7), and the eigenvalues $\{E_i^L(B)\}_{i=1}^{\infty}$ are called the Landau levels. Since $\sigma(-\frac{d^2}{dz^2}) = [0, \infty)$ on $L^2(\mathbb{R})$, it is not hard to check that

$$\sigma(H_0(B)) = \bigcup_{i=1}^{\infty} [E_i^L(B), \infty)$$

by constructing approximate eigenfunctions $\phi_i(x, y)e^{ikz}$, where $H_0^{(2)}(B)\phi_i = E_i^L(B)\phi_i$. We now consider the perturbation V, which we take to be

$$V(x) = -\|x\|^{-1}, \tag{16.11}$$

that is, Coulombic. If V had the special form $-|z|^{-1}$, the effect would be to add a sequence of eigenvalues $\{e_i\}$, with $e_i \to 0^-$, accumulating at each E_j^L. In the case with V Coulombic, (16.11), we find eigenvalues accumulating at E_1^L. Even though these eigenvalues are not analytic in B at $B = 0$, the Rayleigh–Schrödinger expansion predicts their behavior in an asymptotic sense. We will see that this behavior is related to the resonances of $H(iB)$, that is, to the tunneling situation portrayed in Figure 16.4. As $B \to 0$, the lifetime of these resonance states grows, and so small B is the semiclassical regime. In terms of tunneling phenomena, if we apply dilation analyticity to

$$V(x, y, z) = \frac{\alpha^2 B^2}{4}(x^2 + y^2) - \frac{1}{|r|},$$

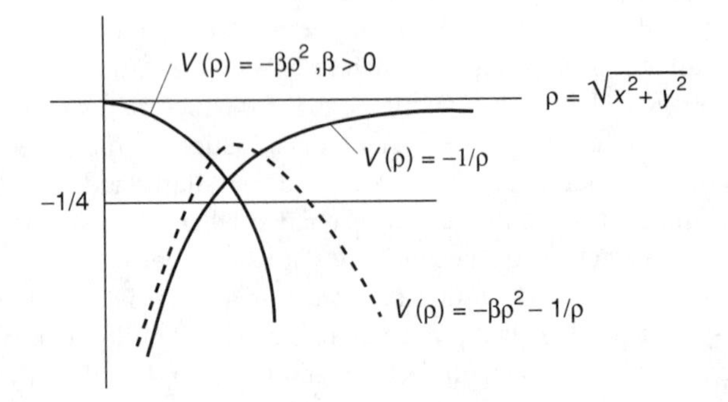

FIGURE 16.4. The potential for the atomic Zeeman effect with a purely imaginary magnetic field.

we obtain

$$V_\theta(x, y, z) = e^{2\theta} \frac{\alpha^2 B^2}{4}(x^2 + y^2) - \frac{e^{-\theta}}{|r|}.$$

Setting $\theta = \frac{i\pi}{2}$, the sign of the oscillator potential becomes negative. At this value, the eigenfunctions of the Coulomb potential can tunnel through the barrier formed by the inverted oscillator potential and escape to infinity. The width of the barrier becomes infinite as $B \to 0$.

16.5 Notes

The theory of resonances in quantum systems is almost as old as quantum mechanics. In Chapter 20, we will discuss the *shape resonance model*, which dates from 1928. Resonances occur in virtually all areas of quantum physics: the theory of atoms and molecules, nuclear and elementary particle physics, and the theory of solids. As mentioned in the Introduction to this chapter, the physical picture of a resonance in these systems is a state that decays very slowly. Resonances are often described as the poles in the meromorphic continuation of the S-matrix or of the resolvent. This relationship between poles and time decay can be seen very easily if one formally writes the inverse Laplace transform representation of the time-evolution group for the Schrödinger operator H:

$$\langle \psi, e^{-itH}\psi \rangle = (2\pi i)^{-1} \lim_{\epsilon \to 0} \text{ Im } \int_0^\infty e^{-itE} \langle \psi, (H - E - i\epsilon)^{-1}\psi \rangle dE,$$

where we have neglected any negative eigenvalues. Suppose that the matrix element of the resolvent can be continued in the complex energy plane so that the contour can be deformed into the fourth quadrant. If the continuation of the resolvent has a pole at $E_0 - i\Gamma$, with $\Gamma > 0$, then by the residue theorem, the coefficient of this term in the integral is

$$e^{-itE_0 - t\Gamma},$$

which decays exponentially in time. This is called the *single-pole approximation* and suggests that the phenomena of resonances can be described by poles of the meromorphic continuation of matrix elements of the resolvent. Mathematical justification of this single-pole approximation is quite difficult. We will discuss some recent results in Section 23.4 (see Gérard and Sigal [GS], Hunziker [Hu3], and Skibsted [Sk1], [Sk2], [Sk3]).

The mathematical theory of resonances seems first to have been established for one-dimensional and spherically symmetric systems. One can look for solutions to the ordinary differential equation with complex energy and certain asymptotic behavior at infinity. The existence of resonances can be reduced to the existence of complex zeros of a Fredholm determinant. For a discussion of the physics of resonances and the one-dimensional and spherically symmetric cases, we refer to the book of R. G. Newton [Ne]. We also refer the reader to the paper of Harrell and Simon [HaSi], which, among other things, discusses resonances for one-dimensional and spherically symmetric systems and gives a good historical overview. The reader can find there a discussion of the Zeeman problem and the Stark effect (see Chapter 23). The situation is more difficult in arbitrary dimensions. Since resonances do not belong to the spectrum of the self-adjoint Schrödinger operator (which is to say, the resonance states are not eigenfunctions of H in the Hilbert space), it is difficult to study them by looking at H alone. One way in which one can infer the existence of resonance states by examining H alone is through the notion of *spectral concentration* mentioned in the Introduction. The notes to Chapter XII of Reed and Simon, Volume IV [RS4] contain references to this idea and some earlier investigations.

Resonances defined in the Aguilar–Balslev–Combes–Simon theory are intrinsic to the Schrödinger operator $H = -\Delta + V$ and the set of analytic vectors. Aspects of the relationship between H and the set of analytic vectors have been explored by Howland [Ho1]. In many situations, resonances can be identified with the poles of the meromorphic continuation of the scattering matrix, which depends only on the pair $\{H_0 = -\Delta, H\}$. Balslev established this result for short-range, dilation analytic potentials in [B1]. Gérard and Martinez [GM2] proved that a suitably defined S-matrix for a family of analytic, two-body, long-range potentials has a meromorphic continuation and that the poles of this continuation coincide with the resonances. Other results on the connection between scattering theory, the poles of the scattering matrix, and resonances can be found in Hagedorn [Ha] and in Jensen [J2].

There are other approaches to spectral deformation. Local distortion methods in momentum space were developed by Babbitt and Balslev [BB] and by Balslev [B2]. Using this method, Jensen [J1] showed that resonances, defined as poles of the meromorphic continuation of the resolvent, coincide with the poles of the meromorphic continuation of the scattering matrix for a class of two-body potentials.

Resonances in quantum mechanics can also be defined without the analyticity assumptions of the Aguilar–Balslev–Combes–Simon theory. They can be defined directly in terms of their physical properties, as done by Lavine [La]. If the two-body potential is exponentially decaying, one can prove that the Green's function has a meromorphic continuation as a bounded operator between exponentially weighted Hilbert spaces. A theory of resonances for Schrödinger operators with potentials that are the sum of an exponentially decaying potential plus a dilation analytic potential was developed by Balslev [B3] and extended by Balslev and Skibsted [BS1]. These authors also explore the relation between the poles of the scattering matrix and resonances for these potentials. Orth [Or] developed a theory of resonances for families of perturbations based on the limiting absorption principal and the Livsic matrix. Gérard and Sigal [GS] developed a theory of resonances in the semiclassical regime directly in terms of the propagation properties of a resonance. They introduce the notion of a *quasiresonance* as a solution to the Schrödinger equation with certain outgoing propagation properties. They also show that such quasiresonances exhibit the expected exponential time-decay behavior (see Section 23.4). We mention the approaches of Davies [Da], the wave equation approach of Lax and Phillips [LP], and the microlocal approach of Helffer and Sjöstrand [HSj4] (we will comment more on the microlocal approach in Chapter 23).

One can also ask if the resonances defined by the Aguilar–Balslev–Combes–Simon theory give rise to the characteristic scattering signatures that physicists associate with resonances. The scattering theory for shape resonance models was studied by Nakamura [N2, N3]. The famous Wigner–Briet formula was proved in the semiclassical regime by Gérard, Martinez, and Robert [GMR].

17

Spectral Deformation Theory

17.1 Introduction to Spectral Deformation

The Aguilar–Balslev–Combes–Simon theory of resonances identifies the resonances of a self-adjoint operator H with the complex eigenvalues of a closed operator $H(\theta)$, which is obtained from H by the method of spectral deformation. In this chapter, we present the general theory of *spectral deformation*. This technique is applicable to many situations in mathematical physics, such as Schrödinger operator theory, quantum field theory, plasma stability theory, and the stability of solutions to certain nonlinear partial differential equations [Si4]. We will discuss the application to Schrödinger operators in Chapter 18.

Originally, spectral deformation theory was formulated for the dilation group. The Schrödinger operators to which it applied had to have dilation analytic potentials. Through the contributions of various researchers, these constraints were relaxed (see the Notes at the end of this chapter). The theory that we discuss here was presented by Hunziker in [Hu2].

The basic idea behind the spectral deformation method is as follows. We consider one-parameter families of diffeomorphisms on $\mathbb{R}^n$ generated by smooth vector fields. The families we choose will admit an extension to smooth maps on a neighborhood of $\mathbb{R}^n$ in $\mathbb{C}^n$ as the parameter θ becomes complex. For real θ, any such family induces a family of unitary operators U_θ on $L^2(\mathbb{R}^n)$. In this chapter, we concentrate on the construction of spectral deformation families, given smooth vector fields on $\mathbb{R}^n$ and a corresponding set of dense, analytic vectors. In Chapter 18, we study how the spectrum of the conjugated Schrödinger operator $H(\theta) \equiv U_\theta H U_\theta^{-1}, \theta \in \mathbb{R}$, deforms as θ becomes complex. We remind the reader that the basic strategy is illustrated in Example 16.3, where the dilation analyticity of the Laplacian is studied.

17.2 Vector Fields and Diffeomorphisms

Let $g: \mathbb{R}^n \to \mathbb{R}^n$ be a smooth mapping (actually, only C^2 is necessary). Consider, for $\theta \in \mathbb{R}$, the related family of maps $\phi_\theta: \mathbb{R}^n \to \mathbb{R}^n$ defined by

$$\phi_\theta(x) = x + \theta g(x). \tag{17.1}$$

Since $\phi_0(x) = x$, we expect ϕ_θ to be invertible for θ sufficiently small.

Examples 17.1.

(1) If $g(x) = x$, then $\phi_\theta(x) = (1+\theta)x$. Comparison with the map $x \to e^\theta x$, for $|\theta|$ small, shows that g is the infinitesimal generator of dilations on $\mathbb{R}^n$. This comparison can be done by taking the Taylor expansion of $e^\theta x$ in θ about $\theta = 0$. This map plays a particularly important role in the theory.

(2) For a fixed vector $\hat{e} \in \mathbb{R}^n$, we define $g(x) = \hat{e}$. We then have $\phi_\theta(x) = x + \theta\hat{e}$. This is simply a one-parameter family of translations on $\mathbb{R}^n$ in the direction $\hat{e}$. Such maps ϕ_θ are invertible and form a one-parameter group of diffeomorphisms on $\mathbb{R}^n$. This family of maps plays a role in the theory of the Stark effect, as we will see in Chapter 23.

To study the maps ϕ_θ, $\theta \in \mathbb{R}$, we need to consider the derivative. We let Df denote the derivative of a map $f: \mathbb{R}^n \to \mathbb{R}^n$. The derivative Df is a linear map on $\mathbb{R}^n$. Associated with this map Df is the $n \times n$ real Jacobian matrix relative to some basis of $\mathbb{R}^n$. We will write

$$Df(x_0) = \left(\frac{\partial f^i}{\partial x^j}(x_0) \right). \tag{17.2}$$

The determinant of this matrix is the Jacobian determinant, which we denote by

$$J_f(x) = \det\left(\frac{\partial f^i}{\partial x^j}(x) \right). \tag{17.3}$$

When we consider a family ϕ_θ, we will write J_θ for J_{ϕ_θ}. Returning to the family ϕ_θ, note that

$$D\phi_\theta(x) = 1 + \theta(Dg)(x), \tag{17.4}$$

where 1 is the $n \times n$ identity matrix. We denote by M_1 the inverse of the sup-norm of the derivative of g,

$$M_1 \equiv \left[\sup_{x \in \mathbb{R}^n} \|Dg(x)\| \right]^{-1}, \tag{17.5}$$

where $\|\cdot\|$ denotes the operator norm on the set of linear transformations on $\mathbb{R}^n$.

Lemma 17.2. *Let $g: \mathbb{R}^n \to \mathbb{R}^n$ be a smooth vector field. Then the map ϕ_θ defined in* (17.1) *is a diffeomorphism of $\mathbb{R}^n$ for $|\theta| < M_1$.*

Proof. It follows from (17.4) that if $|\theta| M_1^{-1} < 1$, then $D\phi_\theta$ is invertible. The inverse is given explicitly by

$$(D\phi_\theta)^{-1} = \sum_n (-1)^n \theta^n (Dg)^n, \tag{17.6}$$

which, for any $x \in \mathbb{R}^n$, is an absolutely convergent series provided $|\theta| < M_1$. Hence, by the inverse function theorem (see, for example, [HiSm]), ϕ_θ is invertible on $\mathbb{R}^n$ provided $|\theta| < M_1$. □

Remark 17.3. Some insight can be gained from the theory of ordinary differential equations (ODE) (see, for example, [HiSm]). Suppose $v: \mathbb{R}^n \to \mathbb{R}^n$ is a vector field with $\| Dv \|_\infty < \infty$. We consider the ODE

$$\frac{d\phi_\theta}{d\theta} = v \circ \phi_\theta, \tag{17.7}$$

with initial condition

$$\phi_0 = 1.$$

Standard theory then implies that there exists a solution ϕ_θ, $\theta \in \mathbb{R}$, to this ODE (17.7) such that for any $x \in \mathbb{R}^n$, $\phi_0(x) = x$ and $\phi_\theta \circ \phi_{\theta'}(x) = \phi_{\theta+\theta'}(x)$, $\theta, \theta' \in \mathbb{R}$. We say that ϕ_θ is the *global flow* generated by v. If we consider the map $\theta \in \mathbb{R} \to \phi_\theta(x) \in \mathbb{R}^n$, for fixed x and for $|\theta|$ small, the Taylor expansion of this map about $\theta = 0$ is

$$\phi_\theta(x) = x + \theta v(x) + \frac{d^2}{dt^2} \phi_t(x) \bigg|_s \theta^2, \tag{17.8}$$

for some $s \in (0, \theta)$. We see that (17.1) is the *infinitesimal* version of the global flow generated by g (if such a global flow exists).

Problem 17.1. Consider the two vector fields given in Examples 17.1. Verify that the global flows exist, and check the assertions in Examples 17.1.

Example 17.4. The theory of *exterior dilations* relative to the ball $B_R(0)$ is based on a smooth map $g: \mathbb{R}^n \to \mathbb{R}^n$ which satisfies

$$g(x) = 0, \qquad |x| < R,$$

$$g(x) = x, \qquad |x| \geq 2R.$$

Problem 17.2. Complete the construction of this vector field, and verify that it generates a global flow.

17.3 Induced Unitary Operators

We now consider the behavior of functions under the action of the maps ϕ_θ generated by a smooth vector field g. For any $f \in \mathcal{S}(\mathbb{R}^n)$, we define a map U_θ on $\mathcal{S}(\mathbb{R}^n)$ by

$$(U_\theta f)(x) = J_\theta(x)^{\frac{1}{2}} f(\phi_\theta(x)), \quad \theta \in \mathbb{R}. \tag{17.9}$$

Proposition 17.5. *The map U_θ defined in (17.9) maps $\mathcal{S}(\mathbb{R}^n)$ into $\mathcal{S}(\mathbb{R}^n)$. For $|\theta| < M_1$ and real, U_θ extends to a unitary operator on $L^2(\mathbb{R}^n)$ and $U_\theta \to 1$ strongly as $\theta \to 0$.*

Proof. We leave it as a problem to check the first part of the proposition. As for the second, we note that for $f \in \mathcal{S}(\mathbb{R}^n)$,

$$\begin{aligned} \|U_\theta f\|^2 &= \int J_\theta(x)|f(\phi_\theta(x))|^2 dx \\ &= \int J_\theta(\phi_\theta^{-1}(y))|f(y)|^2 \det[D\phi_\theta^{-1}(y)]dy, \end{aligned} \tag{17.10}$$

where we took $y \equiv \phi_\theta(x)$. Consider the identity, valid for all $y \in \mathbb{R}^n$,

$$\phi_\theta \cdot \phi_\theta^{-1}(y) = \phi_\theta(\phi_\theta^{-1}(y)) = y. \tag{17.11}$$

Differentiating both sides of (17.11) with the aid of the chain rule, we obtain

$$D\phi_\theta(\phi_\theta^{-1}(y)) \cdot (D\phi_\theta^{-1})(y) = 1. \tag{17.12}$$

Taking the determinant of both sides of (17.12), we obtain

$$J_\theta(\phi_\theta^{-1}(y)) \cdot \det[D\phi_\theta^{-1}(y)] = 1.$$

We can use this formula in the right side of (17.10) to obtain

$$\|U_\theta f\| = \|f\|, \quad \theta \in \mathbb{R}, \tag{17.13}$$

and so U_θ is an isometry. We define a map V_θ on $\mathcal{S}(\mathbb{R}^n)$, which is similar to U_θ, as

$$(V_\theta f)(x) = J_{\phi_\theta^{-1}}(x) f(\phi_\theta^{-1}(x)). \tag{17.14}$$

We leave it as an exercise to check that V_θ is an isometry and that on $\mathcal{S}(\mathbb{R}^n)$,

$$V_\theta U_\theta = U_\theta V_\theta = 1, \tag{17.15}$$

for $|\theta| < M_1$. Hence, by the density of $\mathcal{S}(\mathbb{R}^n)$ in $L^2(\mathbb{R}^n)$ and facts (17.13)–(17.15), we see that U_θ is unitary and that $U_\theta^{-1} = V_\theta$ for $|\theta| < M_1$. □

Problem 17.3. Fill in the details, and complete the proof, of Proposition 17.5 For the last part, the Lebesgue dominated convergence theorem, Theorem A2.7 of Appendix 2, will be useful.

Problem 17.4. Suppose $\theta \in \mathbb{R} \to \phi_\theta$ is a global flow on $\mathbb{R}^n$ as described in Example 17.4. Defining U_θ as in (17.9), show that $\{U_\theta \mid \theta \in \mathbb{R}\}$ forms a strongly continuous, one-parameter unitary group.

Examples 17.6. For the vector fields in Examples 17.1, we have the following unitary groups. The *dilation group* is given in $\mathbb{R}^n$ by

$$(U_\theta f)(x) = e^{\frac{\theta n}{2}} f(e^\theta x). \tag{17.16}$$

The one-parameter *translation group* in $\mathbb{R}^n$ with direction $\hat{e}$ is given by

$$(U_\theta f)(x) = f(x + \theta \hat{e}). \tag{17.17}$$

Note that in both examples, $U_\theta^{-1} = U_{-\theta}$.

17.4 Complex Extensions and Analytic Vectors

As discussed earlier in this chapter, we are interested in deforming $\mathbb{R}^n$ into $\mathbb{C}^n$ by allowing the parameter θ to become complex. We now investigate the manner in which we can extend the operators U_θ, implementing ϕ_θ, from $\theta \in \mathbb{R}$ to $\theta \in \mathbb{C}$, at least for small $|\theta|$. Of course, for $\theta \in \mathbb{C}$, the operators U_θ will no longer be unitary. Formula (17.9) indicates that such an extension will be possible provided $J_\theta(x)^{1/2}$ and f have extensions into some complex neighborhood of $\mathbb{R}^n$. Formula (17.4) shows that $J_\theta(x)^{1/2}$ extends analytically to complex θ, provided $|\theta| < M_1$, since the determinant map preserves the analyticity in θ of the matrix on the right side of (17.4). The condition $|\theta| < M_1$ guarantees that the argument of the function stays away from the branch point.

As to the second problem, the analyticity of f, we need to find a dense set of functions in $L^2(\mathbb{R}^n)$ that are the restrictions to $\mathbb{R}^n$ of functions analytic on a small complex neighborhood of $\mathbb{R}^n$ in $\mathbb{C}^n$, and such that $f \circ \phi_\theta$ remains in $L^2(\mathbb{R}^n)$ as a function of x for $|\theta| < M_1$. For such a function f, the map $\theta \in \mathbb{C}$, $|\theta| < M_1$, to the vector $U_\theta f \in L^2(\mathbb{R}^n)$ will be strongly analytic. We introduce a large class of analytic functions as follows.

Definition 17.7. *Let $\mathcal{A}$ be the linear space of all entire functions $f(z)$ having the property that in any conical region C_ε,*

$$C_\varepsilon \equiv \left\{ z \in \mathbb{C}^n \mid |\mathrm{Im}\, z| \leq (1-\varepsilon)|\mathrm{Re}\, z| \right\},$$

for any $\varepsilon > 0$, we have for any $k \in \mathbb{N}$,

$$\lim_{\substack{|z|\to\infty \\ z \in C_\varepsilon}} |z|^k |f(z)| = 0.$$

We note that $\mathcal{A}$ is not empty since any entire function of the form

$$f(z) = e^{-\alpha z^2} p(z),$$

for $\alpha > 0$ and any polynomial p, belongs to $\mathcal{A}$. We now use the linear space $\mathcal{A}$ to define our set of functions in $L^2(\mathbb{R}^n)$.

Definition 17.8. *The set of analytic vectors in $L^2(\mathbb{R}^n)$ is the set of $\psi \in L^2(\mathbb{R}^n)$ such that $\exists f \in \mathcal{A}$ and $\psi(x) = f(x)$, $x \in \mathbb{R}^n$.*

This definition is a generalization of one coming from the theory of unitary group representations. We recall Stone's theorem (see [RS2]), which establishes a one-to-one correspondence between strongly continuous, one-parameter unitary groups $\{U_\theta \mid \theta \in \mathbb{R}\}$ and self-adjoint operators. Each such group is generated by a self-adjoint operator A, in the sense of Problem 17.5, and conversely. In the finite-dimensional case when A is a symmetric matrix, we have

$$U_\theta = \exp(i\theta A),$$

where we define the exponential function of a matrix by its power series. For a matrix A, this power series converges absolutely. In the infinite-dimensional case, if A is a bounded, self-adjoint operator, then the series again converges in norm. In the general unbounded case, however, the series does not converge in norm and one has to define the right side through the functional calculus. However, the series might converge strongly, at least for certain vectors. A vector $\psi \in L^2(\mathbb{R}^n)$ is said to be *analytic for* A if the power series

$$\sum_{n=0}^{\infty} \frac{\theta^n}{n!} A^n \psi$$

has a nonzero radius of convergence. On such vectors ψ, the map

$$\theta \in \mathbb{R} \to U_\theta \psi$$

can be continued to a small complex neighborhood of the origin.

Problem 17.5. Consider the unitary groups of Examples 17.1. Construct the self-adjoint operators A that generate these groups. Here, the formula

$$\lim_{\theta \to 0} \frac{(U_\theta - I)f}{\theta} = -iAf \tag{17.18}$$

will be useful (think of the matrix case). Prove that A is self-adjoint. Construct the set of analytic vectors for both generators (i.e., give an explicit description of the sets).

There is a theorem of Nelson [RS2] that says that a closed symmetric operator A is self-adjoint if and only if its domain contains a dense set of analytic vectors. Combined with Stone's theorem, we see that in the case that our vector field g generates a global flow, so that the corresponding family U_θ, $\theta \in \mathbb{R}$, is a unitary group, then we have immediately a dense set of analytic vectors. We can use these vectors to extend $U_\theta g, \theta \in \mathbb{R}$, into a small complex neighborhood of zero such that $U_\theta g$ remains in $L^2(\mathbb{R}^n)$ (see [HS2] for this construction). However, in the general case that g induces only a local flow, we cannot appeal to these general theorems. Hence, we have to construct an appropriate set of vectors, as in Definition 17.8. We now explore the properties of $\mathcal{A}$.

Lemma 17.9. *The set of functions in $\mathcal{A}$ restricted to $\mathbb{R}^n$ form a dense, linear subset of $L^2(\mathbb{R}^n)$. Furthermore, for any $f \in \mathcal{A}$, $f(z) \in L^2(\mathbb{R}^n)$ for $z \in \mathcal{C}_\varepsilon$, any $\varepsilon > 0$.*

Problem 17.6. Prove Lemma 17.9. The denseness of the Hermite functions in $L^2(\mathbb{R}^n)$ will be useful.

We now consider the action of a spectral deformation family $\mathcal{U}$ on $\mathcal{A}$. To do this, we must make an additional assumption on the vector field g that generates the family:

(L) $\exists R > 0$ and a constant $c_1 \geq 0$, such that $\|g(x)\| \leq \|x\| + c_1$ $\forall x \in \mathbb{R}^n$ with $\|x\| > R$.

We will always assume this in what follows. Consistent with condition L, we now normalize g so that $M_1 = 1$. Note that the vector fields of Examples 17.1 satisfy both of these conditions. We will see in the next section that the linear growth of g outside $B_R(0)$ greatly facilitates the calculation of the essential spectrum of $H(\theta)$ for $\theta \in \mathbb{C}$. We define a domain D_0 in $\mathbb{C}$ about the origin by

$$D_0 = \left\{ \theta \in \mathbb{C} \mid |\theta| < \frac{1}{\sqrt{2}} \right\}. \tag{17.19}$$

Proposition 17.10. *Let $\mathcal{U}$ be a spectral deformation family associated with a smooth vector field g satisfying condition L and with $M_1 = 1$. Then,*

(i) *the map $(\theta, f) \in D_0 \times \mathcal{A} \to U_\theta f$ is an analytic L^2-valued function;*

(ii) *for any $\theta \in D_0$, $U_\theta \mathcal{A}$ is dense in $L^2(\mathbb{R}^n)$.*

Proof.

(1) We first show that for any $\theta \in D_0 \cap \mathbb{R}$ and $f \in \mathcal{A}$, $U_\theta f$ extends to an analytic L^2-valued function on D_0. For this, we show that $z \equiv \phi_\theta(x) \in \mathcal{C}_\varepsilon \cap \{z \in \mathbb{C}^n \mid \|z\| > R\}$, $\forall \theta \in D_0$. Writing $\theta = \theta_1 + i\theta_2 \in D_0$, we have

$$\|\mathrm{Re}\, z\|^2 = \|x\|^2 + 2\theta_1 x \cdot g(x) + \theta_1^2 \|g(x)\|^2, \tag{17.20}$$

$$\|\mathrm{Im}\, z\|^2 = \theta_2^2 \|g(x)\|^2. \tag{17.21}$$

Using the simple inequality

$$|\theta_1 x \cdot g(x)| \leq \left(\frac{1}{\sqrt{2}} \|x\| \right) (\sqrt{2}|\theta_1| \, \|g\|) \leq \frac{1}{4} \|x\|^2 + \theta_1^2 \|g(x)\|^2, \tag{17.22}$$

and (17.20) and (17.21), we obtain

$$\|\text{Re } z\|^2 - \|\text{Im } z\|^2 \geq \frac{1}{2}\|x\|^2 - |\theta|^2\|g(x)\|^2. \tag{17.23}$$

Next, restricting to $\|x\| > R$, the linearity assumption on g implies that for $|\theta|^2 < \frac{1}{2} - \varepsilon$, $\varepsilon > 0$,

$$\begin{aligned}\|\text{Re } z\|^2 - \|\text{Im } z\|^2 &\geq \frac{1}{2}\|x\|^2 - |\theta|^2(\|x\|^2 + c_2) \\ &\geq \varepsilon\|x\|^2 - c_3 \\ &\geq \varepsilon\|g(x)\|^2 - c_4 \\ &\geq \varepsilon\|\text{Im } z\|^2 - c_4,\end{aligned} \tag{17.24}$$

where c_4 is a finite constant. In the last inequality, we used (17.21). From (17.24), we obtain

$$\|\text{Im } z\|^2 \leq (1+\varepsilon)^{-1}\|\text{Re } z\|^2 + c.$$

Since we have $\|\text{Re } z\| > R$, for any $\varepsilon > 0$, we obtain

$$\|\text{Im } z\| \leq (1-\varepsilon)\|\text{Re } z\|,$$

showing that Ran $\phi_\theta(x)$, $\theta \in D_0$ and $\|x\| > R$, lies in a cone $\mathcal{C}_\varepsilon$. Hence, by the decay assumptions on f, $f \circ \phi_\theta \in L^2(\mathbb{R}^n)$ (the Jacobian factors are uniformly bounded). Finally, to prove strong analyticity of $U_\theta f$, we note that the map

$$\theta \in D_0 \rightarrow \langle \psi, U_\theta f \rangle,$$

for any $\psi \in L^2(\mathbb{R}^n)$, is analytic on D_0. Because weak analyticity implies strong analyticity, the result follows.

Problem 17.7. Prove that weak analyticity implies strong analyticity in a Hilbert space.

(2) To prove that $U_\theta \mathcal{A}$ is dense in $L^2(\mathbb{R}^n)$, for any $\theta \in D_0$, we first note, as above, that the Jacobian factors are uniformly bounded for $\theta \in D_0$. Consequently, it suffices to show that the functions

$$f(\phi_\theta(x)) = f(x + \theta g(x)) \tag{17.25}$$

are dense for $f \in \mathcal{A}$. For any $h \in C_0^\infty(\mathbb{R}^n)$, define a sequence of functions h_k by

$$h_k(x) = \left(\frac{k}{\pi}\right)^{\frac{n}{2}} \int dy J_\theta(y) h(y) e^{-k(x-y-\theta g(y))^2}. \tag{17.26}$$

Since supp h is compact, the functions h_k are entire and decay like e^{-kz^2} for any $\theta \in D_0$ (see Problem 17.8). Hence, for each $k \in \mathbb{N}$, $h_k \in \mathcal{A}$. Fixing $\theta \in D_0$, we consider $h_k \circ \phi_\theta$. We prove that $s - \lim_{k\to\infty} h_k \circ \phi_\theta = h$, which proves the density of $\mathcal{A}$ in $L^2(\mathbb{R}^n)$. Let the exponent in (17.26) for $h_k \circ \phi_\theta(x)$ be designated by $k\psi_\theta(x, y)$, where

$$\psi_\theta(x, y) = \left[(x - y) + \theta(g(x) - g(y))\right]^2.$$

In the same manner as in part (1), we can show that

$$\|\mathrm{Re}\ \psi_\theta\|^2 - \|\mathrm{Im}\ \psi_\theta\|^2 \geq \varepsilon\|x - y\|^2,$$

for some $\varepsilon > 0$, so that

$$\|e^{-k\psi_\theta}\| \leq e^{-\varepsilon k\|x-y\|^2}.$$

We leave it as a problem to show that

$$\left(\frac{k}{\pi}\right)^{\frac{n}{2}} \int dy J_\theta(y) e^{-k\psi_\theta^2} = 1. \tag{17.27}$$

It then follows that

$$\begin{aligned} |h(x) - h_k \circ \phi_\theta(x)| &\leq \left(\frac{k}{\pi}\right)^{\frac{n}{2}} \int dy |J_\theta(y)|\, |h(x) - h(y)| e^{-k|\psi_\theta|^2} \\ &\leq \left(\frac{k}{\pi}\right)^{\frac{n}{2}} \int dy |J_\theta(y)|\, |h(x) - h(y)| e^{-\varepsilon k\|x-y\|^2}. \end{aligned} \tag{17.28}$$

Problem 17.8. First, prove in detail that $h_k \in \mathcal{A}$. Then, verify the identity (17.27). It is useful to compute the integral for $\theta \in D_0 \cap \mathbb{R}$ and then to use analyticity.

(3) To finish the proof, we take the L^2-norm of the expression in (17.28) and obtain an upper bound

$$\left(\frac{k}{\pi}\right)^n \int dx \left| \int dy |h(x) - h(y)| e^{-\varepsilon k\|x-y\|^2} \right|^2.$$

Changing variables from (x, y) to $(u \equiv k^{1/2}x, v \equiv k^{1/2}y)$, we obtain

$$\begin{aligned} & c_0 k^{-\frac{n}{2}} \int du \left| \int dv |h(u) - h(v)| e^{-\varepsilon\|u-v\|^2} \right|^2 \\ &\leq c_0 k^{-\frac{n}{2}} \left\{ \int du \left| \int dv |h(u)| e^{-\varepsilon\|u-v\|^2} + \int dv |h(v)| e^{-\varepsilon\|u-v\|^2} \right|^2 \right\} \\ &\leq c_1 k^{-\frac{n}{2}}, \end{aligned}$$

since h has compact support. This proves that

$$\lim_{k\to\infty} \|h - h_k \circ \phi_\theta\| = 0,$$

and so $U_\theta \mathcal{A}$ is dense in $L^2(\mathbb{R}^n)$, $\theta \in D_0$.

□

17.5 Notes

The theory of dilation analytic potentials is presented in Reed and Simon, Volume IV [RS4]. This theory works well for atoms since the Coulomb potential centered at the origin is dilation analytic. Indeed, one of the first applications by Simon [Sim9] of the resonance theory discussed here was to the helium and other, multielectron, atoms. It was soon realized, however, that the potential for multicentered problems like molecules, which consist of several atoms, is not dilation analytic. Motivated by this and related problems, there were successive modifications of the dilation analytic theory. Simon introduced *exterior dilation analyticity* in [Sim11]. This paper shows that only analyticity near infinity is necessary. The problem of spectral deformation for molecules was also one of the motivating factors of Hunziker's work [Hu2]. The use of flows generated by general vector fields in configuration space was developed by Hislop and Sigal in [HS2]. Hunziker [Hu2] realized that the global flow is not necessary and worked only with the first-order approximation to the flow. This form of the theory is presented in this chapter. Another problem that could not be treated by dilation analyticity is a Schrödinger operator with a nonsmooth, exponentially decaying potential; for example, a potential of compact support. It was realized that this could be treated by spectral deformation methods using flows generated by vector fields in the Fourier transform variable. Since the potential operator V becomes a convolution operator in this representation, there are new technical difficulties. This was solved by Sigal in [Si2], by Cycon [Cy], and, from a pseudodifferential operator perspective, by Nakamura in [N4] and [N5].

18

Spectral Deformation of Schrödinger Operators

The method of spectral deformation in configuration space, developed in the last chapter, is quite general. It has been applied to a variety of problems. Our main application is to the semiclassical theory of shape resonances. For this, we need to study the behavior of Schrödinger operators under spectral deformations. In this chapter, we first study the effect of local deformations on the Laplacian and its spectrum. We then show that the effect of adding a relatively compact potential, which is the restriction of a function analytic in some neighborhood of $\mathbb{R}^n$, does not change the essential spectrum. This shows that the hypotheses of the Aguilar–Balslev–Combes–Simon theory are satisfied and opens the way for a study of resonances.

18.1 The Deformed Family of Schrödinger Operators

We apply the theory of spectral deformation to study Schrödinger operators $H = -\Delta + V$. We assume that $\mathcal{U} \equiv \{U_\theta \mid |\theta| < 1/\sqrt{2}\}$ is a spectral deformation family associated with a vector field g normalized so that $M_1 = 1$. We also assume that g satisfies condition L of Chapter 17. As for H, we assume that it is self-adjoint with domain $D(H) = H^2(\mathbb{R}^n)$. We will impose more conditions on V later. Let D_0 be the disk $D_0 \equiv \{\theta \in \mathbb{C} \mid |\theta| < 1/\sqrt{2}\}$. Consider, for $\theta \in D_0 \cap \mathbb{R}$, the family of unitarily equivalent operators

$$H(\theta) \equiv U_\theta H U_\theta^{-1} = p_\theta^2 + V_\theta, \tag{18.1}$$

where

$$p_\theta^2 = U_\theta p^2 U_\theta^{-1}, \quad p_j \equiv -i \frac{\partial}{\partial x^j}, \tag{18.2}$$

and

$$V_\theta = U_\theta V U_\theta^{-1}. \tag{18.3}$$

The operator V_θ is given by

$$(V_\theta f)(x) = V(x + \theta g(x)) f(x), \tag{18.4}$$

for suitable $f \in L^2(\mathbb{R}^n)$. The operator p_θ^2 can be computed explicitly. We first compute, for any $f \in \mathcal{S}(\mathbb{R}^n)$,

$$\left(U_\theta \frac{\partial}{\partial x^i} U_\theta^{-1} f \right)(x) = J_\theta(x)^{\frac{1}{2}} \left(\frac{\partial}{\partial x^i} U_\theta^{-1} f \right)(\phi_\theta(x)). \tag{18.5}$$

The derivative of the map ϕ_θ is given by

$$J_{ij}(x) \equiv [D\phi_\theta(x)]_{ij} = \frac{\partial \phi_\theta(x)^i}{\partial x^j}.$$

We denote the inverse of the matrix J by J^{-1} and write

$$(J^{-1}(x))_{\ell i} \equiv J(x)^{i\ell}.$$

We will use the standard convention that *repeated indices are always summed over their full range of values*. Using the chain rule, we find

$$\left(\frac{\partial}{\partial x^i} U_\theta^{-1} f \right)(\phi_\theta(x)) = J(x)^{i\ell} \frac{\partial}{\partial x^\ell} \overline{J}_\theta(\phi_\theta(x))^{\frac{1}{2}} f(x). \tag{18.6}$$

Here, we write $\overline{J}_\theta(x)$ for the Jacobian determinant of $\phi_\theta^{-1}(x)$. From equation (17.12), we see that

$$\overline{J}_\theta(\phi_\theta(x)) = J_\theta(x)^{-1},$$

and so from (18.5) and (18.6), we get

$$U_\theta \frac{\partial}{\partial x^i} U_\theta^{-1} = J_\theta^{\frac{1}{2}} J^{\ell i}(x) \frac{\partial}{\partial x^\ell} J_\theta^{-\frac{1}{2}}. \tag{18.7}$$

The adjoint of this operator (θ is real) is given by

$$\left(U_\theta \frac{\partial}{\partial x^i} U_\theta^{-1} \right)^* = -J_\theta^{-\frac{1}{2}} \frac{\partial}{\partial x^\ell} J^{\ell i}(x) J_\theta^{\frac{1}{2}}. \tag{18.8}$$

Multiplying results (18.7) and (18.8) together and suppressing the x-dependence, we obtain, from (18.2),

$$p_\theta^2 \equiv U_\theta p^2 U_\theta^{-1} = J_\theta^{-\frac{1}{2}} p_\ell J^{\ell i} J_\theta J^{mi} p_m J_\theta^{-\frac{1}{2}}. \tag{18.9}$$

A particularly useful form of (18.9) is

$$p_\theta^2 = -\nabla a_\theta \nabla + g_\theta, \tag{18.10}$$

where the matrix-valued function a_θ is

$$[a_\theta(x)]_{ij} = \sum_{m=1}^{n} J^{im}(x) J^{jm}(x), \tag{18.11}$$

and g_θ is the function

$$\begin{aligned} g_\theta(x) &= \frac{1}{4}\left[\frac{\partial}{\partial x^k}(\log J_\theta)\right] J^{km} J^{\ell m}\left[\frac{\partial}{\partial x^\ell}(\log J_\theta)\right] \\ &\quad + \frac{1}{2}\left[\frac{\partial}{\partial x^\ell}, J^{km} J^{\ell m} \frac{\partial}{\partial x^k}(\log J_\theta)\right]. \end{aligned} \tag{18.12}$$

Problem 18.1. Verify these calculations.

The functions $[a_\theta(x)]_{ij}$ and $g_\theta(x)$ are analytic in θ on D_0. They satisfy the following estimates. Since $\left|\frac{\partial g^i}{\partial x^j}\right|$ is bounded, it follows that $J_\theta(x)$ is uniformly bounded in x and $\theta \in D_0$. By expanding the determinant, we see that

$$|J_\theta(x)| \le 1 + c_1|\theta|,$$

for $|\theta|$ sufficiently small. Similarly, from the definitions of a_θ and g_θ, we find

$$[a_\theta] = 1 + [b_\theta],$$

where

$$[b_\theta]_{ij} \le c|\theta|$$

and

$$|g_\theta(x)| \le c|\theta|,$$

for all $|\theta|$ sufficiently small.

Proposition 18.1. *The family of operators p_θ^2, $\theta \in D_0$, defined in (18.2) is a type-A analytic family of operators with domain $D(p_\theta^2) = H^2(\mathbb{R}^n)$.*

Proof. We first verify the constancy of domain. Suppose $u \in H^2(\mathbb{R}^n)$. Since a_θ is smooth, we have

$$p_i a_{ij} p_j u = a_{ij} p_i p_j u + [p_i, a_{ij}] p_j u.$$

It follows from this and the boundedness of a_θ and g_θ that

$$\|p_\theta^2 u\| \le c_1 \|p^2 u\| + c_2 \|u\|,$$

and so $D(p^2) = H^2(\mathbb{R}^n) \subset D(p_\theta^2)$. The proof of the opposite inclusion is more involved. We begin with some matrix estimates. Recall that $a_\theta = (J^{-1})(J^{-1})^T$.

Let $b_\theta \equiv a_\theta^{-1} = J^T J$. Since $J = 1 + \theta(Dg)$, where Dg is the Jacobian matrix of g, we have, for any $\xi \in \mathbb{C}^n$,

$$\langle \xi, b_\theta \xi \rangle = \|\xi\|^2 + 2\theta \operatorname{Re}\langle \xi, (Dg)\xi \rangle + \theta^2 \|(Dg)\xi\|^2. \tag{18.13}$$

Recall that by our normalization, $\|Dg\| \leq 1$. It then follows that $\exists c_0 > 0$ such that

$$|\langle \xi, b_\theta \xi \rangle| \geq (1 - |\theta|)^2 \|\xi\|^2 \geq c_0 \|\xi\|^2. \tag{18.14}$$

Since b_θ is invertible, if we replace ξ by $a_\theta \xi$ in (18.14), we obtain

$$|\langle \xi, a_\theta \xi \rangle| \geq c_0 |\langle \xi, a_\theta^* a_\theta \xi \rangle| \geq c \|\xi\|^2. \tag{18.15}$$

The constant c is strictly positive. For any $u \in H^2(\mathbb{R}^n)$, we can write

$$\begin{aligned} \|p^2 u\|^2 &= \sum_k \langle p_k u, p^2 p_k u \rangle \\ &\leq c \sum_k \left(|\langle p_k u, p_\theta^2 p_k u \rangle| + |\langle p_k u, g_\theta p_k u \rangle| \right), \end{aligned} \tag{18.16}$$

where we used (18.10), (18.15), and the definition of p_θ^2. Commuting p_k to the left in each term of (18.16), we obtain for any $\delta > 0$,

$$|\langle p^2 u, p_\theta^2 u \rangle| \leq \delta \|p^2 u\|^2 + \delta^{-1} \|p_\theta^2 u\| \tag{18.17}$$

and

$$|\langle p^2 u, g_\theta u \rangle| \leq \delta \|p^2 u\| + c\delta^{-1} \|u\|^2, \tag{18.18}$$

plus commutator terms. These commutator terms can be bounded by terms similar to those in (18.17) and (18.18). This leads to the estimate

$$\|p^2 u\|^2 \leq c_1 \|p_\theta^2 u\|^2 + c_2 \|u\|^2,$$

obtained by taking δ sufficiently small. It follows that $D(p_\theta^2) \subset H^2(\mathbb{R}^n)$, completing the proof of the stability of the domain. The type-A analyticity on D_0 follows from the analyticity of a_θ and g_θ and $(p_i a_\theta)$. □

Problem 18.2. Verify estimates (18.17) and (18.18).

18.2 The Spectrum of the Deformed Laplacian

We next want to compute $\sigma_{\text{ess}}(p_\theta^2)$. In the dilation analytic case of Example 16.3, this is simply a half-line from the origin rotated through the angle $-2 \operatorname{Im} \theta$. For the case we have been studying with $g(x) \to x$ as $\|x\| \to \infty$, we obtain a similar result. The other well-known case, when $g(x) \to 1$ as $\|x\| \to \infty$, corresponding to translation analyticity, is treated in Problem 18.9. To compute $\sigma_{\text{ess}}(p_\theta^2)$, with p_θ^2 a closed, non–self-adjoint operator, we use the results of Section 10.3.

Proposition 18.2. *Let p_θ^2 be as defined in* (18.2) *with g satisfying condition L. Then* $\sigma(p_\theta^2) = \sigma_{\text{ess}}(p_\theta^2) = \{z \mid \arg z = -2\arg(1+\theta)\}$, for any $\theta \in D_0$.

Proof. Referring to Section 10.3, we first show that $Z(p_\theta^2) = W(p_\theta^2)$ using condition L on g. Clearly, $C_0^\infty(\mathbb{R}^n)$ is a core for p_θ^2. Defining χ_d as in Theorem 10.12, we compute

$$[p_\theta^2, \chi_d] = [-\nabla a_\theta \nabla, \chi_d] = -\sum_{ij} \left(\partial_i a_{ij}(\partial_j \chi_d) + (\partial_i \chi_d) a_{ij} \partial_j\right).$$

Since $\|\partial_i \chi_d\|_\infty \leq c_0 d^{-1}$, we see that for any $u \in C_0^\infty(\mathbb{R}^n)$,

$$\|[p_\theta^2, \chi_d]u\| \leq c_1 d^{-1} \||\nabla u|\| + c_2 d^{-2} \|u\|.$$

Finally, using the relative, mutual boundedness of p^2 and p_θ^2, we obtain

$$\|[p_\theta^2, \chi_d]\| \leq c d^{-1}(\|p_\theta^2 u\| + \|u\|).$$

It remains to show that $\rho(p_\theta^2) \neq \emptyset$, which we will do by proving that $p_\theta^2 + \lambda$, for $\lambda \in \mathbb{R}$ and sufficiently large, is invertible. Recall from (18.13) that for $\xi \in \mathbb{C}^n$, $\|\xi\| = 1$,

$$\langle \xi, b_\theta \xi \rangle = 1 + 2\theta\eta r + \theta^2 r^2,$$

where $\eta \in \mathbb{R}$ satisfies $-1 \leq \eta \leq 1$ and we have written $r = \|(Dg)\xi\| \leq 1 + c_0$ (see (18.12)). It follows that $\langle \xi, b_\theta \xi \rangle$ lies on the line segment between points $(1 \pm \theta r)^2$. Since $|\text{Im}\,\theta| < 1/\sqrt{2}$, we see that

$$\langle \xi, b_\theta \xi \rangle \geq 1 - \frac{1}{\sqrt{2}}(1 + c_0) \geq -M_0, \tag{18.19}$$

for some $0 \leq M_0 < \infty$, and

$$\langle \xi, b_\theta \xi \rangle \in S_0 \equiv \left\{ z \mid |\arg(z + M_0)| \leq \arg\left[1 + i\frac{\sqrt{2}}{2}(1 + c_0)\right] < \frac{\pi}{2} \right\}, \tag{18.20}$$

for $\theta \in D_0$. As in the proof of Proposition 18.1, these two properties, (18.19) and (18.20), are true for a_θ, so for any $u \in H^2(\mathbb{R}^n)$,

$$\langle u, (-\nabla a_\theta \nabla) u \rangle \;\in\; S_0, \tag{18.21}$$

$$\text{Re}\langle u, (-\nabla a_\theta \nabla) u \rangle \;\geq\; -M_0 > -\infty. \tag{18.22}$$

It follows from (18.12) that g_θ, $\theta \in D_0$, is uniformly bounded. Thus properties (18.21) and (18.22) continue to hold for p_θ^2 with a small shift in the vertex of the sector, which we still call $-M_0$. Consequently, we obtain

$$\text{Re}\langle u, (p_\theta^2 + \lambda) u \rangle \geq (\lambda - M_0)\|u\|^2, \tag{18.23}$$

and by the Schwarz inequality,

$$(\lambda - M_0)\|u\| \leq \|(p_\theta^2 + \lambda)u\|, \tag{18.24}$$

for all $u \in H^2(\mathbb{R}^n)$ and $\theta \in D_0$. It is easy to check that a similar estimate holds for $(p_\theta^2+\lambda)^*$. Consequently, $\text{Ran}(p_\theta^2+\lambda)$ is closed (λ large enough) and $[\text{Ran}(p_\theta^2+\lambda)]^\perp = \ker(p_\theta^2+\lambda)^* = \{0\}$. Hence, $(p_\theta^2+\lambda)^{-1}$ exists as a bounded operator, so $\lambda \in \rho(p_\theta^2)$. Theorem 10.12 now states that $Z(p_\theta^2) = W(p_\theta^2)$. To compute $W(p_\theta^2)$, we note that $g_\theta(x) \to 0$ as $\|x\| \to \infty$ and that $a_\theta(x) \to (1+\theta)^{-2}$ as $\|x\| \to \infty$, which implies (see Problem 18.4) that $W(p_\theta^2) = \{z \mid \arg z = -2\arg(1+\theta)\}$. Finally, since $\mathbb{C}\backslash Z(p_\theta^2)$ consists of one connected component that contains a point of $\rho(p_\theta^2)$ by the above discussion, we conclude that $\sigma_{ess}(p_\theta^2) = W(p_\theta^2)$ by Theorem 10.10. To finish the proof, we recall the Aguilar–Balslev–Combes theorem, Theorem 16.4, which identifies the poles of the meromorphic continuation of matrix elements of $(p^2 - z)^{-1}$, $\text{Im } z > 0$, with $\sigma_d(p_\theta^2)$ in $S_\theta^- \equiv (\mathbb{C}\backslash Z(p_\theta^2)) \cap \{z | \text{Re } z > 0\}$. Examination of the kernel of $(p^2 - z)^{-1}$, as given in Problem 16.1, and its continuation shows that the matrix elements have no poles, so $\sigma_d(p_\theta^2) \cap S_\theta^- = \emptyset$. □

Problem 18.3. Verify the claim that properties (18.21) and (18.22) for $-\nabla a_\theta \nabla$ extend to p_θ^2 with a shift in the vertex of the sector S_0.

Problem 18.4. By constructing Weyl sequences, show that $W(p_\theta^2) = \{z \mid \arg z = -2\arg(1+\theta)\}$.

18.3 Admissible Potentials

We now consider the potential V. We saw that Ran $\phi_\theta(x)$ lies in a cone C_ε for $\theta \in D_0$ and $\|x\| > R$. We call C_ε^R the *truncated cone* defined by $\{z \in \mathbb{C}^n \mid \|\text{Im } z\| \leq (1-\epsilon)\|\text{Re } z\| \text{ and } \|\text{Re } z\| > R\}$. We impose two conditions on V so that (1) the transformed potential $V \circ \phi_\theta$ admits a continuation to $\theta \in D_0$ as a relatively p^2-bounded operator, and (2) we can compute $\sigma_{\text{ess}}(H(\theta))$, given the result of the previous section on $\sigma_{\text{ess}}(p_\theta^2)$.

Definition 18.3. *A real-valued function V on $\mathbb{R}^n$ is an admissible potential for a spectral deformation family $\mathcal{U}$ if*

(V1) V is relatively p^2-compact;

(V2) V is the restriction to $\mathbb{R}^n$ of a function, which we also denote by V, that is analytic on the truncated cone C_ε^R, for any $\varepsilon > 0$ and some $R > 0$ sufficiently large.

Lemma 18.4. *Let V satisfy* (V1) and (V2). *Then $V \circ \phi_\theta$ extends to $\theta \in D_0$ as an analytic, relatively p_θ^2-compact operator.*

Problem 18.5. Prove Lemma 18.4 using condition L on the vector field g and Proposition 18.1. Theorem 9.8 will be useful in proving the relative compactness.

Corollary 18.5. *Let V satisfy* (V1) and (V2). *Then the self-adjoint operator $H(\theta) = p_\theta^2 + V_\theta$, defined for $\theta \in D_0 \cap \mathbb{R}$, extends to an analytic type-A family of operators on D_0 with domain $H^2(\mathbb{R}^n)$.*

18.4 The Spectrum of Deformed Schrödinger Operators

The last results we need concern the spectrum of $H(\theta) = p_\theta^2 + V_\theta$, for an admissible potential V, and the resonances. Recall from Section 16.2 that we expect the essential spectrum of $H(\theta)$ to deform off of the real axis into the lower half-plane for Im $\theta > 0$. In the case of dilation analyticity discussed in Example 16.3, we saw that $\sigma_{\text{ess}}(-e^{2\theta}\Delta) = e^{2i \operatorname{Im} \theta}\mathbb{R}_+$. It follows from Proposition 18.2 that this is typically the case for p_θ^2 when the spectral deformation family $\mathcal{U}$ is generated by a vector field g that approaches x at infinity, in the sense of condition L. We expect that the addition of an admissible potential will not change the essential spectrum.

As for the resonances, we can apply the Aguilar–Balslev–Combes theorem of Section 16.2 to our family of deformed Hamiltonians $H(\theta)$ and the set of analytic vectors $\mathcal{A}$. Note that $\mathcal{A}$ is independent of the specific vector field used, so the resonances of H, obtained as in Section 16.2, depend only on H and the set $\mathcal{A}$.

Theorem 18.6. *Let $\mathcal{U}$ be a spectral deformation family for the Schrödinger operator $H = -\Delta + V$, with V satisfying* (V1) and (V2). *Then for any $\theta \in D_0$,*

$$\sigma_{\text{ess}}(H(\theta)) = \{z \in \mathbb{C} \mid \arg z = -2\arg(1+\theta)\}. \tag{18.25}$$

Let $\mathcal{R}(H)$ be the set of resonances as defined in the Aguilar–Balslev–Combes–Simon theory, and let S_θ^- be the open region in $\mathbb{C}^-$ bounded by $\mathbb{R}^+$ and $\sigma_{\text{ess}}(H(\theta))$, for $\theta \in D_0 \cap \mathbb{C}^+$. Then,

$$\sigma_d(H(\theta)) \cap S_\theta^- \subset \mathcal{R}(H). \tag{18.26}$$

In particular, the resonances in this sector depend only on H and $\mathcal{A}$.

Remarks 18.7. We will not prove (18.25) here completely, but only a weaker version. The proof of (18.25) uses a version of Weyl's theorem, Theorem 14.6, for closed operators. We state this theorem without proof. The proof can be found, for example, in [K].

Theorem 18.8. *Let T be a closed operator on a Hilbert space $\mathcal{H}$ and A be a relatively T-compact operator. Then $\sigma_{\text{ess}}(T) = \sigma_{\text{ess}}(T + A)$.*

Problem 18.6. Suppose T and A satisfy the hypotheses of Theorem 18.8 and that $\sigma_{\text{ess}}(T) = W(T)$. Prove that $\sigma_{\text{ess}}(T + A) = \sigma_{\text{ess}}(T)$.

Proof of Theorem 18.6. Fix $\theta \in D_0$, let $S_\theta \equiv \{z \mid \arg z = -2\arg(1+\theta)\}$, and define $\mathbb{C}_\theta \equiv \mathbb{C} \setminus \overline{S}_\theta$. We will show that $\sigma_{\text{ess}}(H(\theta)) \cap \mathbb{C}_\theta = \emptyset$, and so $\sigma_{\text{ess}}(H(\theta)) \subset \overline{S}_\theta$. This is a weaker version of (18.25) but sufficient for our purposes. The statement (18.25) follows from Proposition 18.2 and Theorem 18.8. By Proposition 18.1, $(p_\theta^2 - z)^{-1}$ is analytic in z on $\mathbb{C}_\theta$. Lemma 18.4 and this fact imply that

$$F_\theta(z) \equiv V_\theta(p_\theta^2 - z)^{-1}$$

is an analytic, compact, operator-valued function on $\mathbb{C}_\theta$. We apply the Fredholm alternative, Theorem 9.12, to study $F_\theta(z)$.

The operator $1 + F_\theta(z)$, $z \in \mathbb{C}_\theta$, is invertible if and only if $-1 \notin \sigma(F_\theta(z))$, which is discrete, except possibly at 0. Suppose $-1 \in \sigma(F_\theta(z_0))$, $z_0 \in \mathbb{C}_\theta$. By analyticity, $\exists r > 0$ such that for $z \in D_r \equiv \{z \mid \|z - z_0\| < r\}$ implies that $\|F_\theta(z) - F_\theta(z_0)\| < 1/2$. Similarly, by Theorem 9.15, there exists a finite-rank operator F such that $\|F_\theta(z_0) - F\| < 1/2$. Consequently, by Theorem A3.30, the operator $1 + F_\theta(z) + F$ is invertible and analytic on D_r. Let us write $1 + F_\theta(z)$ as

$$1 + F_\theta(z) = (1 - G_\theta(z))(1 + F_\theta(z) + F),$$

where the finite-rank, analytic, operator-valued function $G_\theta(z) \equiv F(1 + F_\theta(z) + F)^{-1}$. Because of the canonical form of a finite-rank operator, we can consider $1 - G_\theta(z)$ as an analytic matrix-valued function. Let $D(z) \equiv \det(1 - G_\theta(z))$ be the determinant of this matrix. It follows from the preceding equation that $1 + F_\theta(z)$ is invertible on D_r if and only if $D(z) \neq 0$. By the assumption that $-1 \in \sigma(F_\theta(z_0))$, we know that $D(z_0) = 0$. Since $D(z)$ is an analytic function on D_r, either $D(z) = 0$, $\forall z \in D_r$, or there exists a discrete set of points, with no limit point in $\overline{D_r}$, where $D(z) = 0$.

We now show that the first alternative cannot hold. As a consequence, the set of points in $\mathbb{C}_\theta$ where $1 + F_\theta(z)$ is not invertible is discrete. Since $\mathbb{R}_- \subset \mathbb{C}_\theta$, it follows from (18.23) that

$$\|(p_\theta^2 + \lambda)^{-1}\| \leq c\lambda^{-1}, \quad \lambda > 0.$$

Since V_θ is relatively p_θ^2-compact, for any $\alpha > 0$, $\exists\, \beta(\alpha) > 0$ such that

$$\|V_\theta u\| \leq \alpha\|p_\theta^2 u\| + \beta(\alpha)\|u\|,$$

for any $u \in H^2(\mathbb{R}^n)$. Consequently, for any $\varepsilon > 0$,

$$\|V_\theta(p_\theta^2 + \lambda)^{-1}\| < \varepsilon,$$

for all λ sufficiently large. This shows that $1 + F_\theta(z)$ is invertible for all $z \in \mathbb{C}_\theta$ with Re z sufficiently negative. Using the connectivity of $\mathbb{C}_\theta$, if the first alternative mentioned above held, that $1 + F_\theta(z)$ is not invertible on D_r, then by repeating the above argument, we would arrive at the region where we know $1 + F_\theta(z)$ is invertible, and hence obtain a contradiction.

Therefore, the value -1 cannot be in $\sigma(F_\theta(z))$ except for a discrete, countable set $\mathcal{P}$ of z_i in $\mathbb{C}_\theta$. This means that $(1 + F_\theta(z))^{-1}$ exists on $\mathbb{C}_\theta \setminus \mathcal{P}$. Writing

$$H_\theta - z = (1 + F_\theta(z))(p_\theta^2 - z),$$

we see that $H_\theta - z$ is invertible on $\mathbb{C}_\theta \setminus \mathcal{P}$ since it is the product of two invertible operators. This also shows that $(H_\theta - z)^{-1}$ is meromorphic on $\mathbb{C}_\theta \setminus \mathcal{P}$.

Problem 18.7. Complete the proof of the theorem using the Aguilar–Balslev–Combes theorem, Theorem 16.4, and fill in the details of the argument for the invertibility of $1 + F_\theta(z)$.

This finishes the proof of Theorem 18.6. □

The combination of analyticity and compactness used in the proof of the theorem yields a result of interest in its own right. The following theorem is called the analytic Fredholm theorem in [RS1].

Theorem 18.9. *Let $F(z)$ be a compact, analytic, operator-valued function on a connected, open subset D in $\mathbb{C}$. Then, exactly one of the following holds:*

(1) the operator $1 + F(z)$ is not invertible for any $z \in D$;

(2) the operator $1 + F(z)$ is invertible on $D \backslash S$, where S is a discrete subset of D with no limit points in $\overline{D}$.

Problem 18.8. Prove that the poles of $(H(\theta) - z)^{-1}$ in $\mathbb{C}_\theta$ are of finite order.

Problem 18.9. The purpose of this problem is to study translation analyticity. Let $g(x)$ be a smooth vector field that is bounded with bounded derivatives and such that $g \to 1$ as $\|x\| \to \infty$ (see Examples 17.1). Suppose that V is the restriction to $\mathbb{R}$ of a function $V(z)$ analytic on a strip $\{z \mid |\text{Im } z| < \alpha\}$. Assume that V is relatively p^2-compact. Prove the analogues of Proposition 18.2 in this situation. In particular, $\sigma_{\text{ess}}(H(\theta)) \subset \{z \mid \text{Re } z \in \mathbb{R} \text{ and Im } z = \beta\}$ for $|\beta| < \alpha$.

18.5 Notes

References to the spectral deformation of Schrödinger operators are given in the Notes for Chapter 16. The calculations of this chapter follow the paper of Hunziker [Hu2]. Translation analyticity was introduced by Avron and Herbst [AvHe] to study Stark Hamiltonians. These Schrödinger operators have the form $H = H_0 + V$, where

$$H_0 \equiv -\Delta + F \cdot x$$

is the free Stark Hamiltonian. The constant vector $F \in \mathbb{R}^n$ is the electric field. Let us fix coordinates so that $F = \|F\|\hat{e}_1$. Let U_θ be the unitary implementation of the translation group $x \rightarrow x + \theta$. It is easy to see that, formally, the free Stark Hamiltonian transforms as

$$U_\theta H_0 U_\theta^{-1} = -\Delta + \|F\| x_1 + \|F\| \theta.$$

It can be proved, as outlined in Problem 18.9, that $\sigma_{\text{ess}}(H_0(\theta)) = \{z \mid \text{Im } z = \|F\| \text{ Im } \theta\}$. The main problem with translation analyticity is that the Coulomb potential is not translation analytic. The free Stark Hamiltonian, on the other hand, does not transform well under dilations. In fact, the spectrum of the free Stark Hamiltonian with a *complex* electric field is empty! Herbst showed in [He2] how to resolve these difficulties and use dilation analyticity to define resonances for the free Stark Hamiltonian with a Coulomb potential.

19

The General Theory of Spectral Stability

In Chapter 15, we studied the question of the stability of a discrete eigenvalue λ_0 of an operator T_0 under an *analytic* perturbation T_κ. The main result, Theorem 15.11, states that the family T_κ, for κ in a small complex neighborhood of 0, will have eigenvalues near λ_0 of total algebraic multiplicity equal to that of λ_0. This is what we mean by the *stability* of λ_0 with respect to the family T_κ. With regard to the perturbation theory of discrete eigenvalues, this theorem is quite satisfactory.

There are, however, other interesting situations that we want to study. We describe two main cases here:

(1) **Nonanalytic perturbations.** Let $\mathcal{S}$ be a sector in $\mathbb{C}$ with vertex at 0. Consider a family T_κ, $\kappa \in \mathcal{S}$, of closed operators depending in a controlled manner on the complex parameter $\kappa \in \mathcal{S}$. Suppose that there exists a closed operator T_0 such that $T_\kappa \to T_0$ in some sense. Furthermore, let us suppose that λ_0 is a discrete eigenvalue of T_0. What can we say about $\sigma_d(T_\kappa)$ in a neighborhood of λ_0 for $\kappa \in \mathcal{S}$ near 0?

(2) **Embedded eigenvalues.** Suppose that λ_0 is an eigenvalue of a self-adjoint operator T_0 lying in the essential spectrum of T_0 . How can we study the behavior of such an eigenvalue under perturbations of T_0?

As we will see, both of these situations occur quite naturally in examples in quantum mechanics. The examples involve both eigenvalues and resonances. We will study aspects of nonanalytic perturbation theory in this chapter that follow the works of Vock and Hunziker [VH], Hunziker [Hu4], and Simon [Sim12]. Basic material on asymptotic perturbation theory is given in Kato [K] and in Reed and Simon, Volume IV [RS4]. We study this problem because of its importance for several problems in quantum mechanics, and because we need to generalize the notion of stability of the discrete spectrum in order to treat problems in the semiclassical regime.

Indeed, as the examples in the next section will show, we must broaden the notion of stability to include not only eigenvalues but also resonances. By considering the stability of eigenvalues in nonanalytic situations and for non–self-adjoint operators obtained, for example, by spectral deformation, we will develop a theory in which eigenvalues and resonances are treated equally. We typically encounter two situations for which we wish to study the effects of a perturbation on an eigenvalue: an embedded eigenvalue of the unperturbed operator T_0, described in case (2), and nonanalytic perturbations of isolated eigenvalues for which the eigenvalue disappears as soon as the perturbation is turned on (examples (2) and (3) of the next section). Perturbation theory of embedded eigenvalues is most commonly associated with spectral resonances, as discussed in Chapter 16. In these examples, the key idea is that one first moves away the essential spectrum and then applies perturbation theory to study the stability of the eigenvalue. In general, the perturbed eigenvalue of the spectrally deformed operator will be a resonance of the perturbed, self-adjoint operator. We will illustrate this approach with a simple example in the last section. We will return to the problem of embedded eigenvalues in our discussion of resonances for the Helmholtz resonator in Chapter 23.

Nonanalytic perturbations are associated with the failure of the Rayleigh–Schrödinger perturbation series to converge about $\kappa = 0$. Yet, it is found that the truncated series gives remarkably good information about the perturbed eigenvalue or resonance. This is the topic of *asymptotic expansions* and *Borel summability*. We will not discuss these topics here, but readers may refer to [RS4], [Sim12], and [Hu4].

The main result of this chapter is the stability criteria for families of Schrödinger operators, Theorem 19.12. It should be compared to the estimates in Theorems 5.6 and 5.9, which give conditions for determining the spectrum of a self-adjoint operator.

19.1 Examples of Nonanalytic Perturbations

We give three classic examples of nonanalytic perturbations in quantum mechanics.

(1) The anharmonic oscillator. We consider a one-dimensional Schrödinger operator of the form

$$H_\kappa \equiv p^2 + x^2 + \kappa x^4 \tag{19.1}$$

acting on $L^2(\mathbb{R})$. For $\kappa = 0$, the Hamiltonian H_0 in (19.1) is just the harmonic oscillator Hamiltonian studied in Chapter 11. The spectrum is purely discrete, as follows from Theorem 10.7, and is given in (11.5). Note that the perturbation x^4 is not relatively H_0-bounded. The properties of the perturbed operator depend dramatically on the sign of the coupling constant κ. When

$\kappa > 0$, the potential $V_\kappa(x) \equiv x^2 + \kappa x^4$ is strictly positive and Theorem 10.7 again indicates that the spectrum is purely discrete. When $\kappa < 0$, the potential satisfies $V_\kappa(x) \to -\infty$ as $|x| \to \infty$ and the Hamiltonian is unbounded from below.

Problem 19.1. Prove that the essential spectrum of H_κ, for $\kappa < 0$, is $\mathbb{R}$, by constructing appropriate Zhislin sequences.

Because of this change in spectral type, from discrete spectrum to only essential spectrum as κ passes through 0, it is impossible that the eigenvalues are analytic in κ about $\kappa = 0$. What is true, however, is that the operator family H_κ is analytic *in a sector* $\mathcal{S}$ containing the positive real axis and with vertex at $\kappa = 0$. In fact, the family H_κ is analytic of type-A about any $\kappa_0 \in \mathcal{S}$. One of the amazing aspects of this example is that if one formally computes the Rayleigh–Schrödinger series for the eigenvalues given in Section 15.5, the values obtained are remarkably close to the eigenvalues of H_κ for $\kappa > 0$.

(2) The Stark effect. We consider the Schrödinger operator for a hydrogen atom in a uniform, external, electric field in the x_1-direction. The unperturbed operator is $H_0 \equiv -\Delta - 1/\|x\|$, the hydrogen atom Hamiltonian, acting on $L^2(\mathbb{R}^3)$. The perturbation is $-Fx_1$, where F is the electric field strength and plays the role of the coupling constant. For $F \neq 0$, the Faris–Lavine theorem (see [RS2]) can be used to prove that the operator is essentially self-adjoint on $C_0^\infty(\mathbb{R}^n)$. What are the spectral properties of

$$H(F) \equiv H_0 - Fx_1 = -\Delta - 1/\|x\| - Fx_1, \tag{19.2}$$

for various values of F? When $F = 0$, the spectrum is again well known: It consists of a sequence of infinitely many negative eigenvalues of the form $\{-1/m^2 \mid m \in \mathbb{Z}\}$, accumulating at 0, and essential spectrum $[0, \infty)$. Are the eigenvalues stable with respect to the perturbation? The perturbation is not relatively H_0-bounded, so some strange behavior might be anticipated. For $F \neq 0$, the essential spectrum of $H(F)$ is the entire real line and there are no eigenvalues. Thus, the negative eigenvalues of H_0 cannot be stable with respect to the perturbation in the sense of Chapter 15. We will see in Chapter 23 that they become resonances of $H(F)$. The formal Rayleigh–Schrödinger series gives remarkably good estimates of the shift of the real part of the eigenvalues. The imaginary part of the resonance, however, is exponentially small in F and cannot be computed in perturbation theory.

(3) A shape resonance model. This simple model is discussed by Hunziker in [Hu4]. One again considers the hydrogen atom Hamiltonian $H_0 \equiv -\Delta - 1/\|x\|$, acting on $L^2(\mathbb{R}^3)$. The perturbation is given by

$$V_\kappa(x) \equiv \frac{\kappa\|x\|}{1 + \kappa\|x\|}, \tag{19.3}$$

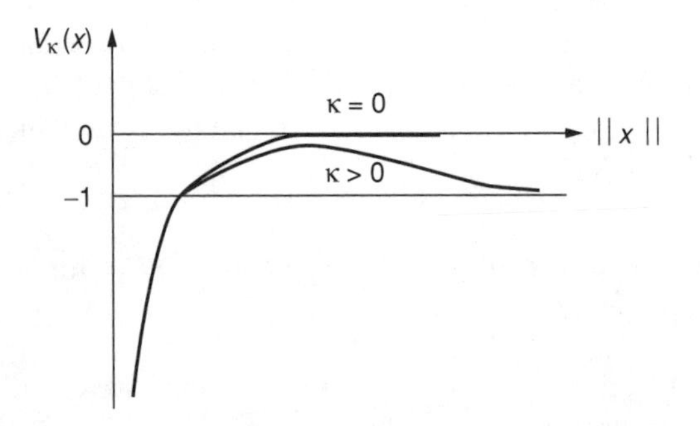

FIGURE 19.1. The radial section of the potential V_κ, for $\kappa > 0$.

for κ real, and we define

$$H_\kappa \equiv H_0 - V_\kappa. \tag{19.4}$$

Since $\lim_{\|x\|\to\infty} V_\kappa(x) = 1$, it is easy to check that $\inf \sigma_{\text{ess}}(H_\kappa) = -1$, for $\kappa \neq 0$. The potential is sketched in Figure 19.1. It is similar to Figure 16.1. We expect that the eigenvalues of H_0 in the interval $(-1, 0)$ will become resonances of H_κ. Once again, the perturbation is nonanalytic, and the notion of stability given in Chapter 15 is inadequate to explain this phenomenon.

We will devote the next few sections to developing a notion of stability applicable to the eigenvalues of the unperturbed operators given in the above examples. In the last section, we will comment on the case of perturbation of eigenvalues embedded in the essential spectrum.

19.2 Strong Resolvent Convergence

In dealing with the general question of eigenvalue stability, we must weaken our notion of the convergence of the perturbed family to the unperturbed operator. In the case of analytic perturbations studied in Chapter 15, we encountered the situation of *norm-resolvent convergence*: For any z in the common resolvent set,

$$\| R_\kappa(z) - R_0(z) \| \to 0 \tag{19.5}$$

as $\kappa \to 0$. This has as a consequence the norm convergence of the projectors, which is the first criterion for stability as given in Definition 15.1. Norm-resolvent convergence is not a *necessary* condition for the norm convergence of the projections. Indeed, in many situations of nonanalytic perturbations, the perturbation is not relatively bounded, so one cannot expect norm-resolvent convergence. We will replace this with *strong resolvent convergence*.

We now consider a family $\{T_\kappa\}$ of closed operators depending on a complex parameter $\kappa \in \mathcal{S} \subset \mathbb{C}$. We assume that $\mathcal{S}$ is a sector of the form

$$\mathcal{S} \equiv \{\kappa \in \mathbb{C} \mid \theta_0 < \arg(\kappa) < \theta_1\},$$

where, for some $\epsilon > 0, -\pi - \epsilon < \theta_0 < \theta_1 < \pi + \epsilon$. We are actually only concerned with the behavior of the family for $\kappa \in \mathcal{S}$ with $|\kappa|$ small. By $\kappa \to 0$, we mean the limit along some sequence in $\mathcal{S}$ approaching 0.

Definition 19.1. *A family of closed operators $\{T_\kappa, \kappa \in \mathcal{S}\}$ converges strongly in the generalized sense to a closed operator T_0 if*

(i) $\exists$ a nonempty, connected subset $\mathcal{A} \subset \mathbb{C}$ such that $\mathcal{A} \subset \rho(T_\kappa) \cap \rho(T_0)$ for all $\kappa \in \mathcal{S}$ and small;

(ii) for some $z \in \mathcal{A}$, we have strong resolvent convergence, that is, for any $u \in \mathcal{H}$,

$$s - \lim_{\kappa \to 0} R_\kappa(z)u = R_0(z)u. \tag{19.6}$$

Problem 19.2. Prove that if $\{T_\kappa\}$ is an analytic type-A family of operators on a neighborhood D about 0, then the family satisfies the conditions of generalized strong convergence.

An immediate question concerns the extension of the strong resolvent convergence from *some* $z \in \mathcal{A}$ to other points of $\mathcal{A}$. In the case of norm-resolvent convergence, (19.5), this can always be done.

Problem 19.3. Prove that norm-resolvent convergence at one point implies it at any other point in the corresponding connected component of the common resolvent sets.

It is clear that the question can be answered in the affirmative whenever there are uniform bounds on the resolvents in some neighborhood of z in $\mathcal{A}$.

Definition 19.2. *Let $\{T_\kappa\}$ be a family of closed operators as in Definition 19.1. The region of boundedness for the family $\{T_\kappa\}$ is a subset $\Delta_b \subset \mathbb{C}$ defined by*

$$\Delta_b \equiv \{z \mid \|R_\kappa(z)\| \leq M, \forall \kappa \in \mathcal{S} \text{ small}\},$$

for some positive constant M.

Problem 19.4. Prove that the region of boundedness is an open subset of $\mathcal{A}$, and use this to extend the strong resolvent convergence condition (19.6) to this set.

Given that a family $\{T_\kappa\}$ converges to an operator T_0 in the strong resolvent sense, what can be said about the relationship between the spectral properties of the family and those of T_0? Under norm-resolvent convergence, an isolated eigenvalue of T_0 cannot spread into an interval of essential spectrum because of the constancy of the dimension of the spectral projections following from (19.5). In general, we say that the spectrum cannot suddenly contract as the parameter approaches 0 (the dimension of the projection cannot go from infinite to finite as $\kappa \to 0$). This shows that there is some stability of the spectrum. In the examples of Section 19.1, we have a change of spectral type which indicates that we cannot have norm-resolvent convergence. Strong resolvent convergence, however, is much weaker. The only general statement that can be made applies to the case when the family $\{T_\kappa\}$ and the limit operator T_0 are self-adjoint [K]. Then, the spectrum of T_κ does not suddenly *expand* at $\kappa = 0$. One way to say this is that $\dim P_\kappa \geq \dim P_0$. In the following sections, we will study the stability of a discrete eigenvalue of T_0 in the general case of a perturbation under the conditions of strong resolvent convergence. This will necessitate additional assumptions, for it is possible that a discrete eigenvalue of T_0 can become embedded in or dissolve into the essential spectrum of T_κ, for any $\kappa \neq 0$, under the conditions of strong resolvent convergence.

19.3 The General Notion of Stability

The notion of stability for a discrete eigenvalue of an operator T_0 with respect to a perturbation T_κ is given by Kato [K]. We extend the notion of *core* introduced in Chapter 8. For a closed operator T with domain $D(T)$, a dense subset $D_0 \subset \mathcal{H}$ is a *core* for T if $T \mid D_0$ is closable with closure T. We first list our assumptions of the family T_κ for which we can define the notion of a stable eigenvalue.

Definition 19.3. *Let T_κ be an analytic family of closed operators for $\kappa \in \mathcal{S}$, a sector in $\mathbb{C}$ defined by*

$$\mathcal{S} \equiv \{\kappa \in \mathbb{C} \ |\theta_0 < \arg(\kappa) < \theta_1\},$$

where, for some $\epsilon > 0$, $-\pi - \epsilon < \theta_0 < \theta_1 < \pi + \epsilon$. Such a family $\{T_\kappa, \mathcal{S}\}$ is called a continuous family of operators if there exists a closed operator T_0, having a common core D_0 with the family T_κ, such that for any $u \in D_0$,

$$s - \lim_{\kappa \to 0} T_\kappa u = T_0 u.$$

Proposition 19.4. *Let T_κ and T_0 be a family of operators as in Definition 19.3. If $\rho(T_0) \cap \Delta_b \neq \emptyset$, then T_κ converges strongly to T_0 in the generalized sense.*

Proof. Let $z_0 \in \rho(T_0) \cap \Delta_b$. By the second resolvent formula, we have

$$R_\kappa(z_0) - R_0(z_0) = R_\kappa(z_0)[T_\kappa - T_0]R_0(z_0). \tag{19.7}$$

Applying this equation to a vector $v \equiv (T_0 - z_0)u$, with $u \in D_0$ so that $R_0(z_0)v \in D_0$, we obtain strong convergence on a dense set. By Problem 19.4, we can extend this to any $z \in \rho(T_0) \cap \Delta_b$. □

For the situation described in Definition 19.3, let us suppose that λ_0 is an isolated eigenvalue of T_0 with finite algebraic multiplicity, so $\lambda_0 \in \sigma_d(T_0)$. Let us recall the definition of the Riesz projections from Chapter 6, equation (6.1). Let Γ be a simple closed curve about λ_0 in the resolvent set of T_0. Then we define

$$P_{\lambda_0} \equiv \frac{1}{2\pi i} \oint_{\Gamma_{\lambda_0}} R_0(z)dz. \tag{19.8}$$

Recall that, in general, the projection P_{λ_0} is not orthogonal. We define P_κ similarly with $R_\kappa(z)$ in place of $R_0(z)$ in (19.8), when it makes sense.

Definition 19.5. *An eigenvalue $\lambda_0 \in \sigma_d(T_0)$ is stable with respect to the family of perturbations T_κ, as in Definition 19.4, if the following two conditions hold:*

(1) $\exists\, \delta > 0$ such that the annular region $A_\delta(\lambda_0) \equiv \{z \mid 0 < |z - \lambda_0| < \delta\}$ is in the boundedness set Δ_b of the family T_κ, for $\kappa \in S$ and small, and in the resolvent set of T_0.

(2) Let P_κ be the operator constructed as in (19.8) with T_0 replaced by T_κ. Then we require that

$$\lim_{\kappa \to 0} \|P_\kappa - P_{\lambda_0}\| = 0. \tag{19.9}$$

It follows from this definition that for all κ small, the operators H_κ have discrete spectra inside $A_\delta(\lambda_0)$ with total algebraic multiplicity equal to that of λ_0.

Problem 19.5. Suppose that the continuous family $\{T_\kappa S\}$ with limit operator T_0 satisfies the *first* condition of Definition 19.5. Prove that the Riesz projector P_κ converges *strongly* to P_0.

19.4 A Criterion for Stability

We will now concentrate on a special class of families of operators T_κ which arise from Schrödinger operators. We do this in order to utilize local compactness results and the techniques of geometric spectral analysis introduced in Chapter 10. Much of what is presented here can be extended to families of operators that are *local* perturbations of elliptic differential operators. The main difficulties in proving a stability result arise in locating the resolvent set and estimating the resolvent of T_κ. The resolvent set might be empty or might be the entire complex plane. Also, we noticed earlier that the *region of boundedness* plays a crucial role in the theory. As Problems 19.4 and 19.5 show, it is important to have information about this region to prove strong convergence of the projections and in order to prove the first condition of stability. Recall that our previous estimates on the resolvent of an operator, Corollary 5.7 and Theorem 5.8, required that the operator be self-adjoint. There are no such estimates for a general, closed operator. Thus, we have to put conditions on the family T_κ so that we can control the resolvent set and the region of boundedness Δ_b. Finally, as Problem 19.5 shows, projections for a continuous family of operators will converge strongly. More information is needed for the second condition of stability.

In order to motivate the additional conditions on a continuous family of operators sufficient for stability, we need another concept. One of the basic tools for the spectral analysis of closed operators is the *numerical range*.

Definition 19.6. *Let T be an operator with domain $D(T)$. The numerical range of T, denoted by $\Theta(T)$, is the subset of $\mathbb{C}$ defined by*

$$\Theta(T) \equiv \{\langle u, Tu\rangle \mid u \in D(T) \text{ and } \|u\| = 1\}. \tag{19.10}$$

An important fact about the numerical range of T is that it is a *convex* subset of $\mathbb{C}$. The proof is not relevant to the discussion here, so we refer to the book by Stone [St]. It is easier to check the following properties of the numerical range.

Problem 19.6. Let T be a closed operator, let $\Theta(T)$ be its numerical range, and define an open set $\Delta(T) \equiv \mathbb{C} \setminus \overline{\Theta(T)}$. For any $z \in \Delta(T)$, the operator $T - z$ is injective and has closed range.

The following proposition is an immediate consequence of Problem 19.6.

Proposition 19.7. *Let T be a closed operator. If $Ran(T - z)$ is dense in the Hilbert space for any $z \in \Delta(T)$, then $\Delta(T) \subset \rho(T)$ and*

$$\|(T - z)^{-1}\| \leq 1/\{\text{dist}(z, \overline{\Theta(T)}\,)\}.$$

Having introduced the concept of the numerical range, the following conditions on a potential are more natural.

Definition 19.8. *A family of complex-valued functions $V_\kappa \equiv V_\kappa^1 + V_\kappa^2 \in L^2_{\text{loc}}(\mathbb{R}^n)$ is called a continuous family of potentials with sector $\mathcal{S}$ if*

(i) V_κ^1 is relatively $p^2 \equiv -\Delta$-bounded in the sense that for any $\alpha > 0$, there exists a constant $\beta(\alpha) > 0$, such that for all $u \in C_0^\infty(\mathbb{R}^n)$,

$$\|V_\kappa^1(u)\| \leq \alpha\, \|p^2 u\| + \beta(\alpha)\, \|u\|, \tag{19.11}$$

uniformly for $\kappa \in \mathcal{S}$;

(ii) V_κ^2 is multiplication by a complex function in $L^\infty_{\text{loc}}(\mathbb{R}^n)$ such that the range of the function lies in a half-plane bounded away from the negative real axis,

$$0 \leq \text{Re}(e^{-i\gamma_\kappa} V_\kappa^2), \tag{19.12}$$

for some real γ_κ with $|\gamma_\kappa| < \pi/2 - \epsilon$, for some ϵ independent of κ and for all $\kappa \in \mathcal{S}$;

(iii) the potential V_κ converges strongly to V_0 as $\kappa \to 0$ within $\mathcal{S}$ on $C_0^\infty(\mathbb{R}^n)$.

We use continuous families of potentials to construct families of Schrödinger operators on $\mathcal{S}$. Let $H_\kappa \equiv p^2 + V_\kappa$ be initially defined on $C_0^\infty(\mathbb{R}^n)$.

Lemma 19.9. *Let $\{H_\kappa, \mathcal{S}\}$ be the densely defined family of operators introduced above for a continuous family of potentials V_κ.*

(1) For any $u \in C_0^\infty(\mathbb{R}^n)$ with $\|u\| = 1$, we have

$$\langle u, p^2 u\rangle \leq a\mathrm{Re}\left\{e^{-i\gamma_\kappa}\langle u, H_\kappa u\rangle\right\} + b, \tag{19.13}$$

for finite, positive constants a and b independent of κ.

(2) The numerical range of H_κ lies in the half-plane Π:

$$\Theta(H_\kappa) \subset \Pi \equiv \{z \in \mathbb{C} \mid -\epsilon < \arg\left(z - \frac{b}{a}\right) < \pi - \epsilon\},$$

where the constants a and b are defined in (19.13) and ϵ is given in Definition 19.8.

(3) The operators H_κ, $\kappa \in S$ are closable. For any $z \in \mathbb{C} \setminus \overline{\Pi}$, the operator $(H_\kappa - z)$ has dense range, so (denoting the closure by H_κ also) $\sigma(H_\kappa) \subset \overline{\Pi}$.

Proof. We will prove the first two statements and refer to [Hu4] for the proof of part (3). Let $H_\kappa = p^2 + V_\kappa$ on domain $C_0^\infty(\mathbb{R}^n)$. It is simple to check that

$$\mathrm{Re}\ V_\kappa^2 = \mathrm{Re}\ H_\kappa - p^2 - \mathrm{Re}\ V_\kappa^1 \tag{19.14}$$

and

$$\mathrm{Im}\ V_\kappa^2 = \mathrm{Im}\ H_\kappa - \mathrm{Im}\ V_\kappa^1. \tag{19.15}$$

In what follows, we write u for any element of $C_0^\infty(\mathbb{R}^n)$ with $\|u\| = 1$. Substituting these identities into the condition (19.12) on V_κ^2, we obtain

$$\mathrm{Re}\langle e^{-i\gamma_\kappa} H_\kappa\rangle_u \geq \cos\gamma_\kappa \langle (p^2 - |V_\kappa^1|)\rangle_u, \tag{19.16}$$

as the reader can easily verify using the fact that

$$\mathrm{Re}(e^{-i\gamma_\kappa} V_\kappa^1) \leq |V_\kappa^1|.$$

We now choose δ such that $0 < \delta < \cos\gamma_\kappa$. From the relative boundedness condition (19.11) on V_κ^1, we have

$$\|V_\kappa^1 u\| \leq (\alpha\delta^{-1})\|\delta p^2 u\| + \beta(\alpha). \tag{19.17}$$

From (19.17) and the Kato–Rellich theorem, Theorem 13.5 and Problem 13.5, it follows that the operator $\delta p^2 - |V_\kappa^1|$ is semibounded with lower bound $-\beta(\alpha)\delta(\delta - \alpha)^{-1}$, for $\alpha < \delta$. Substituting this into (19.16) leads to the inequality

$$\mathrm{Re}\langle e^{-i\gamma_\kappa} H_\kappa\rangle_u \geq (\cos\gamma_\kappa - \delta)\langle p^2\rangle_u - \beta(\alpha)\delta(\delta - \alpha)^{-1}. \tag{19.18}$$

Since $|\gamma_\kappa| < \pi/2 - \epsilon$, for some fixed $\epsilon > 0$, it follows that $(\cos\gamma_\kappa - \delta) > \sin\epsilon - \delta$. Hence, result (19.13) follows from both (19.18) and this observation with suitable a and b independent of κ.

The statements about the numerical range are proved by observing that for any $u \in C_0^\infty(\mathbb{R}^n)$ with $\|u\| = 1$, condition (19.13) implies that

$$0 \leq a \operatorname{Re}\langle (e^{-i\gamma_\kappa} H_\kappa) \rangle_u + b. \tag{19.19}$$

In particular, the numerical range lies in a half-plane. □

Let us recall that for self-adjoint Schrödinger operator H with $C_0^\infty(\mathbb{R}^n)$ as a core, the condition

$$\|(H - z)u\| \geq \epsilon \|u\|,$$

for all $u \in C_0^\infty(\mathbb{R}^n)$, implies that $z \in \rho(H)$. For non–self-adjoint operators, this is no longer true in general, and we need additional information on the adjoint H^* of H. This is the motivation for the following definition.

Definition 19.10. *A family of Schrödinger operators of the form*

$$H_\kappa \equiv -\Delta + V_\kappa, \tag{19.20}$$

where V_κ a continuous family of potentials with sector $\mathcal{S}$, is called a continuous family of Schrödinger operators with sector $\mathcal{S}$ if

(i) the adjoint of H_κ, denoted H_κ^*, is the complex conjugate of H_κ, that is, if J is the complex conjugation operator, then

$$H_\kappa^* = J H_\kappa J; \tag{19.21}$$

(ii) the family H_κ is an analytic family of operators on any open region in the sector $\mathcal{S}$.

Examples 19.11.

(1) Consider an admissible potential V as in Definition 18.3. Let g be the vector field for the corresponding spectral deformation family. The Schrödinger operator is

$$H_\theta = p_\theta^2 + V_\theta, \tag{19.22}$$

on the sector of analyticity of V. Recall that

$$V_\theta(x) \equiv V(\phi_\theta(x)). \tag{19.23}$$

From (18.10), we can write

$$H_\theta = p^2 + V_\theta^1, \tag{19.24}$$

with the potential V_θ^1 satisfying condition (19.11). It is easy to check that this is a continuous family of Schrödinger operators.

(2) The anharmonic oscillator H_κ, given in (19.1), is a continuous family of Schrödinger operators on the sector $\mathcal{S} = \{z \mid |\arg z| < \pi/2 - \epsilon\}$. In this case, $V_\theta^1 = 0$ and one can check condition (19.12).

Problem 19.7. Show that a continuous family of Schrödinger operators is a continuous family of operators in the sense of Definition 19.3 and hence converges in the generalized strong sense to H_0, as in Definition 19.1.

We are now in position to state the main result on continuous families of Schrödinger operators due to Vock and Hunziker [VH] (see also Hunziker [Hu4]).

Theorem 19.12. *Let H_κ, $\kappa \in \mathcal{S}$, be a continuous family of Schrödinger operators as in Definition 19.10. Suppose that λ_0 is a discrete eigenvalue of H_0. The eigenvalue λ_0 is stable if, for all $u \in C_0^\infty(\|x\| > n)$, for some $n > 0$, there is an $\epsilon > 0$ such that*

$$\|(H_\kappa - \lambda_0)u\| \geq \epsilon \|u\| > 0. \tag{19.25}$$

19.5 Proof of the Stability Criteria

We now proceed with the proof of the stability criteria, Theorem 19.12, for a continuous family of Schrödinger operators. Let H_κ be a continuous family of Schrödinger operators on a sector $\mathcal{S}$ satisfying the conditions of Definition 19.10. We suppose that λ_0 is an isolated eigenvalue of H_0 with multiplicity m_0. We need two preliminary lemmas. The first concerns local compactness of a continuous family, and the second gives some important statements that are equivalent to inequality (19.25).

Lemma 19.13. *Let $F \in C_0^\infty(\mathbb{R}^n)$. Let $\kappa \to 0$ be a sequence in $\mathcal{S}$, and choose a sequence $u_\kappa \in D(H_\kappa)$ satisfying*

$$\|u_\kappa\| + \|H_\kappa u_\kappa\| \leq M < \infty, \tag{19.26}$$

for some constant M. Then if $w - \lim_{\kappa \to 0} u_\kappa = 0$, we have

$$\lim_{\kappa \to 0} \|F u_\kappa\| = 0 \tag{19.27}$$

and

$$\lim_{\kappa \to 0} \|[H_\kappa, F]u_\kappa\| = 0. \tag{19.28}$$

Proof. Let us define $|p| \equiv \sqrt{-\Delta}$. From the proof of Lemma 19.9, it follows that for any $u \in C_0^\infty(\mathbb{R}^n)$,

$$\| \, |p|u \, \|^2 \leq \langle p^2 \rangle_u \leq a \operatorname{Re}(e^{-i\gamma_\kappa} H_\kappa) + b\|u\|^2. \tag{19.29}$$

Applying the Schwarz inequality to the right side of (19.29), we obtain

$$\| \, |p|u \, \| \leq \left(\frac{a}{2\delta}\right) \|H_\kappa u\| + \left(b + \frac{a\delta}{2}\right) \|u\|, \tag{19.30}$$

for any $\delta > 0$. This shows that $|p|$ is relatively H_κ-bounded with an arbitrarily small relative bound. In Problem 10.2, it was established that the operator $|p| \equiv \sqrt{-\Delta}$ is locally compact. From this fact and the relative boundedness, it follows that H_κ is locally compact. This first statement, (19.27), follows from local compactness and the boundedness property of the sequence. As for the second condition, the commutator is

$$[H_\kappa, F] = 2i\partial_j F \cdot p_j - \Delta F. \tag{19.31}$$

The above derivation can be done with p_k, $k = 1, \ldots, n$, in place of $|p|$. This shows that the commutator is relatively H_κ-bounded with an arbitrarily small relative bound. This proves that the sequence $\{H_\kappa F u_\kappa\}$ is uniformly bounded. Let $G \in C_0^\infty(\mathbb{R}^n)$ satisfy $FG = F$. Then for any $\alpha > 0$ and any $u \in C_0^\infty(\mathbb{R}^n)$, we have

$$\|[H_\kappa, F]u_\kappa\| \leq \alpha \|H_\kappa G u_\kappa\| + \beta \|G u_\kappa\|. \tag{19.32}$$

The first term is unformly bounded in κ. Since the second term converges to zero and α is arbitrarily small, the result follows. □

Lemma 19.14. *The following three conditions are equivalent:*

(1) the number $z \in \Delta_b$;

(2) for all $u \in C_0^\infty(\mathbb{R}^n)$ and $\forall\, \kappa \in \mathcal{S}$ with $|\kappa|$ small,

$$\|(H_\kappa - z)u\| \geq \epsilon \|u\| > 0; \tag{19.33}$$

(3) the number $z \in \rho(H_0)$ and $\forall\, u \in C_0^\infty(\|x\| > n)$ and $\forall\, \kappa \in \mathcal{S}$ with $|\kappa|$ small,

$$\|(H_\kappa - z)u\| \geq \epsilon \|u\| > 0. \tag{19.34}$$

Proof. It is clear that (1) $\Rightarrow$ (2) and that (1) $\Rightarrow$ (3). We first prove that (2) $\Rightarrow$ (1) and then that (3) $\Rightarrow$ (1). From (19.14) and condition (19.33), it follows that for all $u \in C_0^\infty(\mathbb{R}^n)$,

$$\|(H_\kappa^* - \bar{z})u\| \geq \epsilon \|u\|. \tag{19.35}$$

This implies that $\mathrm{Ran}(H_\kappa - z) = L^2(\mathbb{R}^n)$ and that $\|R_\kappa(z)\| \leq \epsilon^{-1}$, which proves $z \in \Delta_b$. It remains to show that (3) $\Rightarrow$ (1). Suppose, to the contrary, that (3) holds but $z \notin \Delta_b$. Then, there exists a sequence $\{u_\kappa\} \subset C_0^\infty(\mathbb{R}^n)$, with $\|u_\kappa\| = 1$, such that

$$\|(H_\kappa - z)u_\kappa\| \to 0, \tag{19.36}$$

as $\kappa \to 0$. We can extract a subsequence that converges weakly to, say, u. For any $v \in D(H_\kappa^*)$, it follows that

$$0 = \lim_{\kappa \to 0} \langle (H_\kappa^* - \bar{z})v, u_\kappa \rangle = \langle (H_0^* - \bar{z})v, u \rangle, \tag{19.37}$$

and so $u \in D(H_0)$ and $(H_0 - z)u = 0$. Since $z \in \rho(H_0)$, this means that $u = 0$. Hence, the sequence $\{u_\kappa\}$ satisfies $w - \lim_{\kappa \to 0} u_\kappa = 0$ and $s - \lim_{\kappa \to 0}(H_\kappa - z)u_\kappa = 0$. By Lemma 19.13, this implies that for any $F \in C_0^\infty(\mathbb{R}^n)$,

$$\|(H_\kappa - z)(1 - F)u_\kappa\| = 0, \tag{19.38}$$

whereas $\|(1 - F)u_\kappa\| = 1$. Choosing F so that $F|B_n(0) = 1$, this contradicts inequality (19.34). □

Proof of Theorem 19.12. Since $\lambda_0 \in \sigma_d(H_0)$, we can choose $\delta > 0$ such that the annular region

$$\mathcal{A}_\delta \equiv \{z \in \mathbb{C} \mid 0 < |z - \lambda_0| < \delta\} \subset \rho(H_0). \tag{19.39}$$

Conditions (19.25) and (19.39) imply that for all $u \in C_0^\infty(\|x\| > n)$ and for all $z \in \mathcal{A}_\delta$, we have

$$\|(H_\kappa - z)u\| \geq \left(\frac{\epsilon}{2}\right) ||u|| > 0. \tag{19.40}$$

This condition also holds for the pair $\{H_\kappa^*, \bar{z}\}$. By the local compactness lemma, Lemma 19.13, and the fact that $C_0^\infty(\mathbb{R}^n)$ is a common core for the family and the adjoint family, we conclude that $\operatorname{Ran}(H_\kappa - z)$ is dense and closed, that is, equal to $L^2(\mathbb{R}^n)$, for any $z \in \mathcal{A}_\delta$. Inequality (19.40) therefore implies that

$$\|R_\kappa(z)\| \leq \frac{2}{\epsilon}, \tag{19.41}$$

for all $z \in \mathcal{A}_\delta$. This means that $\mathcal{A}_\delta \subset \Delta_b$, for all κ sufficiently small.

It follows from this and Problem 19.5 that the Riesz projectors satisfy

$$\lim_{\kappa \to 0} P_\kappa u = P_0 u$$

and

$$\lim_{\kappa \to 0} P_\kappa^* u = P_0^* u,$$

for any $u \in C_0^\infty(\mathbb{R}^n)$. The first of these conditions implies the principle of nonexpansion of the spectrum, that is,

$$\dim P_\kappa \geq \dim P_0.$$

By a lemma of Kato (see Lemma 1.24 of [K]), the above two conditions and the condition that

$$\dim P_\kappa \leq \dim P_0, \tag{19.42}$$

for all κ sufficiently small, are sufficient to prove part (2) of the stability criteria, that is, the norm convergence of the projections. Hence, we turn to the proof of (19.42).

Suppose that inequality (19.42) is false. Then there exists a sequence u_κ such that

$$P_\kappa u_\kappa = u_\kappa \quad \text{and} \quad P_0 u_\kappa = 0, \tag{19.43}$$

for some sequence $\kappa \to 0$.

Problem 19.8. Prove that there is a subsequence $\{u_{\kappa(i)}\}$ of $\{u_\kappa\}$ which converges weakly to zero. (*Hint*: Use the strong convergence of the projections proved in Problem 19.5.)

We now use the lower-bound inequality (19.34) to obtain a contradiction. Since $\mathcal{A}_\delta \subset \Delta_b$, there exists a neighborhood of $\kappa = 0$ in S on which $\| R_\kappa(z) \|$ is uniformly bounded in κ and z, and so we have

$$\| H_\kappa u_\kappa \| \leq \| H_\kappa P_\kappa \| \leq C_0. \tag{19.44}$$

The local compactness lemma, Lemma 19.13, and the uniform boundedness of the resolvent imply that for any $F \in C_0^\infty(\mathbb{R}^n)$, we have

$$\lim_{\kappa \to 0} \|(1 - F)u_\kappa\| = 1. \tag{19.45}$$

Taking $F = 1$ for $\|x\| < n$, inequality (19.34) implies that

$$\|(H_\kappa - z)(1 - F)u\| \geq \left(\frac{\epsilon}{2}\right) ||(1 - F)u|| > 0. \tag{19.46}$$

Extending this to $u \in D(H_\kappa)$ and using the invertibility of $(H_\kappa - z)$ for $z \in \mathcal{A}_\delta$, it follows from (19.46) that

$$\begin{aligned} \left(\frac{\epsilon}{2}\right) ||(1 - F)R_\kappa(z)u_\kappa\| &\leq \|(H_\kappa - z)(1 - F)R_\kappa(z)u_\kappa\| & (19.47) \\ &\leq \|(1 - F)u_\kappa\| + \| [R_\kappa(z), F]u_\kappa\|. & (19.48) \end{aligned}$$

Let $\Gamma \subset \mathcal{A}_\delta$ be a simple closed contour of radius $r < \delta$. Integrating both sides of inequality (19.48) about Γ, we obtain

$$\left(\frac{\epsilon}{2}\right) ||(1 - F)u_\kappa\| \leq r\|(1 - F)u_\kappa\| + 2\pi^{-1}C_0 \oint_\Gamma |dz| \, \|[H_\kappa, F]R_\kappa(z)u_\kappa\|, \tag{19.49}$$

where we used the uniform boundedness of $\|R_\kappa(z)\|$ on Γ and for all κ small. Since $[H_\kappa, F]R_\kappa(z)$ is a compact operator that is uniformly bounded on Γ and $w - \lim_{\kappa \to 0} u_\kappa = 0$, it follows that the integrand on the right side of (19.49) converges to zero. Since we have chosen $r < \delta < \epsilon/2$, we obtain a false inequality from (19.49) unless $\|(1 - F)u_\kappa\| \to 0$. This, however, contradicts (19.45). □

19.6 Geometric Techniques and Applications to Stability

Theorem 19.12 reduces the question of eigenvalue stability to the calculation of a lower bound on the norm of the operator $H_\kappa - \lambda_0$ acting on smooth states supported in the complement of an arbitrarily large region, say $B_R(0)$. Quite often, we will have local information about the potential on subsets of such a region. We can utilize this by using a *partition of unity* in the complement of $B_R(0)$ and estimating each piece separately.

The use of partitions of unity for such estimations plays an important role in geometric spectral analysis. We will give some general formulas in this section. These will be used in our discussion of resonances in Chapters 20–23.

For a function $j \in C^2(\mathbb{R}^n)$, we define a first-order differential operator $W(j)$ by

$$W(j) \equiv [\Delta, j] = \nabla \cdot \nabla j - \nabla j \cdot \nabla. \tag{19.50}$$

Note that this operator is localized to the support of ∇j. We will use two types of partitions of unity $\{j_i\}$: a standard partition for which $\sum_{i=1}^N j_i = 1$, and sometimes it will be convenient to choose the functions such that $\sum_{i=1}^N {j_i}^2 = 1$.

In this section, we will work formally and assume that $C_0^\infty(\mathbb{R}^n)$ is a core for the Schrödinger operator $H = -\Delta + V$.

(1) Lower-bound inequality. Let $\{j_i\}_{i=1}^N$ be a partition of unity with $\sum_{i=1}^N j_i^2 = 1$. Then, for $H = -\Delta + V$, we have

$$\|(H - z)u\|^2 \geq \frac{1}{2} \sum_{i=1}^N \left(\|(H - z)j_i u\|^2 - \|W_i u\|^2 \right), \tag{19.51}$$

for $u \in D(H)$ and any $z \in \mathbb{C}$. We have written $W_i = W(j_i)$. If this identity is to be applied to (19.18), for example, the partition of unity is chosen to be one on the exterior of some ball.

Problem 19.9. Prove this inequality.

We can improve this estimate for families of continuous Schrödinger operators (or a family depending on some parameter) using local compactness. Let us consider a finite partition of unity $\{j_{i,\kappa}\}$, for the exterior of $B_n(0)$, where n is as in Theorem 19.12. Suppose that, in addition, the functions satisfy

$$\partial^\alpha j_{i,\kappa} \to 0, \tag{19.52}$$

as $\|x\| \to \infty$ and $\kappa \to 0$, for multi-indices α, with $|\alpha| = 1$ and 2. The proof of the following lemma can be found in [Hu4]; one direction follows immediately from (19.25), and the other direction uses the local compactness lemma, Lemma 19.13.

Lemma 19.15. *Inequality (19.18) holds if and only if*

$$\|(H_\kappa - \lambda_0) j_{i,\kappa} u\| \geq \epsilon || j_{i,\kappa} u|| > 0, \tag{19.53}$$

for some $\epsilon > 0$ *and for all* $j_{i,\kappa}$.

(2) IMS localization formula. Suppose that $\{j_i\}_{i=1}^N$ is a set of C^2-functions such that $\sum_{i=1}^N j_i^2 = 1$. The *IMS localization formula* (see [Si1]) for a Hamiltonian H is

$$H = \sum_{i=1}^N (j_i H j_i - |\nabla j_i|^2). \tag{19.54}$$

Problem 19.10. Prove the IMS formula by computing $[H, [H, j_i]]$.

(3) Geometric resolvent formulas. *Geometric resolvent formulas*, that are similar to the second resolvent formula, provide a powerful tool for comparing the resolvents of operators that are the same when acting on functions localized to certain regions of $\mathbb{R}^n$, but differ in other regions where the resolvents can be controlled. We will see many examples in the next chapters. The heart of geometric perturbation theory method is to estimate $H = -\Delta + V$ by simpler Hamiltonians H_i, $i = 1, \dots, N$, $H_i \equiv -\Delta + V_i$, with V_i differing from V in a region of $\mathbb{R}^n$ that is classically forbidden for the interval of energies I we are considering. For an interval I, we define $\text{CFR}(I) \equiv \bigcap_{E \in I} \text{CFR}(E)$. Typically, the V_i's are obtained from V as follows. Let $\{j_i\}_{i=1}^N$ be a partition of unity for $\mathbb{R}^n$ such that $j_i \in C^2$, $j_i \geq 0$, and $\sum_{i=1}^N j_i = 1$. The operators $\{H_i\}_{i=1}^N$ are self-adjoint Schrödinger operators on $L^2(\mathbb{R}^n)$ with potentials V_i having the property that

$$V_i \mid \text{supp } j_i = V, \tag{19.55}$$

so that $\text{supp}(V - V_i) \subset \text{CFR}(I)$. Each V_i is extended to $\mathbb{R}^n$ in a suitable manner. Each H_i is simple in the sense that the resolvent $R_i(z) \equiv (H_i - z)^{-1}$ can be analyzed. We relate $R_i(z)$ to $R(z) \equiv (H - z)^{-1}$ by the *geometric resolvent equation*. As above, let W_i be the first-order differential operator defined by

$$W_i \equiv [\Delta, j_i]. \tag{19.56}$$

We assume that W_i is relatively H_i-bounded.

Lemma 19.16. *Let H and $\{H_i\}$ be constructed as above using a partition of unity $\{j_i\}_{i=1}^N$. Then, for all z in the intersection of the resolvent sets of H and each H_i,*

$$R(z) - \sum_{i=1}^N j_i R_i(z) = \sum_{i=1}^N R(z) W_i R_i(z). \tag{19.57}$$

Problem 19.11. Prove Lemma 19.15.

19.7 Example: A Simple Shape Resonance Model

We illustrate the ideas of this chapter with the shape resonance model of Section 6.2. This model is due to Hunziker [Hu4]. Our discussion will also serve as an introduction to the theory of shape resonances in the semiclassical regime, which will be discussed in Chapter 20. We are concerned with a family of Schrödinger operators, H_κ, $\kappa \in \mathbb{R}$, acting on $L^2(\mathbb{R}^3)$, of the form

$$H_\kappa = -\Delta - \frac{1}{\|x\|} - \frac{\kappa \|x\|}{1 + \kappa \|x\|}. \tag{19.58}$$

The potential is sketched in Figure 19.1. When $\kappa = 0$, this is the hydrogen atom Hamiltonian. There is discrete spectrum in the interval $(-1, 0)$, and the essential spectrum is the half-line $[0, \infty)$. For $\kappa \neq 0$, the essential spectrum shifts to the half-line $[-1, \infty)$. Theorem 16.1 can be used to show that the Schrödinger operator

$$\tilde{H}_\kappa \equiv H_\kappa + 1,$$

has no positive eigenvalues so that H_κ has no eigenvalues greater than -1. We prove that the eigenvalues of H_0 in the interval $(-1, 0)$ become resonances of H_κ, $\kappa \in \mathbb{R}$.

The first step is to deform spectrally the family of operators H_κ. Because of the simple form of the potential, we can use dilation analyticity. For $\theta \in \mathbb{R}$, the dilation transformation $x \mapsto e^\theta x$ generates the spectral deformation family. The unitarily equivalent family of Schrödinger operators, for $\theta \in \mathbb{R}$, is

$$H_\kappa^\theta = e^{-2\theta} p^2 - \frac{e^{-\theta}}{\|x\|} - \frac{e^\theta \kappa \|x\|}{1 + e^\theta \kappa \|x\|}. \tag{19.59}$$

For fixed κ, this is an analytic type-A family of Schrödinger operators with respect to θ for $|\arg \theta| < \pi/2$. Moreover, using the results of Chapter 18, it is easy to verify that

$$\sigma_{\text{ess}}(H_0^\theta) = \{z \in \mathbb{C} \mid \arg z = -2i \text{ Im } \theta\} \tag{19.60}$$

and, for $\kappa \in \mathbb{R}$,

$$\sigma_{\text{ess}}(H_\kappa^\theta) = \{-1 + z \mid z \in \mathbb{C} \text{ and } \arg z = -2i \text{ Im } \theta\}. \tag{19.61}$$

It also follows from the Aguilar–Balslev–Combes theorem that the discrete spectrum of H_0 in the interval $(-1, 0)$ remains the discrete spectrum of H_0^θ. We want to show that these eigenvalues are stable with respect to the perturbation H_κ^θ.

We now fix $0 < \text{Im } \theta < \pi/2$ and study the Schrödinger operators H_κ^θ as functions of κ. The family H_κ^θ, up to a factor of $e^{-2\theta}$, is a *continuous family of Schrödinger operators*, in the sense of Definition 19.10, with sector $\mathcal{S} = \{\kappa \in \mathbb{C} \mid \epsilon < \arg \kappa + \text{Im } \theta < \pi - \epsilon\}$, for any $\epsilon > 0$. The form of the sector can be verified by examining the denominator of the perturbing potential. Because of this, we can apply Theorem 19.12 to establish the stability of a discrete eigenvalue of H_0^θ in the interval $(-1, 0)$. In order to do this, we use the geometric method and Lemma 19.15. It is important to note that the stability estimate (19.25) cannot be obtained simply by a numerical range argument. This is because the numerical range, being a convex set, includes the entire sector $\{-1 + z \mid z \in \mathbb{C} \text{ and } \arg z = -2i \text{ Im}\,\theta\}$. Consequently, the resolvent estimate of Proposition 19.7 gives no useful information in a neighborhood of any eigenvalue in the interval $(-1, 0)$. This is the typical situation one encounters in dealing with resonances. Instead, we have the following result.

Lemma 19.17. *The set of λ for which the stability estimate (19.25) fails to hold is the range of the function Φ defined by*

$$(\kappa, p, x) \mapsto e^{-2\theta} p^2 - \frac{e^{\theta} \kappa \|x\|}{1 + e^{\theta} \kappa \|x\|},$$

on $\mathcal{S} \times \mathbb{R}^3 \times \mathbb{R}^3$.

As the reader can check, the complement of this set in $\mathbb{C}$ includes a neighborhood of the interval $(-1 + \epsilon, -\epsilon)$, so that the stability estimate (19.25) holds for the eigenvalues in this region. Consequently, by Theorem 19.12, these eigenvalues are stable.

Sketch of the proof of Lemma 19.17. Since the Coulomb potential contribution to V_κ^θ is a Kato potential, we do not expect it to contribute to the determination of the stability set. Now, if λ is in the range of the map Φ, we can use the previous comment to construct a sequence $\kappa_n \to 0$ in the sector $\mathcal{S}$ and corresponding functions $u_{\kappa_n} \in C_0^\infty(\|x\| > n)$ such that

$$\lim_{n \to \infty} (H_{\kappa_n}^\theta - \lambda) u_{\kappa_n} = 0.$$

Conversely, we construct a finite partition of unity $J_{\alpha,\kappa}$ of the range of the map Φ, and corresponding operators $H_{\kappa,\alpha}^\theta$, such that

$$\|(H_\kappa^\theta - H_{\kappa,\alpha}^\theta) J_{\alpha,\kappa} u\| \leq (|e^\theta n|^{-1} + \delta) \| J_{\alpha,\kappa} u \|, \tag{19.62}$$

for any $u \in C_0^\infty(\|x\| > n)$ and $\delta > 0$ small. The operators $H_{\kappa,\alpha}^\theta$ have numerical range contained in the range of Φ. If λ is not in the range of Φ, it then follows from Lemma 19.15 and (19.62) that the stability criterion (19.53) holds for each α, for suitable n and δ. To construct the partition of unity, let $\phi \equiv \operatorname{Im} \theta + \arg \kappa$, and define a function

$$V(\phi, s) \equiv \frac{e^{i\phi} s}{1 + e^{i\phi} s},$$

for $0 \leq s$ and $\epsilon < \phi < \pi - \epsilon$, as above. We first choose a finite partition of unity J_α of $\mathbb{R}^+$ as follows. For any $\delta > 0$, choose finitely many points s_α so that

$$|V(\phi, s) - V(\phi, s_\alpha)| < \delta,$$

for all $\epsilon < \phi < \pi - \epsilon$ and for all s in supp J_α. The partition $\{J_{\alpha,\kappa}\}$ is defined by $J_{\alpha,\kappa}(x) \equiv J_\alpha(|\kappa e^\theta| \, \|x\|)$. We then define the operators $H_{\kappa,\alpha}^\theta \equiv e^{-2\theta} p^2 - V(\phi, s_\alpha)$. The reader can verify the claims made above. □

20

Theory of Quantum Resonances II: The Shape Resonance Model

20.1 Introduction: The Gamow Model of Alpha Decay

The shape resonance model was developed by Gamow [Ga] and by Gurney and Condon [GC] to describe the decay of an unstable atomic nucleus by alpha-particle emission. The idea is very simple. The atomic nucleus is modeled by a potential barrier of finite width which traps the alpha particle. A typical situation is shown in Figure 16.1. According to quantum theory, the wave function for the alpha particle, initially localized in the potential well between the barriers, will oscillate between the barriers. However, because the barriers have finite thickness, as measured by the Agmon metric, the wave function penetrates the barriers, and hence the particle has a nonzero probability of escaping to infinity. In fact, the probability that the particle will escape to infinity in infinite time is 1. To say this another way, there are typically no bound states for this model, and consequently a decay condition such as (16.1) holds: The wave function will eventually leave every bounded region. The classical limit of this model is clear: The barriers are infinitely high, so the distance in the Agmon metric across the barrier is infinite. This forces the wave function to vanish in the classically forbidden region, and the alpha particle is in a bound state. As we will show by a rescaling of the Schrödinger operator, this is equivalent to taking Planck's constant to zero. The semiclassical regime, therefore, is described by very large potential barriers relative to the energy of the alpha particle. We describe the alpha particle as a quantum resonance of a Hamiltonian $H(\lambda) = -\Delta + \lambda^2 V$, where V has the "shape" of a confining potential barrier of finite thickness determined by λ. We will work in the large λ regime.

The lifetime of the resonance is controlled by the characteristics of this barrier given by V and by the semiclassical parameter λ. The lifetime is the inverse of the imaginary part of the resonance $E - i\Gamma$, $\Gamma > 0$. In this and the next two chapters, we will prove the existence of resonances in the large λ regime and give an upper bound on the resonance width $\Gamma(\lambda)$.

Let us point out some differences between this problem and the shape resonance model of Section 19.1. The perturbation in the model we study is formally around $\lambda = \infty$. The low-lying eigenvalues of the problem of a potential well with infinitely high potential barriers will play the role of the unperturbed eigenvalues. We will show that these are stable under the perturbation given by large, but finite, λ. This Hamiltonian has no bound states, so one must first apply the method of spectral deformation to it to move away the essential spectrum. The goal is now to show that this spectrally deformed Hamiltonian has an eigenvalue close to a low-lying eigenvalue of the unperturbed problem. For this, we need a stability estimate similar to the one appearing in Theorem 19.12. This will be obtained using localization methods as in Section 19.6. In the potential well region, the spectrally deformed Hamiltonian is close to the unperturbed model. In the classically forbidden region, we will use tunneling estimates to control the perturbation. In the exterior region, where the energy is above the potential, we need to use *quantum nontrapping methods* to control the resolvent. These estimates, of interest in their own right, are discussed in the next chapter.

20.2 The Shape Resonance Model

The model we study is described by a potential V on $\mathbb{R}^n$ having the following *geometric* and *complex analytic* properties. We first describe the geometric properties.

(V1) The potential $V \in C^2(\mathbb{R}^n)$ is a real-valued, bounded function such that for some $\epsilon > 0$ and for any multi-index $\alpha = (\alpha_1, \ldots, \alpha_n)$, $\alpha_i \in \mathbb{N} \cup \{0\}$, with $|\alpha| = \sum_{i=1}^n \alpha_i$ and $\partial^\alpha = \frac{\partial^{\alpha_1}}{\partial x_1^{\alpha_1}} \cdots \frac{\partial^{\alpha_n}}{\partial x_n^{\alpha_n}}$, the potential V satisfies

$$|\partial^\alpha V(x)| \leq c_\alpha (1 + \|x\|)^{-|\alpha| - \epsilon}, \tag{20.1}$$

for $|\alpha| \leq 0, 1, 2$ and finite constants $c_\alpha > 0$.

(V2) The potential V has a single, nondegenerate minimum at x_0 such that $V(x_0) > 0$.

The nondegeneracy of V at x_0 means that the $n \times n$ matrix of second partial derivatives of V at x_0,

$$A \equiv \left[\frac{\partial^2 V}{\partial x_i \partial x_j}(x_0) \right], \tag{20.2}$$

is a positive definite matrix. By a simple translation, we can assume that $x_0 = 0$. The results derived here can be extended to the case of several nondegenerate minima. One can also allow the potential V to have local singularities (see [HS2]).

Conditions (V1) and (V2) have the following geometric consequences. For any $E > V(0)$, we define the *classically forbidden region* for V at energy E, denoted by $\mathrm{CFR}(E)$ as in Section 16.1, as the subset of $\mathbb{R}^n$ given by

$$\mathrm{CFR}(E) \equiv \{x \mid V(x) > E\}. \tag{20.3}$$

Condition (V1) guarantees that this set has compact closure, for, if not, there exists a sequence $\{x_n\}$, with $\|x_n\| \to \infty$, such that $\lim_{n\to\infty} V(x_n) \geq E$, which contradicts (V1). If $E > \sup_{x\in\mathbb{R}^n} V(x)$, then $\mathrm{CFR}(E) = \emptyset$. In the following text, we will always assume that $E > V(0)$ is chosen such that $\mathrm{CFR}(E) \neq \emptyset$ and that $\mathrm{Int}\,\mathrm{CFR}(E) \neq \emptyset$. This is actually no additional restriction, since we will work with low-lying eigenvalues of the potential well.

For such an $E > V(0)$, the complement of $\mathrm{CFR}(E)$ in $\mathbb{R}^n$ consists of two, disjoint, closed regions. One region is compact and contains the origin:

$$W(E) \equiv \left\{x \mid V(x) \leq E \text{ and } x \text{ is path connected to } 0\right\}. \tag{20.4}$$

We call $W(E)$ the *potential well* for V and E. The orbits $x(t)$ of a classical particle with energy $E' \leq E$ and initial conditions (x_0, p_0) such that $x_0 \in W(E')$ and $\frac{1}{2}p_0^2 + V(x_0) = E'$ remain in $W(E)$ for all time.

Problem 20.1. Consider the classical equations of motion:

$$\begin{cases} \dfrac{d}{dt}x(t) = p(t), \\[2ex] \dfrac{d}{dt}p(t) = -\nabla V(x(t)). \end{cases}$$

Given initial conditions (x_0, p_0) so that $x_0 \in W(E')$ and $E' = \frac{1}{2}p_0^2 + V(x_0)$, as above, prove that $x(t) \in W(E')$ for all time. (*Hint*: Use the fact that $p(t)$ vanishes when $V(x(t)) = E'$.)

The other region, which we call the exterior region, $\mathcal{E}(E)$, is unbounded. A classical particle with energy E and initial conditions (x_0, p_0), such that $x_0 \in \mathcal{E}(E)$ and $\frac{1}{2}p_0^2 + V(x_0) = E$, will move out to infinity (see the discussion in Section 21.2). We define *classical turning surfaces* relative to E as follows:

$$S^-(E) \equiv \partial W(E), \tag{20.5}$$
$$S^+(E) \equiv \partial \mathcal{E}(E). \tag{20.6}$$

These surfaces are disjoint. Each is compact and connected with

$$\partial\mathrm{CFR}(E) = S^-(E) \cup S^+(E). \tag{20.7}$$

These sets and surfaces are illustrated in Figure 20.1.

The name "shape resonance" comes from the fact that it is the geometry of the potential well $W(E)$ that generates the resonances. As we will illustrate below, the resonances are associated with the quantization of the closed orbits of the

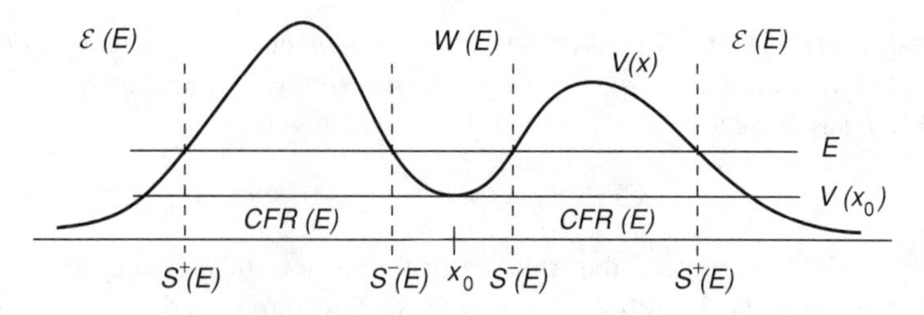

FIGURE 20.1. A shape resonance potential: the potential well $W(E)$, the classically forbidden region CFR(E), the exterior region $\mathcal{E}(E)$, and the classical turning surfaces $S^{\pm}(E)$.

well. Although it is believed that resonances should exist for $H = -\Delta + V$—a Schrödinger operator with potential V satisfying (V1) and (V2) and some form of (V3) and (V4) ahead—this has not yet been proven except in the semiclassical regime, which we now describe.

20.3 The Semiclassical Regime and Scaling

The semiclassical Hamiltonian is

$$H(h) = -h^2\Delta + V, \tag{20.8}$$

where h is a small, nonnegative parameter (see Section 11.1 for a discussion of the semiclassical regime). Factoring h^2 from $H(h)$ and defining $\lambda \equiv h^{-1}$, we obtain

$$H(h) = h^2(-\Delta + \lambda^2 V) = h^2 \tilde{H}(\lambda). \tag{20.9}$$

We now consider $\tilde{H}(\lambda)$. Intuitively, as $\lambda \to \infty$, the potential barrier becomes very large, decoupling the well $W(E)$ from the exterior $\mathcal{E}(E)$. Since, however, the bottom of the well $\lambda^2 V(0)$ also increases with λ, this situation is difficult to work with unless $V(0) = 0$. We do another rescaling to achieve this. Let U_λ, $\lambda \in \mathbb{R}^+$, be the transformation on $L^2(\mathbb{R}^n)$ defined by

$$(U_\lambda f)(x) \equiv \lambda^{\frac{n}{2}} f\left(\lambda^{\frac{1}{2}} x\right). \tag{20.10}$$

The change of independent variable implemented by U_λ is $x \mapsto y \equiv \lambda^{\frac{1}{2}} x$. This is just the dilation group studied in Chapter 11 and in Example 16.3. We find formally that

$$U_\lambda \tilde{H}(\lambda) U_\lambda^{-1} = \lambda\left(-\Delta_y + \lambda V\left(\lambda^{-\frac{1}{2}} y\right)\right). \tag{20.11}$$

Problem 20.2. Prove that for $\lambda > 0$, the map U_λ defined in (20.10) is unitary on $L^2(\mathbb{R}^n)$, and verify (20.11).

We adjust the energy scale so that the minimum potential energy at $x_0 = 0$ is zero by defining a new potential,

$$V(x;\lambda) \equiv \lambda\left(V\left(\lambda^{-\frac{1}{2}}x\right) - V(0)\right). \tag{20.12}$$

We define a new Hamiltonian on $L^2(\mathbb{R}^n)$ by

$$H(\lambda) \equiv -\Delta + V(x;\lambda). \tag{20.13}$$

Then, we have $V(0;\lambda) = 0$, and the limit of the potential at infinity is $-\lambda V(0) < 0$, which diverges toward negative infinity for large λ. It is easy to check that $H(\lambda)$ is self-adjoint on the same domain as $H(h)$. Furthermore, these two Hamiltonians are equivalent in that the existence of resonances for $H(\lambda)$, λ large, implies the existence of resonances for $H(h)$, h small. Summarizing the transformations performed above, the two operators are related by the formula

$$H(h) = \frac{1}{\lambda}U_\lambda^{-1}H(\lambda)U_\lambda + V(0).$$

We will work with $H(\lambda)$ from now on. We will write V for the unscaled potential and V_λ for $V(x;\lambda)$, given in (20.12).

In addition to fixing $V_\lambda(0) = 0$, the main advantage of working with $H(\lambda)$ is the following. The Taylor expansion for $V(x;\lambda)$ about $x_0 = 0$ is

$$V(x;\lambda) = \frac{1}{2}\langle x, Ax\rangle + \lambda^{-\frac{1}{2}}\mathcal{O}(\|x\|^3). \tag{20.14}$$

We used (V1) and (V2), and the matrix A is defined in (20.2). The notation $\mathcal{O}$ means

$$h(x) = \mathcal{O}(\|x\|^p) \quad \text{if} \quad |h(x)|(\|x\| + 1)^{-p} < c_0 < \infty \text{ for all } x \in \mathbb{R}^n.$$

For all $\|x\| \approx 0$ and λ large, the potential $V(x;\lambda)$ is essentially a harmonic oscillator potential (recall Chapter 11). Setting $V_0(x) \equiv \frac{1}{2}\langle x, Ax\rangle$, we define a reference Hamiltonian by

$$K \equiv -\Delta + V_0(x), \tag{20.15}$$

which is completely solvable and independent of λ.

Problem 20.3. Find $\sigma(K)$ and the eigenfunctions of K (see Section 11.2). (*Hint*: Use the facts that the matrix A can be diagonalized and that one can separate variables in Cartesian coordinates, thus reducing the problem to one dimension. Then, look for solutions of the form $p(x)e^{-\alpha x^2}$, for polynomials p and constants α proportional to the eigenvalues of A.)

The nondegeneracy condition implies that on any compact set $\mathcal{K} \subset \mathbb{R}^n$ containing the origin, $V(x;\lambda)$ can be uniformly approximated by $V_0(x)$ as $\lambda \to \infty$ since

$$\sup_{x\in\mathcal{K}} |V(x;\lambda) - V_0(x)| \leq c(\mathcal{K})\lambda^{-\frac{1}{2}}. \tag{20.16}$$

This suggests that the resonances of $H(\lambda)$ will be close to the low-lying eigenvalues of K in the large λ regime. We will prove this *spectral stability* result ahead, making precise the meaning of "close to." With reference to Section 20.1, the operator K is the unperturbed operator at $\lambda = \infty$. We will prove the existence of resonances for $H(\lambda)$ by proving the stability of the low-lying eigenvalues of K with respect to the perturbation of "lowering the infinite potential barriers."

Let us fix an eigenvalue $e_0 \in \sigma(K)$. We will need to consider all the geometric objects of Section 20.2, the classically forbidden regions, potential well, and so on, defined relative to the potential V_λ given in (20.12). These objects will now depend on λ. We will exhibit this dependence explicitly to indicate that the potential V_λ is being used in the definition. Any notation without an explicit λ-dependence refers to the potential V. Clearly, from conditions (V1) and (V2) and the definition of V_λ, there exists a $\lambda_0 > 0$ such that for all $\lambda > \lambda_0$, Int CFR$(e_0;\lambda) \neq \emptyset$ and the potential well $W(e_0;\lambda)$ is a connected set. The assumptions then imply that this is true for all $e_i \leq e_0$. As in (20.5) and (20.6), we define the classical turning surfaces relative to e_0 and $V(x;\lambda)$. To indicate the λ-dependence, we write

$$S^-(e_0;\lambda) \equiv \partial W(e_0;\lambda), \tag{20.17}$$

$$S^+(e_0;\lambda) \equiv \partial \mathcal{E}(e_0;\lambda). \tag{20.18}$$

For all large λ, these surfaces are disjoint, compact, and connected. We also note that if $x(\lambda) \in S^\pm(e_0;\lambda)$, then

$$x(\lambda) = \lambda^{\frac{1}{2}} y + \mathcal{O}(\lambda^{-1}), \tag{20.19}$$

where y satisfies $V(y) = V(0)$ and is therefore λ-independent. This follows from the fact that

$$\begin{aligned} S^\pm(e_0;\lambda) &= \left\{x \mid \lambda\left(V\left(\lambda^{-\frac{1}{2}}x\right) - V(0)\right) = e_0\right\} \\ &= \lambda^{\frac{1}{2}}\left\{y \mid V(y) = V(0) + \lambda^{-1}e_0\right\} \end{aligned}$$

and from an estimation of the gradient of $V(y)$ for $y \approx y_0$, where $V(y_0) = V(0)$. We remark that if we know, in addition to (V1) and (V2), that $\partial_i V(x_0) \neq 0$ for each $x_0 \in S^\pm(e_0;\lambda)$ and some $i = 1, \ldots, n$, then one can use the implicit function theorem to verify that $S^\pm(e_0;\lambda)$ are C^2-surfaces. (However, we don't need this here.) Finally, it is clear that $S^\pm(e_0,\lambda)$ vary continuously in λ and e (at least for λ large and e near e_0).

Given the turning surface $S^+(e_0;\lambda)$, we denote by Int $S^+(e_0;\lambda)$ the open, connected interior of the surface which contains the origin, and we define

$$\text{Ext } S^+(e_0;\lambda) = \text{Int}(\mathbb{R}^d \setminus \text{Int } S^+(e;\lambda)). \tag{20.20}$$

For $\delta > 0$ small, to be fixed below, we define two energies relative to e_0:

$$e' \equiv e_0 + \lambda^\delta, \tag{20.21}$$

$$e \equiv e_0 + 2\lambda^\delta. \tag{20.22}$$

The assumptions on V imply that for all large λ,

$$S^{\pm}(e';\lambda) \subset \text{Int } S^{+}(e_0;\lambda) \tag{20.23}$$

and

$$S^{\pm}(e;\lambda) \subset \text{Int } S^{+}(e';\lambda). \tag{20.24}$$

20.4 Analyticity Conditions on the Potential

We now give the two analyticity assumptions on V. These are two conditions on V restricted to the exterior region Ext $S^{+}(V(0) + e_0 + \varepsilon)$, for some $\varepsilon > 0$. We let $E_0 \equiv e_0 + V(0) + \varepsilon$. The first condition, (V3), is that $V \mid \text{Ext } S^{+}(E_0)$ is the restriction of a function analytic in a truncated cone in $\mathbb{C}^n$ containing Ext $S^{+}(E_0)$. The second requires that V is a *nontrapping potential* in Ext $S^{+}(E_0)$. We will show in Chapter 21 that this nontrapping condition implies that $V \mid \text{Ext } S^{+}(E_0)$ does not produce resonances near e_0. We will discuss this condition in detail in Chapter 21 and give a simple geometric condition on V which implies the estimate (V4).

For a given $\delta_1, \delta_2 > 0$, we set $\tilde{\delta} \equiv (\delta_1, \delta_2)$ and define a truncated cone $\Gamma_{\tilde{\delta}}$ with respect to $S^{+}(E_0)$ by

$$\Gamma_{\tilde{\delta}} \equiv \left\{ z \in \mathbb{C}^n \mid |\text{Im } z| \leq \delta_1 \,\text{dist}(S^{+}(E_0), \text{Re } z) + \delta_2 \right\}, \tag{20.25}$$

where $E_0 = V(0) + e_0 + \varepsilon$, for some $\varepsilon > 0$. Our third condition on V is

(V3) The potential $V \mid \text{Ext } S^{+}(E_0)$ is the restriction of a function, also denoted by V, which is analytic in $\Gamma_{\tilde{\delta}}$, for some $\delta_1, \delta_2 > 0$.

To formulate the final condition on V, we note that the geometric conditions (V1) and (V2) and the analyticity condition (V3) enable us to construct a vector field relative to V in the region Ext $S^{+}(E_0)$ and satisfying condition L. We will use this vector field to construct a spectral deformation family for $H(\lambda)$ as in Chapters 16–18.

Let $S \subset \mathbb{R}^n$ be a C^k-, $k \geq 1$, compact, connected surface with $0 \in \text{Int } S$ so that $\mathbb{R}^n = \overline{\text{Int } S} \cup \text{Ext } S$.

Definition 20.1. *A C^2 vector field g on $\mathbb{R}^n$ is said to be exterior to the surface S if*

(i) $g \mid \overline{\text{Int } S} = 0$ *and* $g \mid \text{Ext } S$ *is nonvanishing;*

(ii) $\sup_{x \in \text{Ext } S} \|Dg(x)\| \equiv M_1 < \infty$;

(iii) $\exists R > 0$ *and* $c_R > 0$, *finite, such that*

$$\|g(x)\| \leq c_R \|x\|,$$

for all x, $\|x\| > R$.

Proposition 20.2. *Let V satisfy* (V1)–(V3). *Then there exists a vector field v exterior to $S^+(e';\lambda)$, with $M_1 = 1$ and satisfying condition L, so that the potential V is an admissible potential for the spectral deformation family $\mathcal{U}$ generated by v. The spectrally deformed family of Schrödinger operators,*

$$H(\lambda,\theta) \equiv p_\theta^2 + V(\phi_\theta(x);\lambda), \tag{20.26}$$

is an analytic type-A family on some disk D_a, $0 < a \leq 1/\sqrt{2}$, where a is determined by $\delta_1, \delta_2 > 0$.

Before proving this proposition, let us state the final condition on V.

(V4) Let $H(\lambda,\theta)$, $\theta \in D_a$, be any spectral deformation family for $H(\lambda)$ constructed as in (20.26). Then $\exists\, b > 0$ such that for $0 < \operatorname{Im}\theta < a \operatorname{Im}\theta < a$, and for any function $u \in C_0^\infty(\operatorname{Ext} S^+(e;\lambda))$,

$$\|(H(\lambda,\theta) - e_0)u\| \geq b|\operatorname{Im}\theta| \left(\|u\| + \sum_{i=1}^{n} \|\eta\partial_i u\| \right), \tag{20.27}$$

where $\eta \not\equiv 0$ is any function supported in $\operatorname{Ext} S^+(e;\lambda) \setminus \operatorname{Ext} S^+(e';\lambda)$.

In Chapter 21, we will formulate the so-called nontrapping condition on a potential V and verify that this condition implies the lower bound (20.27). It will then be clear that there are many potentials satisfying (V1)–(V3) for which (V4) holds.

Proof of Proposition 20.2.

(1) Construction of the vector field. For any $e \in \mathbb{R}^+$, let $\phi_e \in C^\infty(\mathbb{R})$ be such that $\phi_e(s) = 0$ for $s \geq e$ and ϕ_e is a bounded, nonnegative function for $s < e$ with $\lim_{s\to-\infty} \phi_e(s) = 1$. We can take, for example,

$$\phi_e(s) = \begin{cases} 0, & s > e, \\ e^{-\varepsilon(e-s)^{-1}}, & s \leq e, \end{cases} \tag{20.28}$$

for any $\varepsilon > 0$. We define a function $\Phi_e(x)$ by

$$\Phi_e(x) \equiv \begin{cases} \phi_e(V(x)), & x \in \operatorname{Ext} S^-(e), \\ 0 & \text{otherwise}, \end{cases} \tag{20.29}$$

and a vector field

$$v_e(x) \equiv \Phi_e(x)x. \tag{20.30}$$

Problem 20.4. Show that $v_e(x)$ is a vector field exterior to the surface $S^+(e)$.

(2) By Lemma 17.2, v_e defines a diffeomorphism on $\mathbb{R}^n$ by

$$\phi_\theta(x) \equiv x + \theta v_e(x), \tag{20.31}$$

for $|\theta| < M_1$, where M_1 is defined in (17.5) and M_1^{-1} appears in condition (ii) of Definition 20.1. By simple computation, it is clear that we can assume that $M_1 = 1$ by choosing the constant $c_R \leq 1$. With this choice, condition L is also satisfied. This family of diffeomorphisms (20.31) has two properties that are easily verified for all λ sufficiently large and $e < E_0$:

(1) ϕ_θ maps Ext $S^+(e)$ onto itself;

(2) for $\theta \in D$, Ran $\phi_\theta(x) \subset \Gamma_{\delta_1,\delta_2}(S^+(e))$, for some $\delta_1, \delta_2 > 0$, depending on c_R and M_1 of conditions (ii) and (iii) in Definition 20.1.

The proposition now follows from these results, Theorem 18.6 and the discussion in that chapter, and the following problem. It is important to note that the potential V_λ is not relatively p^2-compact, but the shifted potential $V_\lambda + \lambda V(0) = \lambda V(\lambda^{-1/2}\cdot)$ is relatively p^2-compact. By Theorem 18.6, $\sigma_{\text{ess}}(H(\lambda;\theta)) = \{z - \lambda V(0) \in \mathbb{C} \mid \arg z = -2 \arg (1+\theta)\}$. □

Problem 20.5. Verify that (V3) implies that, for all λ sufficiently large, $V_\lambda|$Ext $S^+(e';\lambda)$ has an analytic continuation to the truncated cone

$$\Gamma_{\tilde{\delta},\lambda} \equiv \left\{z \in \mathbb{C}^n \mid |\text{Im } z| \leq \delta_1 \text{dist}(S^+(e';\lambda), \text{Re } z) + \delta_2\right\}. \tag{20.32}$$

20.5 Spectral Stability for Shape Resonances: The Main Results

We now make precise the manner in which the eigenvalues of K are close to the resonances of $H(\lambda)$. By the Aguilar–Balslev–Combes–Simon theory, the resonances of $H(\lambda)$ are eigenvalues of the spectrally deformed family $H(\lambda,\theta)$, Im $\theta > 0$, which is specified in Proposition 20.2. The stability theory for eigenvalues under type-A analytic perturbations is presented in Chapter 15. Extensions of this theory to nonanalytic perturbations are discussed in Chapter 19. There are two main ideas in Chapter 19: (1) The notion of eigenvalue stability has to be extended to allow the possibility that an eigenvalue might turn into a resonance under perturbation; and (2) this general notion of eigenvalue stability can be derived from a lower-bound estimate on the Schrödinger operator, as given in Theorem 19.12. Such a *stability estimate* will play an important role in this and the following chapters. In order to extend the ideas of eigenvalue stability given in Chapter 19, we simply identify the unperturbed operator $H(\infty)$ with K given in (20.15) and the perturbed family with $H(\lambda,\theta)$ given in (20.26). With this change, Definition 19.5 applies to an eigenvalue of K and for large λ.

Let $\{e_i\}_{i=1}^{\infty}$ be the eigenvalues of K listed in increasing order including multiplicity. Our intuition tells us that the low-lying eigenvalues of K (in a sense to be made precise ahead) are close to the eigenvalues of $H(\lambda, \theta)$, for $\theta \in D_a^+$ and λ large. The main result of this chapter is that such eigenvalues e_i are indeed stable, in the sense of Definition 19.5, with respect to the perturbation $H(\lambda, \theta)$, about $\lambda = \infty$.

Theorem 20.3. *Let $\tilde{e}_i$ be a distinct eigenvalue of K such that for some $\varepsilon > 0$ and any $e_i = \tilde{e}_i$, $d_A^{e_i}(S^-(e_i), S^+(e_i)) > \varepsilon$. Assume that V satisfies* (V1)–(V4). *Then $\tilde{e}_i$ is stable with respect to the perturbation $H(\lambda, \theta)$, for $\theta \in D_a^+$ and all λ sufficiently large.*

An immediate consequence of this theorem, and Theorem 16.1 on the absence of positive embedded eigenvalues for $H(\lambda)$ (as commented upon in the proof of Proposition 20.2, one applies the result to $H(\lambda) + \lambda V(0)$), is the existence of resonances for $H(\lambda)$ for λ sufficiently large.

Theorem 20.4. *Let $\{\tilde{e}_i\}_{i=1}^{\infty}$ be the distinct eigenvalues of K with multiplicities $\{m_i\}_{i=1}^{\infty}$, $m_i \geq 1$. Suppose $\tilde{e}_\ell$ satisfies the hypotheses of Theorem 20.3. Then $\exists \lambda_0 > 0$ such that $H(\lambda)$, for $\lambda > \lambda_0$, has distinct resonances $\{z_{k,\ell}(\lambda)\}_{k=1}^{N}$ with $1 \leq N \leq m_\ell$ and*

(i) $\lim_{\lambda \to \infty} \mathrm{Re}\, z_{k,\ell}(\lambda) = \tilde{e}_\ell$;

(ii) $\mathrm{Im}\, z_{k,\ell}(\lambda) < 0$;

(iii) *the total algebraic multiplicities of the $z_{k,\ell}(\lambda) = m_\ell$.*

Problem 20.6. Prove Theorem 20.4 given Theorems 16.4, 18.6, and 20.3, and the appropriately modified definition of stability, Definition 19.5.

In Chapter 22, we will prove an upper bound on $|\mathrm{Im}\, z_{k,\ell}(\lambda)|$. This has the familiar tunneling form $e^{-\lambda d}$, where the distance d approaches $d_A^{\tilde{e}_\ell}(S^-(\tilde{e}_\ell), S^+(\tilde{e}_\ell))$, the Agmon distance across the barrier at energy $\tilde{e}_\ell$ as $\lambda \to \infty$.

The proof of Theorem 20.3 will be given in the next two sections. The keys to the stability estimates needed for the proof are *geometric perturbation theory* and *localization formulas* developed in Chapter 19. We would like to compare the operators K and $H(\lambda, \theta)$. However, away from a small neighborhood of the origin, these two operators are not close in any usual sense. This can be compensated for by the introduction of two approximate operators, $H_0(\lambda)$ and $H_1(\lambda)$, which are close to $H(\lambda, \theta)$ in the region around the origin and in the exterior, respectively. These operators are constructed by means of an appropriately chosen partition of unity $\{j_0, j_1\}$, which localizes to the well and the exterior and such that the gradients are supported in the classically forbidden region. We then relate the resolvents of these operators through the localization formula of Lemma 19.16.

In this and the next chapter, we will need various partitions of unity. We will use the following construction. For $E \in \mathbb{R}^+$, let ξ_E be a smooth function such that

$0 \leq \xi_E \leq 1$ and

$$\xi_E(s) = \begin{cases} 1, & s \geq E, \\ 0, & s \leq E - \frac{1}{4}\lambda^{\delta}. \end{cases} \tag{20.33}$$

For our present work, we set $E = e$, as given in (20.21), and introduce a cut-off function

$$j_{0,e}(x) \equiv \xi_e(V(x;\lambda)). \tag{20.34}$$

To obtain a partition of unity, we set

$$j_{1,e}(x) \equiv 1 - j_{0,e}(x).$$

We then have

$$j_{0,e}(x) + j_{1,e}(x) = 1.$$

We will occasionally need partitions of unity satisfying

$$\sum_{i=0}^{1} j_{i,e}^2 = 1,$$

and we discuss this construction in Section 20.7. When the energy e is fixed, we will write j_i for $j_{i,e}$.

The partition of unity $\{j_0, j_1\}$ constructed above satisfies

$$\begin{aligned} \text{supp } j_0 &\subset \text{Int } S^+(e';\lambda); \\ j_0 \mid \text{Int } S^+(e;\lambda) &= 1. \end{aligned}$$

Note that supp $j_1 \subset \text{Ext } S^+(e;\lambda)$, and for $i = 0, 1$,

$$\text{supp}|\nabla j_i| \subset \text{Int } S^+(e';\lambda) \setminus \text{Int } S^+(e;\lambda).$$

Finally, from the definition of V_λ and j_i, we have

$$\|\nabla j_i\|_\infty = \mathcal{O}(\lambda^{\frac{1}{2}-\delta}). \tag{20.35}$$

Consequently, we will always take $\delta > 1/2$.

The two approximate Schrödinger operators have potentials constructed as follows. We choose V_i, $i = 0, 1$, to satisfy

$$V_i \mid \text{supp } j_i = V,$$

and we extend V_i continuously to $\mathbb{R}^n$ as follows:

$$V_0(x,\lambda) \equiv \begin{cases} V(x,\lambda), & x \in \text{Int } S^+(e';\lambda), \\ \geq e_0 + \frac{1}{2}\lambda^{\delta}, & x \in \text{Ext } S^+(e';\lambda), \end{cases} \tag{20.36}$$

and

$$V_1(x;\lambda) \equiv \begin{cases} \geq e, & x \in \text{Int } S^+(e;\lambda), \\ V(x,\lambda), & x \in \text{Ext } S^+(e;\lambda), \end{cases} \tag{20.37}$$

for $\delta > 1/2$. We define the two approximate Hamiltonians on $L^2(\mathbb{R}^n)$ by

$$H_i \equiv -\Delta + V_i(x, \lambda), \tag{20.38}$$

for $i = 0$ and 1. Since H_1 is constructed with V_1 equaling V outside $S^+(e; \lambda)$, the potential V_1 satisfies condition (V3) and we can apply the same spectral deformation family to H_1 as to $H(\lambda)$. We will write $H_1(\lambda, \theta)$ for the resulting family. Moreover, the estimate of condition (V4) applies to the deformed family $H_1(\lambda, \theta)$.

We will apply formula (19.57) to $H(\lambda, \theta)$, $H_0(\lambda)$, and $H_1(\lambda, \theta)$ for $\theta \in D_a^+$. The *geometric resolvent formula* in this situation takes the following form. Let W_i be the first-order differential operator defined by

$$W_i \equiv [\Delta, j_i]. \tag{20.39}$$

By the conditions on the potentials, the localized operator W_i is relatively H_i-bounded. For all z in the intersection of the resolvent sets of $H(\lambda)$, $H_0(\lambda)$, and $H_1(\lambda, \theta)$ for $\theta \in D_a^+$, we have

$$R(z) - \sum_{i=0}^{1} j_i R_i(z) = \sum_{i=0}^{1} R(z) W_i R_i(z), \tag{20.40}$$

where we write $R(z) \equiv (H(\lambda, \theta) - z)^{-1}$, $R_0(z) \equiv (H_0(\lambda) - z)^{-1}$, and $R_1(z) \equiv (H_1(\lambda, \theta) - z)^{-1}$.

The first step in the proof of Theorem 20.2 is to demonstrate that the low-lying eigenvalues of K (in the sense of Theorem 20.2) are stable with respect to the perturbation $H_0(\lambda)$. This is similar to the discussion of Chapter 11, and we present the results in the next section. Second, given that $e_{0(i)}(\lambda) \in \sigma(H_0(\lambda))$ satisfies $e_{0(i)}(\lambda) \to e_0 \in \sigma(K)$, we consider a contour Γ_0 inside the annular region

$$A(e_0) \equiv \{z \mid a_0 < |z - e_0| < a_1\}, \tag{20.41}$$

for $0 < a_0 < a_1$, chosen such that $A(e_0) \cap \sigma(K) = \{e_0\}$ (we will specify a_1 more carefully below). Then, for all λ large, $\{e_{0(i)}(\lambda)\} \subset \{z \mid |z - e_0| < a_0\}$. We must show that $H(\lambda, \theta), \theta \in D_a^+$, has an eigenvalue inside Γ_0 for all λ large. By Chapter 6 this is equivalent to showing that the Riesz projector

$$P(\lambda, \theta) \equiv \frac{1}{2\pi i} \oint_{\Gamma_0} (H(\lambda, \theta) - z)^{-1} dz \tag{20.42}$$

exists and is nonzero. For this, we use the geometric resolvent formula (20.40). Integrating both sides of that equation around Γ_0, we see that

$$\|P(\lambda, \theta) - j_0 P_0(\lambda)\| \leq \frac{1}{2\pi} \sum_{i=0}^{1} \oint_{\Gamma_0} |dz| \, \|R(z) W_i R_i(z)\|, \tag{20.43}$$

where $P_0(\lambda)$ is the projection for $H_0(\lambda)$ and the eigenvalue cluster $\{e_{0(i)}(\lambda)\}$,

$$P_0(\lambda) \equiv \frac{1}{2\pi i} \oint_{\Gamma_0} (H_0(\lambda) - z)^{-1} dz. \tag{20.44}$$

We will see that as a consequence of condition (V4), the resolvent $R_1(z)$ is analytic on and inside the contour Γ_0, so its contribution to the left side of (20.40) vanishes upon integration. So our task is to show that the right side of (20.43) exists for $z \in A(e_0)$ and is less than 1 for all λ large. Existence of the right side of (20.43) is clear from the analyticity of $R_1(z)$ and the fact that we know about $R_0(z)$ because of its relation to $(K-z)^{-1}$. Hence it remains to show that the norm is less than 1. This is why we must work in the semiclassical regime of large λ (this will also allow us to remove the j_0 on the left side of (20.43) by the exponential decay estimates of Chapter 3). We also use the fact that $\text{supp}|\nabla j_i| \subset \text{CFR}([e_0 - \eta, e_0 + \eta])$ for suitable $\eta > 0$.

20.6 The Proof of Spectral Stability for Shape Resonances

The proof of spectral stability consists of two parts. The first is the stability of the low-lying eigenvalues of $K \equiv -\Delta + \frac{1}{2}\langle x, Ax\rangle$ with respect to the perturbation $H_0(\lambda)$ for λ large. We accomplished this in Chapter 11, and we recall the main results here. Taking into account the different λ-scaling used here (see (11.9)), we have the following result.

Theorem 20.5. *Let $\Sigma \equiv \inf \sigma_{ess}(H_0(\lambda))$. Fix $n \in \mathbb{N}$. For λ sufficiently large, $H_0(\lambda)$ has at least n eigenvalues $\{e_i(\lambda)\}$ below Σ, and these satisfy*

$$e_i(\lambda) = e_i + \mathcal{O}\left(\lambda^{-\frac{1}{5}}\right), \tag{20.45}$$

(the error is not uniform in i), where $\{e_i\} \subset \sigma(K)$ are ordered in increasing fashion including multiplicity.

For an eigenvalue $\tilde{e}_k \in \sigma(K)$, with $\tilde{e}_k < \Sigma$, let $e_{k(i)} \in \sigma(H_0(\lambda))$ be the cluster of eigenvalues converging to $\tilde{e}_i$ as $\lambda \to \infty$. Recall that the total geometric multiplicity m_k of this cluster is equal to the multiplicity of $\tilde{e}_k$. The second part of the stability proof is to establish the stability of this cluster of eigenvalues $e_{k(i)}(\lambda)$, which converge to $\tilde{e}_k \in \sigma(K)$, with respect to the perturbation $H(\lambda, \theta)$, $\theta \in D_a^+$. We will refer to this finite cluster of eigenvalues of $H_0(\lambda)$ as $\tilde{e}_i(\lambda)$. Each element satisfies (20.45).

We next specify the constants a_1 and a_2 appearing in the definition of the annular region $A(e_0)$ in (20.41). Let $b > 0$ be the constant appearing in (V4). We choose a_1 so that

$$0 < a_1 \leq \frac{1}{2}(\text{Im}\,\theta)b, \tag{20.46}$$

for $\theta \in D_a^+$, and set $a_0 = \frac{1}{2}a_1$. An immediate corollary of Theorem 20.5 is an estimate on $R_0(z) \equiv (H_0(\lambda) - z)^{-1}$, for $z \in A(e_0)$.

Corollary 20.6. *For all λ sufficiently large, and for all $z \in A(e_0)$,*

$$\begin{aligned} \|R_0(z)\| &\leq 4a_1^{-1}, \\ \|\eta \partial_i R_0(z)\| &\leq c, \end{aligned}$$

for some finite constant $c > 0$ and any function η with $\operatorname{supp} \eta \subset \operatorname{Int} S^+(e'; \lambda) \setminus \operatorname{Int} S^+(e; \lambda)$.

Problem 20.7. Prove Corollary 20.6.

We want to apply formula (20.40) to study $R(z) \equiv (H(\lambda, \theta) - z)^{-1}, z \in A(e_0)$, for λ large. We will show that a consequence of (V4) and Corollary 20.6 is that $R(z)$ is analytic in $A(e_0)$. We will prove the following a priori estimate on $R(z)$.

Theorem 20.7. *Assume conditions* (V1)–(V4), *and fix* $\operatorname{Im} \theta \in D_a^+$. *Then for all λ sufficiently large $\exists\, 0 < c_1 < \infty$ such that $\forall z \in A(e_0)$,*

$$\|(H(\lambda, \theta) - z)^{-1}\| < c_1. \tag{20.47}$$

We defer the proof of these results until the next section. As an intermediate step in the proof, we need to know that $R_1(z) \equiv (H_1(\lambda, \theta) - z)^{-1}$ satisfies an a priori estimate similar to (20.47). The proof of the following theorem is similar to that of Theorem 20.7 and will also be given in the next section.

Theorem 20.8. *Assume conditions* (V1)–(V4), *and fix $\theta \in D_a^+$. Then for all λ sufficiently large, $R_1(z)$ is analytic inside $\{z \mid |z - e_0| = a_1\}$ and satisfies the estimates*

$$\begin{aligned} \|R_1(z)\| &\leq c_1(\operatorname{Im} \theta), \\ \|\eta \partial_i R_1(z)\| &\leq c_2(\operatorname{Im} \theta), \end{aligned}$$

where η is as in Corollary 20.6.

Given Corollary 20.6 and Theorems 20.7 and 20.8, proof of the spectral stability theorem, Theorem 20.3, is rather easy.

Proof of Theorem 20.3. Let Γ_0 be a simple closed contour lying in $A(e_0)$. By Corollary 20.6 and Theorems 20.7 and 20.8, Γ_0 lies in the resolvent sets of $H(\theta, \lambda)$, $H_0(\lambda)$, and $H_1(\lambda, \theta)$. Hence for $z \in \Gamma_0$, the geometric resolvent formula (20.40) is valid. Recall from (20.44) and (20.42) that $P_0(\lambda)$ and $P(\lambda, \theta)$ are the corresponding Riesz projectors for Γ and $H_0(\lambda)$ and $H(\lambda, \theta)$, respectively. Let P_0 be the Riesz projection defined as in (20.44) with K in place of $H_0(\lambda)$. We apply (20.40) to $H_0(\lambda)$, $H_1(\lambda, \theta)$, and $H(\lambda, \theta)$, since we already know from Theorem 20.5 (see Chapter 11 for the details) that

$$\lim_{\lambda \to \infty} \|P_0 - P_0(\lambda)\| = 0. \tag{20.48}$$

By Theorem 20.8, we obtain

$$\oint_\Gamma R_1(z) dz = 0,$$

and so an integration of formula (20.40) yields (20.43):

$$\|P(\lambda, \theta) - j_0 P_0(\lambda)\| \leq \frac{1}{2\pi} \sum_{i=0}^{1} \oint_{\Gamma_0} d|z| \, \|R(z) W_i R_i(z)\|. \tag{20.49}$$

We estimate the right side of (20.49). Theorem 20.7 states that $\|R(z)\| \leq c$ on Γ. Recalling from (20.35) that ∇j_i satisfies

$$\|\nabla j_i\|_\infty = \mathcal{O}(\lambda^{\frac{1}{2}-\delta}), \tag{20.50}$$

for $\delta > \frac{1}{2}$, the estimates of Theorem 20.8 and Corollary 20.6 imply, for $z \in \Gamma_0$,

$$\|W_0 R_0(z)\| \leq c_0 \lambda^{\frac{1}{2}-\delta} \tag{20.51}$$

and

$$\|W_1 R_1(z)\| \leq c_1 \lambda^{\frac{1}{2}-\delta}. \tag{20.52}$$

Since the diameter of $A(e_0)$ is independent of λ, these estimates give an upper bound for the right side of (20.49),

$$\|P(\lambda, \theta) - j_0 P_0(\lambda)\| \leq c_0 \lambda^{\frac{1}{2}-\delta}, \tag{20.53}$$

for $\delta > 1/2$. We now use the semiclassical result that the eigenfunctions of $H_0(\lambda)$ corresponding to $\tilde{e}_0(\lambda)$ decay exponentially in $\operatorname{supp}(1 - j_0)$ to conclude that

$$\lim_{\lambda \to \infty} \|(1 - j_0) P_0(\lambda)\| = 0. \tag{20.54}$$

This follows from an Agmon-type argument as presented in Chapter 3, and the proof is relegated to Problem 20.8. Consequently, for the contour Γ_0,

$$\|P(\lambda, \theta) - P_0(\lambda)\| \leq c_0 \lambda^{-\varepsilon}, \tag{20.55}$$

for some $\varepsilon > 0$, and because $P_0(\lambda) \neq 0$, we have shown that $P(\lambda, \theta) \neq 0$. This implies by Lemma 15.4 that $\sigma(H(\lambda, \theta)) \cap \operatorname{Int} \Gamma_0 \neq \emptyset$, proving the stability result. □

Problem 20.8. Since Ran $P_0(\lambda)$ is finite-dimensional, use the Agmon estimates on eigenfunctions of $H_0(\lambda)$ derived in Chapter 3 to verify (20.54).

20.7 Resolvent Estimates for $H_1(\lambda, \theta)$ and $H(\lambda, \theta)$

We give the proof of Theorem 20.7. The proof of Theorem 20.8 is similar and is left as Problem 20.11. Condition (V4) controls $(H(\lambda, \theta) - z)$ on functions u supported on the exterior of $S^+(e, \lambda)$. We use a partition of unity and the geometric resolvent equation (19.51) to extend this estimate to arbitrary $u \in C_0^\infty(\mathbb{R}^n)$. Recall the definitions, $e' \equiv e_0 + \lambda^\delta$ and $e \equiv e_0 + 2\lambda^\delta$, for $\delta > 1/2$, given in (20.21) and (20.22).

For $E \in \mathbb{R}^+$, let ξ_E be a smooth function as defined in (20.33). Setting $E = e$, defined above, we introduce a cut-off function

$$j_{0,e}(x) \equiv \xi_e(V_1(x;\lambda)), \tag{20.56}$$

with $V_1(x;\lambda)$ given in (20.37). We note that

$$V_0 j_{0,e} = V j_{0,e}. \tag{20.57}$$

In order to use the geometric resolvent formula (19.51), we need a partition of unity satisfying

$$\sum_{i=0}^{1} j_{i,e}^2 = 1. \tag{20.58}$$

We obtain this by setting

$$j_{1,e}(x) = \left(1 - j_{0,e}^2(x)\right)^{\frac{1}{2}}. \tag{20.59}$$

It is left as Problem 20.9 to show that there exist smooth functions $\xi_E(s)$ so that $j_{i,e} \in C^2(\mathbb{R}^n)$. As in (20.35), we have for $1 \le i, j \le n$,

$$|\partial_i^\alpha \partial_j^\beta j_{k,e}(x)| = \mathcal{O}(\lambda^{\frac{1}{2}-\delta}), \tag{20.60}$$

where α and β are nonnegative integers satisfying $\alpha+\beta = 2$ and $k = 0, 1$. Analogous to (20.57), we have

$$V_1 j_{1,e} = V j_{1,e}. \tag{20.61}$$

Problem 20.9. Show how to construct a smooth function $\xi_E(s)$, as in (20.56), so that $(1 - (\xi_E(s))^2)^{1/2}$ is twice continuously differentiable. Using such a function, verify claims (20.57), (20.60), and (20.61).

In what follows, we consider e fixed as above and write j_i for $j_{i,e}$.

Proof of Theorem 20.7. We first prove that for all λ large enough, $\exists\, c_0 > 0$ such that for any $u \in C_0^\infty(\mathbb{R}^n)$ and for all $z \in A(e_0)$,

$$\|(H(\lambda,\theta) - z)u\| \ge c_0 \|u\|,$$

$\theta \in D_a^+$. From (19.51) and Problem 19.7, we have the geometric resolvent equation

$$\|(H(\lambda,\theta) - z)u\|^2 \ge \frac{1}{2} \sum_{i=0}^{1} \|(H(\lambda,\theta) - z) j_i u\|^2 - \mathcal{R}(u), \tag{20.62}$$

where the remainder $R(u)$ is given by

$$\mathcal{R}(u) \equiv \sum_{i=0}^{1} \|[\Delta, j_i]u\|^2. \tag{20.63}$$

We first estimate the main terms of (20.62). Because of the choice of j_0, we have

$$(H(\lambda, \theta) - z) j_0 u = (H_0(\lambda) - z) j_0 u. \tag{20.64}$$

It follows from Corollary 20.6 that $\exists c_1 > 0$ such that for all λ sufficiently large,

$$\|(H(\lambda, \theta) - z) j_0 u\| \geq c_1 \| j_0 u\|. \tag{20.65}$$

Next, condition (V4) and the definition of $A(e_0)$, (20.41), imply that

$$\|(H(\lambda, \theta) - z) j_1 u\| \geq \frac{1}{2} (\text{Im } \theta) b \left(\| j_1 u\| + \sum_{i=1}^{n} \|\eta \partial_i u\| \right), \tag{20.66}$$

for some η with supp $\eta \subset \text{Int } S^+(e'; \lambda) \setminus \text{Int } S^+(e, \lambda)$ and all $z \in A(e_0)$. Estimates (20.65) and (20.66) give a lower bound for the main term on the right in (20.62).

As for the remainder term $\mathcal{R}(u)$, (20.63), we expand the commutator

$$[\Delta, j_i] u = (2(\nabla j_i) \cdot \nabla + (\Delta j_i)) u.$$

Estimate (20.60) on $|\nabla j_i|$ leads to

$$\sum_{i=0}^{1} \|[-\Delta, j_i] u\|^2 \leq \sum_{i=0}^{1} c_2 \lambda^{2(\frac{1}{2} - \delta)} \left(\sum_{k=1}^{n} \|\eta_1 \partial_k u\|^2 + \|\eta_1 u\|^2 \right), \tag{20.67}$$

where η_1 is a function supported on a set slightly larger than supp$|\nabla j_i|$. Combining estimates (20.65), (20.66), and (20.67), we find that the right side of (20.62) is bounded from below as

$$\begin{aligned} \|(H(\lambda, \theta) - z) u\|^2 \geq\ & (c_0 - c_1 \lambda^{1-2\delta}) \|u\|^2 \\ & + (c_2 |\text{Im } \theta|^2 b^2 - c_3 \lambda^{1-2\delta}) \sum_{k=1}^{n} \|\eta_1 \partial_k u\|^2, \end{aligned} \tag{20.68}$$

where c_0 depends on b and $|\text{Im } \theta|$, and c_1, c_2, and c_3 are finite, positive constants. Since $\delta > 1/2$, we find from (20.68) that for all large λ,

$$\|(H(\lambda, \theta) - z) u\| \geq c_0 \|u\|. \tag{20.69}$$

Recall that $C_0^\infty(\mathbb{R}^n)$ is a core for $H(\lambda, \theta)$. This follows from the fact that (V1) and (V2) imply that $C_0^\infty(\mathbb{R}^n)$ is a core for $H(\lambda)$ and the type-A analyticity. Hence, estimate (20.69) shows that ker$(H(\lambda, \theta) - z)$ has no nontrivial elements. The estimate also shows that Ran$(H(\lambda, \theta) - z)$ is closed as $(H(\lambda, \theta) - z)$ is a closed operator. If $g \in [\text{Ran}(H(\lambda, \theta) - z)]^\perp$, then $g \in \ker(H(\lambda, \theta)^* - \overline{z})$. We leave it to Problem 20.10 to show that an estimate similar to (20.69) holds for $H(\lambda, \theta)^*$. Consequently, $(H(\lambda, \theta) - z)$ is invertible with a bounded inverse. This proves the theorem. □

Problem 20.10. Prove that for $u \in D(H)$ and $z \in A(e_0)$,

$$\|(H(\lambda, \theta)^* - z)u\| \geq c_0 \|u\|,$$

by replacing the argument used in the proof of Theorem 20.7. Note that it is necessary to consider Theorem 18.6 for $\sigma_{\text{ess}}(H(\lambda, \theta)^*)$.

It remains to prove Theorem 20.8 on $R_1(z)$. We outline the necessary modifications to the proof of Theorem 20.7 in Problem 20.11.

Problem 20.11. Prove Theorem 20.8. (*Hints*: The main modification occurs in the estimate of $(H_1(\lambda, \theta) - z)$ on supp j_0. For $H = -\Delta + V$, use the Schwarz inequality to establish

$$\|(H - E) j_0 u\| \geq \|j_0 u\|^{-1} \langle j_0 u, (H - E) j_0 u \rangle.$$

Apply this to $H_1(\lambda, \theta)$, taking into account the simplifications of $H_1(\lambda, \theta)$ on supp j_0. Finally, to estimate $\|\eta \partial_k R_1(z)\|$, retain the kinetic energy term on the right side of (20.62), and use the simple inequality

$$\|f\|^2 + \|g\|^2 \geq 2\|f\| \, \|g\|$$

to obtain

$$\|\partial_k j_0 u\|^2 \|j_0 u\|^{-1} \geq a \|\partial_k j_0 u\| - a^2 \|j_0 u\|.$$

Apply this inequality by making a judicious λ-dependent choice of a.)

20.8 Notes

Recent papers on the semiclassical theory of shape resonances are by Ashbaugh and Harrell [AH], Combes, Duclos, Klein, and Seiler [CDKS], Helffer and Sjöstrand [HSj4], Hislop and Sigal [HS1, HS2], and Sigal [Si5]. Ashbaugh and Harrell consider one-dimensional and spherically symmetric shape resonance potentials of compact support in the large λ regime. They prove the existence of resonances and compute upper bounds on the width. The other papers consider the multi-dimensional case and the methods are similar to those presented here. Another approach to the semiclassical theory of resonances was developed by Helffer and Sjöstrand in [HSj4]. We will comment more on this in Chapter 23. For the relation between the Helffer–Sjöstrand definition of resonances and the Aguilar–Balslev–Combes–Simon theory, we refer to the paper of Helffer and Martinez [HeM]. Using techniques similar to those discussed here, the existence of shape resonances in the semiclassical regime was proved by Combes, Duclos, Klein, and Seiler in [CDKS]. Shape resonances potentials of compact support were studied by Nakamura [N4, N5]. The question of the order of the pole at a resonance in the semiclassical regime was studied by Kaidi and Rouleux [KR].

The theory of resonances for generalized semiclassical regimes (see Section 16.4) has been applied to a variety of problems. Among the problems in the quantum mechanics of two-body systems, resonances in the Stark effect were studied by Graffi and Grecchi [GrGr], Harrell and Simon [HaSi], Herbst [He3], Sigal [Si5], and Titchmarsh [T2]. The Zeeman effect and resonances in magnetic fields are studied in Briet [Br], Helffer and Sjöstrand [HSj4] and Wang [W2]. Resonances in the Born–Oppenheimer model for molecules were studied by Martinez [M3, M4] and in a model of molecular dissociation by Klein [Kl2].

We mention some other applications of the theory. Resonances in quantum mechanical systems often correspond to bounded orbits in a corresponding classical system. This idea has been applied to various geometrical situations. Resonances for the Laplace–Beltrami operator on an $\mathbb{R}^n$ with certain spherically symmetric Riemannian metrics, such that there exist families of trapped geodesics, were studied by DeBièvre and Hislop [DeBH]. Barrier top resonances were studied by Briet, Combes, and Duclos in [BCD2]. This is a case of resonances generated by closed, hyperbolic trajectories, which were analyzed by Gérard and Sjöstrand in [GSj]. The resonances for a Schrödinger operator with a periodic potential were studied by Gérard [G2] and Klopp [Klp]. Gérard also considered resonances in the scattering of atoms off surfaces of semi-infinite periodic structures in Gérard [G1]. We refer to Reed and Simon, Volume IV [RS4] for a discussion of resonances in the scattering of an electron off a helium atom.

21

Quantum Nontrapping Estimates

Let H_κ be a family of Schrödinger operators for $\kappa \in \mathcal{S}$, some sector in $\mathbb{C}$, having 0 as a limit point. Suppose that λ_0 is an unperturbed eigenvalue of H_κ. Lower-bound estimates on $(H_\kappa - \lambda_0)u$, for functions u supported in the complement of some compact region, play an essential role in proving the stability of the eigenvalue with respect to the perturbation H_κ. We have explored some numerical range and localization methods for establishing such estimates in Chapter 19. In our discussion of shape resonances, we assumed a condition, (V4), which would be rather difficult to verify for a given potential. In this chapter, we formulate a *nontrapping condition* directly on the potential, relative to a given vector field used for the spectral deformation, and show that the stability estimate (V4) can be derived from this condition on V. The nontrapping condition implies that the potential outside a compact region does not produce any resonances in a neighborhood of the unperturbed eigenvalue λ_0.

21.1 Introduction to Quantum Nontrapping

In order to motivate the quantum nontrapping condition associated with Hunziker's method of spectral deformation, let us recall a very simple form of the virial theorem (see [RS4]). Consider a potential $V \in C^\infty(\mathbb{R}^n)$ satisfying for $\alpha = (\alpha_1, \ldots, \alpha_n)$, $|\alpha| = 0, 1$,

$$|\partial^\alpha V(x)| \leq c(1 + \|x\|)^{-1}. \tag{21.1}$$

We suppose that V has a positive *virial*,

$$-x \cdot \nabla V(x) \geq c_0 > 0, \quad \forall x \in \mathbb{R}^n. \tag{21.2}$$

This condition states that the force on the particle is everywhere positive so that potential is everywhere repulsive. It is intuitively clear that this implies that $H \equiv -\Delta + V$ has no bound states. We will show this under one additional assumption (which can be removed; see Problem 21.2). Let us suppose that H has an eigenvalue E with an eigenfunction $\psi_E \in L^2(\mathbb{R}^n)$. We assume that $x \cdot \nabla \psi_E \in L^2(\mathbb{R}^n)$. This amounts to requiring some decay on ψ_E and $\partial_i \psi_E$, since the smoothness is assured by elliptic regularity, Theorem 3.8. We define a skew-adjoint operator A by

$$A \equiv \frac{1}{2}(x \cdot \nabla + \nabla \cdot x). \tag{21.3}$$

Problem 21.1. Prove that iA is self-adjoint. Moreover, prove that under hypothesis (21.1), $D(H) = H^2(\mathbb{R}^n)$ and $D(H) \cap D(A)$ is dense in $L^2(\mathbb{R}^n)$.

Formally, we have that the commutator between H and A is

$$[H, A] = -2\Delta - x \cdot \nabla V, \tag{21.4}$$

which, in fact, holds on an appropriate dense domain by Problem 21.1. We can compute the quadratic form (since $\psi_E \in H^2(\mathbb{R}^n)$)

$$\langle \psi_E, [H, A]\psi_E \rangle = 2\langle \psi_E, (-\Delta)\psi_E \rangle + \langle \psi_E, (-x \cdot \nabla V)\psi_E \rangle, \tag{21.5}$$

and, as $-\Delta \geq 0$, we obtain a lower bound from (21.2),

$$\langle \psi_E, [H, A]\psi_E \rangle \geq c_0 \|\psi_E\|^2 > 0. \tag{21.6}$$

On the other hand, since $[H, A] = [H - E, A]$, and we are assuming that $\psi_E \in D(A)$, the left side of (21.6) vanishes:

$$\begin{aligned} \langle \psi_E, [H, A]\psi_E \rangle &= \langle \psi_E, [H - E, A]\psi_E \rangle \\ &= \langle (H - E)\psi_E, A\psi_E \rangle + \langle A\psi_E, (H - E)\psi_E \rangle \\ &= 0. \end{aligned}$$

Hence, H can have no bound states if V satisfies (21.2).

Problem 21.2. The purpose of this problem is to show that the assumption $\psi_E \in D(A)$ can be removed (see [CFKS]). For $\lambda \in \mathbb{R}$, define $R_\lambda \equiv \lambda(A + \lambda)^{-1}$, which is bounded for $\lambda \neq 0$ by Problem 21.1. Prove that $R_\lambda \colon H^2(\mathbb{R}^n) \to H^2(\mathbb{R}^n)$ by showing that for $\phi \in H^2(\mathbb{R}^n)$, $\lim_{\varepsilon \to 0}(1 - \Delta)(1 - \varepsilon\Delta)^{-1} R_\lambda \phi$ exists and is in $L^2(\mathbb{R}^n)$. To see this, let $H_\varepsilon \equiv (1 - \Delta)(1 - \varepsilon\Delta)^{-1}$, and write

$$H_\varepsilon R_\lambda \phi = R_\lambda H_\varepsilon \phi + (iA + \lambda)^{-1}[iA, H_\varepsilon] R_\lambda \phi.$$

Show that the second term is uniformly bounded for all large $|\lambda|$. Since $H^2(\mathbb{R}^n)^* = H^{-2}(\mathbb{R}^n)$, conclude that $R_\lambda \colon H^{-2}(\mathbb{R}^n) \to H^{-2}(\mathbb{R}^n)$. Finally, show that $s-$

$\lim_{\lambda\to\infty} R_\lambda = I$ in $H^{\pm 2}(\mathbb{R}^n)$, by noting that $D(A) \cap H^2(\mathbb{R}^n)$ is dense and that for $\phi \in D(A) \cap H^2(\mathbb{R}^n)$,

$$(R_\lambda - 1)\phi = i\lambda^{-1} R_\lambda A\phi.$$

We can now write the left side of (21.6) as

$$\lim_{\lambda\to\infty} \langle R_\lambda \psi_E, [H, iA] R_\lambda \psi_E \rangle = \lim_{\lambda\to\infty} \langle \psi_E, [H, AR_\lambda] \psi_E \rangle = 0,$$

since AR_λ is uniformly bounded.

We seek a condition on V, similar to the virial condition (21.2), which will guarantee the absence of resonances, in the sense of (V4), originating from V in the exterior region and lying in a neighborhood of e_0. The global repulsive condition (21.2) is far too strong. The shape resonance potentials of Chapter 20 do not satisfy this condition. However, at the exterior classical turning surfaces $S^+(e_0)$, we might expect the potential to be locally repulsive. Away from those exterior surfaces, we have the positivity condition $(e_0 - V) > \delta > 0$. With the form of a shape resonance potential in mind, we formulate a nontrapping condition on the potential which combines this local repulsive character of the potential near the exterior of the well $W(e_0)$ with the positivity of the energy $(e_0 - V(x))$ for $\|x\|$ sufficiently large.

Definition 21.1. *A potential V is nontrapping at energy E on an open subset $\Omega \in \mathbb{R}^n$ if $\exists\, \varepsilon_0 > 0$ and a smooth (C^1 suffices) vector field v defined on Ω such that $\forall u \in C_0^\infty(\Omega)$,*

$$\left\langle u, \left[-v \cdot \nabla V + 2\frac{\langle p(Dv)p\rangle_u}{\langle p^2\rangle_u}(E - V)\right] u \right\rangle \geq \varepsilon_0 \|u\|^2, \tag{21.7}$$

where $\langle A\rangle_u \equiv \langle u, Au\rangle$, $p^2 \equiv -\Delta$, and Dv is the differential (Jacobian matrix) of v.

We refer to (21.7) as the *nontrapping (NT) condition*. Let us note some simplifications that might occur. First, suppose that the Jacobian of the vector field v is strictly positive on Ω, that is, there exists $\delta > 0$ such that $Dv > \delta$ on Ω. Then the quantity in square brackets in (21.7) can be written as the matrix element in the state u of the *generalized virial* $S_E(x; V, v)$ given by

$$S_E(x; V, v) \equiv -v \cdot \nabla V + 2\delta(E - V). \tag{21.8}$$

Similarly, the NT condition simplifies in one dimension. It suffices to have the condition

$$S_E(x; V, v) \equiv 2v'(x)(E - V(x)) - v(x)V'(x) \geq \varepsilon_0, \tag{21.9}$$

which is similar to the classical case, as we will discuss in the next section. The construction of a vector field satisfying (21.7) is not always easy. We will discuss the NT condition and vector fields exterior to a surface in Section 21.4. In one dimension, however, we can solve the differential inequality (21.9) on a half-line $[a_0, \infty)$ and obtain a formula for $v(x)$:

$$v(x) = \frac{1}{2}[E - V(x)]^{-\frac{1}{2}} \int_{a_0}^{x} a(s)[E - V(s)]^{-\frac{1}{2}} ds, \tag{21.10}$$

where $a(s) > \varepsilon_0$ is an unspecified function that can be chosen according to a particular problem. In resonance problems, the vector field v is used to generate the spectral deformation group for the Schrödinger operator.

Problem 21.3. Show that $v(x)$ in (21.10) satisfies

$$S_E(x; V; v) = a(x) \text{ on } [a_0, \infty).$$

21.2 The Classical Nontrapping Condition

Let us briefly discuss the case of nontrapping potentials in classical mechanics. This will provide some insight into the quantum mechanical situation and the NT condition (21.7). Suppose that we are given a Hamiltonian function

$$h(x, \xi) = \frac{1}{2}\xi^2 + V(x) \tag{21.11}$$

on the phase space $\mathbb{R}^{2n}$. The corresponding equations of motion for the Hamiltonian flow $(x(t; y, \eta), \xi(t; y, \eta))$ are

$$\dot{x}(t; y, \eta) = \xi(t; y, \eta), \tag{21.12}$$

$$\dot{\xi}(t; y, \eta) = -(\nabla_x V)(x(t; y, \eta)), \tag{21.13}$$

with the initial conditions $x(0; y, \eta) = y$ and $\xi(0; y, \eta) = \eta$. If we fix the energy $E \geq 0$, then for given initial conditions (y, η) with $h(y, \eta) = E$, the flow $(x(t; y, \eta), \xi(t; y, \eta))$ remains on the energy surface $\{(x, \xi) \mid h(x, \xi) = E\} \equiv S_E$ for all time. We are interested in the situation when the surface S_E is unbounded. The bounded case was dealt with in Problem 20.1.

Definition 21.2. *An energy $E \geq 0$ is nontrapping for the Hamiltonian system* (21.12) and (21.13) *if, for any $R \gg 1$ large enough, $\exists\, T \equiv T(R)$ such that*

$$\|x(t; y, \eta)\| > R,$$

for all $|t| > T$ when $|y| < R$ and $E = \frac{1}{2}\|\eta\|^2 + V(y)$.

This time-dependent condition indicates that V will not produce bounded orbits from initial conditions on the energy surface S_E and with $\|y\| < R$. We would like a time-independent condition directly on V which guarantees that V is nontrapping in the sense of Definition 21.2.

To formulate such a condition, we first note that a repulsive potential is nontrapping, for if the force $F \equiv -x \cdot \nabla V \geq 0$, then we immediately have $\|\ddot{x}(t)\| \geq 0$, which shows that $\|x(t)\| \geq c_0 t + c_1$. On the other hand, if $E > V$, then $\|\xi\| = [2(E - V(x))]^{1/2} > 0$, and so by Hamilton's equations,

$$\|\dot{x}(t)\| = [2(E - V(x(t)))]^{\frac{1}{2}},$$

provided the right side is strictly positive. It follows that in this case the potential is nontrapping at energy E. If, however, there is a surface on which $E = V(x)$, then we may find a bounded orbit for some initial conditions.

Problem 21.4. Consider a potential on $\mathbb{R}^1$,

$$V(x) = e^{-x^2},$$

and show that every positive energy $E \neq 1$ is nontrapping. Explore the Hamiltonian flow for $E = 1$. Is $E = 1$ NT?

We can now write a condition on V that allows a local repulsive character of V to compensate for the vanishing of $E - V$ and, conversely, that allows the local positivity of $E - V$ to dominate the possible local attractivity of V. The quantity we study is

$$S_E(x; V) \equiv 2(E - V(x)) - x \cdot \nabla V(x), \tag{21.14}$$

which is a generalization of the virial (21.2). Note also the relationship between this quantity and the commutator (21.4). If we define a classical analogue of the operator A by $a(x, \xi) \equiv x \cdot \xi$, then the Poisson bracket between the Hamiltonian h in (21.11) and a is

$$\begin{aligned} \{h(x, \xi), a(x, \xi)\} &= \sum_{j=1}^{n} \left(\frac{\partial h}{\partial \xi_j} \frac{\partial a}{\partial x_j} - \frac{\partial h}{\partial x_j} \frac{\partial a}{\partial \xi_j} \right) \\ &= S_E(x; V), \end{aligned} \tag{21.15}$$

provided we restrict to the energy surface S_E.

We have the following simple proposition.

Proposition 21.3. *Suppose $V \in C^1$ and an energy $E > 0$ is such that $S_E(x, V) > c_0 > 0$, for all $x \in \mathbb{R}^n$ and some $c_0 > 0$. Then for initial conditions (y, η) such that $h(y, \eta) = E$, the flow satisfies*

$$\|x(t; y, \eta)\| \geq c_1 t + c_2,$$

for some $c_1 > 0$, some c_2, and all $t \in \mathbb{R}$.

Proof. Let $a = x \cdot \xi$. The Poisson bracket of a with h is

$$\begin{aligned} \{h, a\} &= \sum_{i=1}^{n} (\partial_{\xi_i} h \cdot \partial_{x_i} a - \partial_{\xi_i} a \cdot \partial_{x_i} h) \\ &= \|\xi\|^2 - x \cdot \nabla V \\ &= 2\left(\frac{1}{2}\|\xi\|^2 + V\right) - 2V - x \cdot \nabla V. \end{aligned} \tag{21.16}$$

If we evaluate (21.16) along a trajectory on the energy surface S_E, we find

$$\{h, a\} = 2(E - V) - x \cdot \nabla V. \tag{21.17}$$

By Hamilton's equations and the NT condition, we obtain

$$\frac{da}{dt}(x(t; y, \eta), \xi(t; y, \eta)) = \{h, a\} \geq c_0 > 0, \tag{21.18}$$

so that

$$x \cdot \xi \geq c_0 t + c_1.$$

Again, by Hamilton's equation, $\xi = \dot{x}$, and so

$$\frac{1}{2} \frac{d}{dt} \|x(t; y, \eta)\|^2 \geq c_0 t + c_1,$$

which allows us to verify the condition of Definition 21.2. □

The classical NT condition shows that a trajectory on a nontrapping energy surface S_E will eventually escape to infinity. No bounded orbits can be created by the potential with initial conditions lying on S_E. To compare this with the quantum NT condition of Definition 21.1, let us examine the classical case in more detail. For this, it is convenient to work in one dimension. We can generalize the virial (21.14) by replacing x by a general vector field v. We then obtain the NT condition

$$-vV' + 2(E - V)v' \geq \varepsilon_0. \tag{21.19}$$

This is close to the quantum condition (21.7) and suffices in the quantum one-dimensional situation (see (21.9)). In analogy with the proof of Proposition 21.3, we can define an observable a_v by

$$a_v \equiv v(x) \cdot \xi, \tag{21.20}$$

with v the vector field appearing in (21.19). The Poisson bracket of a_v and h is

$$\begin{aligned} \frac{da_v}{dt} &= \{h, a_v\} = \xi^2 v' - V'v \\ &= 2(h(x, \xi) - V(x))v'(x) - V'(x)v(x). \end{aligned}$$

Restricting to an energy surface $h(x, \xi) = E$, we see that

$$\frac{da_v}{dt} = \{h, a_v\}\Big|_{S_E} = S_E(x, V) \geq \varepsilon_0. \tag{21.21}$$

Hence, a_v is a classical observable that increases along the trajectories on the surfaces S_E. The existence of such an observable implies that there are no bounded trajectories on S_E.

21.3 The Nontrapping Resolvent Estimate

In this section, we prove that a nontrapping condition (21.7) on a potential V implies an a priori lower bound on the spectrally deformed Hamiltonian $H(\lambda, \theta)$ in the semiclassical regime. This reduces the verification of condition (V4) for shape resonance potentials to showing that they are nontrapping. We discuss this in the next section.

Let $e_0 \in \sigma(K)$ as in Chapter 20. We defined in (20.21) and (20.22) the energies $e' \equiv e_0 + \lambda^\delta$ and $e \equiv e_0 + 2\lambda^\delta$, for $\delta > \frac{1}{2}$. We assume that $V(x;\lambda)$ is nontrapping on Ext $S^+(e;\lambda)$ with vector field v constructed in Proposition 20.2. Let $H(\lambda, \theta)$ be the corresponding spectral deformation family. We can take $\theta = i\beta$, $0 < \beta < a$, and write $H_\beta(\lambda) \equiv H(\lambda, i\beta)$, for simplicity. We prove the following nontrapping estimate.

Theorem 21.4. *Let $V(x;\lambda)$ be nontrapping at energy e in* Ext $S^+(e;\lambda)$ *for all λ large. Then $\exists\, c_0 > 0$ such that $\forall u \in C_0^\infty(\text{Ext } S^+(e;\lambda))$, $\forall \lambda$ large and sufficiently small $0 < \beta < a$,*

$$\|(H_\beta(\lambda) - e_0)u\| \geq \beta c_0 \left(\|u\| + \sum_{i=1}^{n} \|\eta \partial_i u\| \right), \tag{21.22}$$

for some function η supported in Ext $S^+(e;\lambda) \setminus$ Ext $S^+(e';\lambda)$.

We remark that this is stable for all z in $A(e_0)$ as defined in (20.41) and (20.46).

Problem 21.5. Verify that Theorem 21.4 implies the corresponding result with e_0 replaced by $z \in A(e_0)$.

As a preliminary to the proof of Theorem 21.4, we introduce another partition of unity for the region Ext $S^+(e;\lambda)$; see Figure 21.1. We define energies F' and F by

$$F' \equiv e_0 + \frac{1}{2}\lambda^\delta, \tag{21.23}$$

$$F \equiv e_0 + \frac{3}{4}\lambda^\delta, \tag{21.24}$$

respectively, for the same δ as in the definition of e' and e. For all λ sufficiently large, we have

$$e_0 < F' < F < e' < e, \tag{21.25}$$

which implies a similar containment for the interior of the turning surfaces,

$$\text{Int } S^+(e_0;\lambda) \supset \text{Int } S^+(F';\lambda) \supset \text{Int } S^+(F;\lambda) \supset \text{Int } S^+(e';\lambda). \tag{21.26}$$

We introduce a partition of unity relative to F in the same way as in equations (20.33)–(20.35). For any $E \in \mathbb{R}$, let ξ_E be a smooth function satisfying $0 \leq \xi_E \leq 1$, and

$$\xi_E(s) = \begin{cases} 1, & E < s, \\ 0, & s < E - \frac{1}{4}\lambda^\delta, \end{cases} \tag{21.27}$$

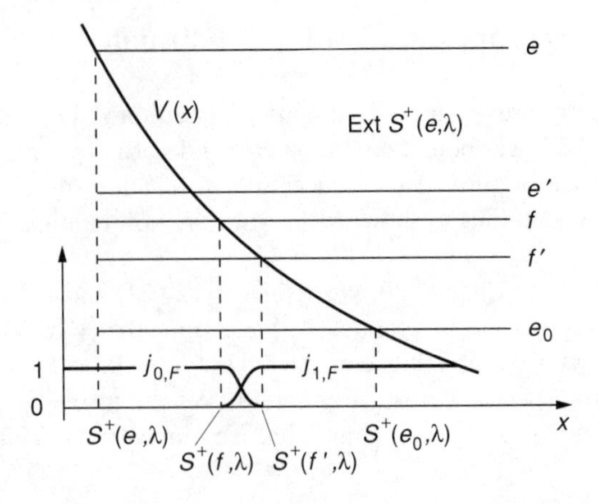

FIGURE 21.1. Geometry and partition of unity for the proof of the nontrapping estimate for a shape resonance potential.

as in (20.33). Taking $E = F$, we define $j_{0,F}$ on Ext $S^+(e;\lambda)$ by

$$j_{0,F}(x) \equiv \xi_F(V(x;\lambda)). \tag{21.28}$$

Then $j_{0,F} \mid \text{Ext } S^+(e;\lambda) \setminus \text{Ext } S^+(F;\lambda) = 1$, $\text{supp}|\nabla j_{0,F}| \subset \text{Int } S^+(F';\lambda) \setminus \text{Int } S^+(F;\lambda)$, and

$$\|\nabla j_{0,F}\|_\infty = \mathcal{O}(\lambda^{\frac{1}{2}-\delta}), \tag{21.29}$$

as in (20.35). Similar to that construction, we set

$$j_{1,F}(x) \equiv \left(1 - j_{0,F}^2(x)\right)^{\frac{1}{2}}$$

so that

$$\sum_{i=0}^{1} j_{i,F}^2(x) = 1 \ \text{ on Ext } S^+(F;\lambda). \tag{21.30}$$

The consequences of Problem 20.9 hold for $\{j_{i,F}\}_{i=0}^{1}$. We take the parameter δ to be the same as in Chapter 20 and $\delta > 1/2$.

Proof of Theorem 21.4.

(1) General strategy. Let $u \in C_0^\infty(\text{Ext } S^+(e;\lambda))$. For each $i = 0, 1$, we obtain, by the Schwarz inequality,

$$\|(H_\beta(\lambda) - e_0)u\| \, \|j_{i,F}^2 u\| \geq \text{Re}\langle u, (H_\beta(\lambda) - e_0) j_{i,F}^2 u\rangle, \tag{21.31}$$

and, for any $\phi \in \mathbb{R}$,

$$\begin{aligned} \|(H_\beta(\lambda) - e_0)u\| \, \|j_{i,F}^2 u\| &\geq \text{Im}\langle u, (H_\beta(\lambda) - e_0) j_{i,F}^2 u\rangle \\ &\geq -\text{Im}\langle e^{i\phi} u, (H_\beta(\lambda) - e_0) j_{i,F}^2 u\rangle. \end{aligned} \tag{21.32}$$

We add (21.31) for $i = 0$ and (21.32) for $i = 1$ to obtain

$$\sum_{i=0}^{1} \|(H_\beta(\lambda) - e_0)u\| \, \|j_{i,F}^2 u\| \geq \operatorname{Re}\langle u, (H_\beta(\lambda) - e_0) j_{0,F}^2 u\rangle - \operatorname{Im}\langle e^{i\phi} u, (H_\beta(\lambda) - e_0) j_{i,F}^2 u\rangle. \tag{21.33}$$

We note that the simple inequality,

$$\sum_{i=0}^{1} \|j_{i,F}^2 u\| \leq \sum_{i=0}^{1} \sup_{w \neq 0} |\langle w, j_{i,F}^2 u\rangle| \, \|w\|^{-1} \leq 2\|u\|,$$

gives an upper bound on the left side of (21.33). It follows that

$$\begin{aligned} 2\|(H_\beta(\lambda) - e_0)u\| \, \|u\| \geq \; & \operatorname{Re}\langle j_{0,F} u, (H_\beta(\lambda) - e_0) j_{0,F} u\rangle \\ & - \operatorname{Im}\langle e^{i\phi} j_{1,F} u, (H_\beta(\lambda) - e_0) j_{1,F} u\rangle \\ & - \sum_{k=0}^{1} |\langle u, [p_{i\beta}^2, j_{k,F}] j_{k,F} u\rangle|. \end{aligned} \tag{21.34}$$

We now turn to the estimation of each of the three terms on the right in (21.34).

(2) Exterior estimate. We first consider the $j_{1,F}$ term in (21.34). Since the smooth function $j_{1,F}u$ is supported in Ext $S^+(F;\lambda)$, we replace it by $w \in C_0^\infty(\text{Ext } S^+(F;\lambda))$, for notational simplicity. Because of the support properties expressed in (21.27) and (21.28), we can apply the nontrapping condition (21.7) on Ext $S^+(F;\lambda)$ and at energy e_0. We choose the real phase ϕ so that

$$\tan\phi = \frac{\operatorname{Im}\langle p_\ell J^{\ell i} J^{mi} p_m\rangle_w}{\operatorname{Re}\langle p_\ell J^{\ell i} J^{mi} p_m\rangle_w}, \tag{21.35}$$

using the notation of (18.9) and the summation convention: Repeated indices are summed. The kinetic energy contribution from $p_{i\beta}^2$ is

$$-\operatorname{Im} e^{-i\phi}\langle w, p_{i\beta}^2 w\rangle = \cos\phi\big[\tan\phi \operatorname{Re}\langle w, p_{i\beta}^2 w\rangle - \operatorname{Im}\langle w, p_{i\beta}^2 w\rangle\big]. \tag{21.36}$$

Recall from (18.10) and (18.11) that

$$p_{i\beta}^2 = p_\ell a_{i\beta,\ell m} p_m + g_{i\beta}.$$

The contribution of the matrix $a_{i\beta}$-term to (21.36) is

$$\tan\phi \operatorname{Re}\langle w, p_\ell a_{i\beta,\ell m} p_m w\rangle - \operatorname{Im}\langle w, p_\ell a_{i\beta,\ell m} p_m w\rangle = 0,$$

in light of (21.35). Since g_θ is bounded, we obtain

$$-\operatorname{Im} e^{-i\phi}\langle w, p_{i\beta}^2 w\rangle \geq -\beta c_0 \|w\|^2, \tag{21.37}$$

for some $c_0 > 0$. The potential energy contribution is estimated using the NT condition (21.7). We have

$$-\mathrm{Im}\, e^{-i\phi}\langle w, (V_{i\beta}(x;\lambda) - e_0)w\rangle = \cos\phi\big[\mathrm{Im}\langle e_0 - V_{i\beta}\rangle_w - \tan\phi\,\mathrm{Re}\langle e_0 - V_{i\beta}\rangle_w\big]. \quad (21.38)$$

We expand $V_{i\beta}(x;\lambda)$ about $\beta = 0$:

$$\begin{aligned} V_{i\beta}(x;\lambda) &= V(x;\lambda) + i\lambda^{\frac{1}{2}}\beta(v\cdot\nabla V)\left(\lambda^{-\frac{1}{2}}x\right) \\ &\quad + \mathcal{O}(\beta^2\partial_i\partial_j V(\lambda^{-\frac{1}{2}}x)v_i(x)v_j(x)), \end{aligned} \quad (21.39)$$

and obtain for the first term on the right in (21.38),

$$\begin{aligned} \mathrm{Im}\langle e_0 - V_{i\beta}\rangle_w &= -\lambda^{\frac{1}{2}}\beta\left\langle v\cdot\nabla V\left(\lambda^{-\frac{1}{2}}x\right)\right\rangle_w \\ &\quad + \mathcal{O}\left(\beta^2\left\langle\partial_i\partial_j V\left(\lambda^{-\frac{1}{2}}x\right)v_iv_j\right\rangle_w\right). \end{aligned} \quad (21.40)$$

To evaluate the real part in (21.38), consider $\tan\phi$. From the facts that the Jacobian is given by

$$J = 1 + i\beta Dv,$$

and that

$$\|Dv\|_\infty \le c_0,$$

we see that for $\beta < c_0^{-1}$, J is invertible. After scaling by $x \to \lambda^{-\frac{1}{2}}x$, we obtain

$$J^{-1} = 1 - i\beta\lambda^{-\frac{1}{2}}(Dv) + \mathcal{O}(\beta^2\lambda^{-1}).$$

This allows us to write

$$\mathrm{Im}\langle p_\ell J^{\ell i}J^{mi}p_m\rangle_w = -2\beta\lambda^{-\frac{1}{2}}\langle p\cdot Dv\cdot p\rangle_w + \mathcal{O}(\beta^2\lambda^{-1})\langle p^2\rangle_w \quad (21.41)$$

and

$$\mathrm{Re}\langle p_\ell J^{\ell i}J^{mi}p_m\rangle_w = \langle p^2\rangle_w(1 + \mathcal{O}(\beta^2\lambda^{-1})). \quad (21.42)$$

Consequently, (21.41) and (21.42) result in the estimate

$$\tan\phi = -2\beta\lambda^{-\frac{1}{2}}\frac{\langle p\cdot Dv\cdot p\rangle_w}{\langle p^2\rangle_w} + \mathcal{O}(\beta^2\lambda^{-1}), \quad (21.43)$$

for $\beta > 0$ small and λ large. Combining (21.43) and (21.46), we obtain

$$\begin{aligned} &\mathrm{Im}\langle e_0 - V_{i\beta}\rangle_w - \tan\phi\,\mathrm{Re}\langle e_0 - V_{i\beta}\rangle_w \\ &\quad = -\lambda^{\frac{1}{2}}\beta\left\langle v\cdot\nabla V\left(\lambda^{-\frac{1}{2}}x\right)\right\rangle_w + \left[2\beta\lambda^{-\frac{1}{2}}\frac{\langle p\cdot Dv\cdot p\rangle_w}{\langle p^2\rangle_w}\right. \\ &\qquad \left. + \mathcal{O}(\beta^2\lambda^{-1})\right]\mathrm{Re}\langle e_0 - V_{i\beta}\rangle_w + \mathcal{O}(\beta^2)\|w\|^2 \end{aligned}$$

$$= \lambda^{\frac{1}{2}}\beta\langle -v\cdot\nabla V\left(\lambda^{-\frac{1}{2}}x\right) + 2\frac{\langle p\cdot Dv\cdot p\rangle_w}{\langle p^2\rangle_w}[e_0 + V(0) - V\left(\lambda^{-\frac{1}{2}}x\right)]\rangle_w + \mathcal{O}(\beta^2)\|w\|^2$$
$$\geq (\varepsilon_0\beta - e_0\beta^2)\|w\|^2, \tag{21.44}$$

where we used the NT condition (21.7) (rescaled) to bound the first term. Estimates (21.40) and (21.44) yield for the exterior term (with $w = j_{1,F}u$)

$$-\text{Im}\, e^{i\phi}\langle j_{1,F}u, (H_\beta(\lambda) - e_0)j_{1,F}u\rangle \geq (e_0\beta - c_1\beta^2)\|j_{1,F}u\|^2, \tag{21.45}$$

for some $c_0, c_1 > 0$, all $\beta > 0$ small, and λ sufficiently large. (Note that (21.43) implies $\cos\phi \geq c_0 > 0$ for large λ.)

(3) Interior estimate. We now treat the $j_{0,F}$-term in (21.34). The kinetic energy condition is

$$\text{Re}\langle u, j_{0,F}p_{i\beta}^2 j_{0,F}u\rangle = \text{Re}\langle j_{0,F}u, p_\ell a_{i\beta,\ell m}p_m j_{0,F}u\rangle + \text{Re}\langle u, j_{0,F}^2 g_{i\beta}u\rangle. \tag{21.46}$$

From the estimates obtained after Problem 18.1, we can bound (21.46) from below by

$$\sum_{\ell=1}^n (1 - c_0\beta)\|j_{0,F}\partial_\ell u\|^2 - c_1\lambda^{\frac{1}{2}-\delta}\|u\|^2 - c_2\beta\lambda^{1-2\delta}\|j_{0,F}u\|^2, \tag{21.47}$$

for $c_0, c_1, c_2 > 0$ and $\delta > 1/2$. As for the potential energy term, we note that

$$V(x;\lambda) \mid [\text{Int}\, S^+(F';\lambda) \setminus \text{Int}\, S^+(e;\lambda)] \geq F',$$

which holds for λ sufficiently large. Using this fact and the expansion (21.39) about $\beta = 0$, we find

$$\text{Re}\langle u, (V_{i\beta}(x;\lambda) - e_0)j_{0,F}^2 u\rangle \geq \left(\frac{1}{2}\lambda^\delta - c_0\beta\right)\|j_{0,F}u\|^2. \tag{21.48}$$

Combining (21.47) and (21.48), we find

$$\text{Re}\langle j_{0,F}u, (H_\beta(\lambda) - e_0)j_{0,F}u\rangle \geq \left(\frac{1}{2}\lambda^\delta - c_0\beta - c_1\beta\lambda^{1-2\delta}\right)\|j_{0,F}u\|^2 - c_2\lambda^{\frac{1}{2}-\delta}\|u\|^2 + \sum_{\ell=1}^n (1 - c_0\beta)\|j_{0,F}\partial_\ell u\|^2,$$
$$\geq c_0\beta\|j_{0,F}u\|^2 + \frac{9}{10}\sum_{\ell=1}^n \|j_{0,F}\partial_\ell u\|^2 - c_1\lambda^{\frac{1}{2}-\delta}\|u\|^2, \tag{21.49}$$

for all λ large and $\beta > 0$ small.

(4) Error terms. Finally, we consider the $k = 0$ term in the sum (21.34). Writing $p_{i\beta}^2$ as in (18.10), we define

$$A_\ell \equiv a_{i\beta,\ell m}(p_m j_{0,F})$$

and

$$\tilde{A}_\ell \equiv a_{i\beta,m\ell}(p_m j_{0,F}).$$

The commutator in (21.341) for $k = 0$ is

$$[p_{i\beta}^2, j_{0,F}]j_{0,F} = 2A_\ell(p_\ell j_{0,F}) + (A_\ell + \tilde{A}_\ell)j_{0,F}p_\ell,$$

from which we obtain

$$\begin{aligned} |\langle u, [p_{i\beta}^2, j_{0,F}]j_{0,F}u\rangle| &\leq \sum_{\ell=1}^{n} c_0\lambda^{\frac{1}{2}-\delta}\|j_{0,F}\partial_\ell u\|\,\|u\| \\ &\quad + c_1\lambda^{1-2\delta}\|u\|, \end{aligned} \tag{21.50}$$

taking into account (21.29). As for the $k = 1$ contribution to the sum, we use an identity as in part (2) of Problem 19.9 and (21.30) to write

$$\begin{aligned} \langle u, [p_{i\beta}^2, j_{1,F}]j_{1,F}u\rangle &= -\frac{1}{2}\langle u, [j_{0,F}^2, p_{i\beta}^2]u\rangle \\ &\quad +\frac{1}{2}\langle u, [j_{1,F}, [j_{1,F}, p_{i\beta}^2]]u\rangle. \end{aligned} \tag{21.51}$$

The double commutator term is bounded above by $c_0\lambda^{1-2\delta}\|u\|^2$, and the term involving $j_{0,F}$ is bounded as in (21.50). Consequently, we obtain

$$\sum_{k=0}^{1} |\langle u, [p_{i\beta}^2, j_{k,F}]j_{k,F}u\rangle| \leq c_0\lambda^{\frac{1}{2}-\delta}\|j_{0,F}\nabla u\|\,\|u\| + c_1\lambda^{1-2\delta}\|u\|^2. \tag{21.52}$$

(5) Summary. We now collect results (21.45), (21.49), and (21.52) to obtain a bound for the left side of (21.34):

$$\begin{aligned} \|(H_\beta(\lambda) - e_0)u\| &\geq c_0\beta(\|j_{0,F}u\|^2 + \|j_{1,F}u\|^2)\|u\|^{-1} \\ &\quad - c_1\lambda^{1-2\delta}\|u\| + \frac{9}{10}\sum_{\ell=1}^{n}\|j_{0,F}\partial_\ell u\|^2\|u\|^{-1} \\ &\quad - c_2\lambda^{\frac{1}{2}-\delta}\sum_{\ell=1}^{n}\|j_{0,F}\partial_\ell u\| \\ &\geq c_0\beta\|u\| + R. \end{aligned} \tag{21.53}$$

We estimate the remainder R using the inequality of Problem 20.11. For any $a > 0$,

$$\|j_{0,F}\partial_\ell u\|^2\|u\|^{-1} \geq a\|j_{0,F}\partial_\ell u\| - a^2\|u\|,$$

and, choosing $a = 2c_0\beta$, we obtain for λ sufficiently large,

$$\|(H_\beta(\lambda) - e_0)u\| \geq c_0\beta \left(\|u\| + \sum_{\ell=1}^{n} \|j_{0,F}\partial_\ell u\| \right), \tag{21.54}$$

proving the theorem. □

21.4 Some Examples of Nontrapping Potentials

We want to explain how to verify the nontrapping condition for some classes of potentials. One basic strategy, explained earlier, is to assume that the potential is repulsive in a neighborhood of the classical turning surface. In this case, the positivity of $-v \cdot \nabla V$ can be used to control the term containing $(E - V)$ as a factor, which is small. On the exterior of this region, it is this factor, $(E - V)$, which guarantees the nontrapping condition. One implements this strategy using a partition of unity. Let us consider a positive shape resonance potential with a single, nondegenerate minimum at the origin and such that for $\|x\| > R$, for some $R > 0$,

$$V(x) = f(x)\|x\|^{-\rho}, \tag{21.55}$$

where $\rho > 1$ and the positive, bounded function f is smooth with bounded derivatives. In addition, f must satisfy some local conditions given below. The behavior of V inside $B_R(0)$ is rather arbitrary, subject to the conditions mentioned above, and there can be local singularities. Let us consider a positive energy level $E > 0$. We can choose E so that in a neighborhood of the classical turning surface $S^+(E)$, the potential V has the form (21.55). We will denote by δ_i various strictly positive constants whose precise value we will adjust below. Let v_E be a vector field exterior to the surface $S^+(E + 2\delta_1)$, as in Definition 20.1,

$$v_E(x) = \phi_{E+2\delta_1}(V(x))x, \tag{21.56}$$

where the form of the function $\phi_{E+2\delta_1}$ will be chosen below. We indicate the proof that such a potential V is nontrapping on Ext $S^+(E + \delta_1)$ with respect to the vector field v_E.

We can find δ_2 such that $E - V > \delta_3 > 0$ for $x \in$ Ext $S^+(E - 2\delta_2)$ because of the smoothness of V. Let us introduce a partition of unity $j_1^2 + j_2^2 = 1$ for Ext $S^+(E + 2\delta_1)$ so that

$$j_1 \mid \text{Ext} S^+(E - 2\delta_2) = 1$$

and

$$j_1 \mid \text{Int} S^+(E - \delta_2) = 0;$$

see Figure 21.2.

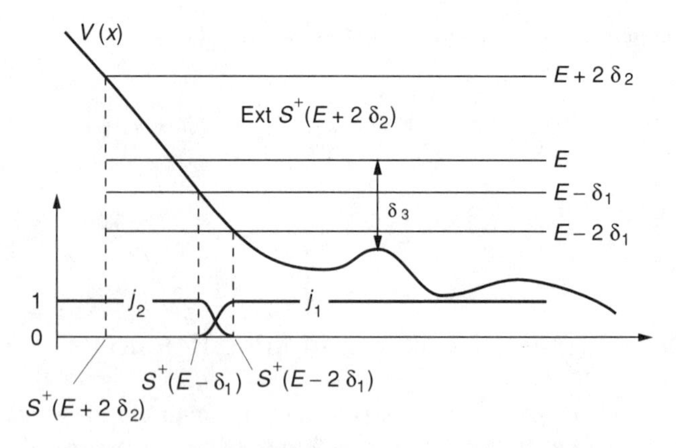

FIGURE 21.2. Geometry and partition of unity used for the verification of the nontrapping condition.

Computing the first term in (21.7), we obtain

$$\mathcal{V}_1 \equiv -v_E \cdot \nabla V = \phi_{E+2\delta_1}(V(x))\,[\rho f - x \cdot \nabla f]\,\|x\|^{-\rho}, \tag{21.57}$$

and for the second term, we obtain

$$\mathcal{V}_2 \equiv 2(E - V(x))\langle p_i[\phi'_{E+2\delta_1}\partial_i V x_j + \phi_{E+2\delta_1}\delta_{ij}]p_j\rangle\langle p^2\rangle^{-1}. \tag{21.58}$$

We now estimate from below $\mathcal{V}_i$ on supp j_k. We choose $\phi_{E+2\delta_1}$ as in the proof of Proposition 20.2 and satisfying $\phi_{E+2\delta_1} > \delta_4$ on $s < E + \delta_1$, and $\phi'_{E+2\delta_1} < 1/\delta_5$, on $s < E + \delta_1$. It then follows that for any $u \in C_0^\infty(\text{Ext } S^+(E + \delta_1))$, we have

$$\langle u, p_i[\phi'_{E+2\delta_1}\partial_i V x_j + \phi_{E+2\delta_1}\delta_{ij}]p_j u\rangle\langle p^2\rangle_u^{-1} > \left(\delta_4 - \frac{C_0}{\delta_5}\right), \tag{21.59}$$

since $\partial_i V x_j$ is uniformly bounded on this exterior region.

The function f must satisfy the condition that $(\rho - (x \cdot \nabla f)f^{-1}) > \kappa > 0$ on the region where the gradients of j_i are supported. This condition is not too restrictive, for if we recall that near $S^+(E)$ we have that $\|x\| \sim [f/E]^{1/\rho}$, and thus the condition is implied by $(\rho - [f/E]^{1/\rho}[\,\|\nabla f\|f^{-1}]) > \kappa > 0$. On the support of $j_1 \cap \text{Ext } S^+(E + \delta_1)$, it follows from (21.57) that

$$\begin{aligned} \mathcal{V}_1 &= \phi_{E+2\delta_1}(V(x))V(x)[\rho - x \cdot (\nabla f)f^{-1}] \\ &\geq [E - 2\delta_2]\kappa, \end{aligned} \tag{21.60}$$

and from (21.58) and (21.59) that

$$\mathcal{V}_2(x) > -2\delta_1[C_1 + \delta_4], \tag{21.61}$$

where C_1 is independent of δ_i. It follows from these expressions that by taking δ_1 small, if necessary, there exists an $\epsilon_1 > 0$ such that

$$\langle u, j_1^2[\mathcal{V}_1 + \mathcal{V}_2]u\rangle > \epsilon_1\|j_1 u\|^2. \tag{21.62}$$

The term involving $E - V$ should dominate on the support of j_2. From (21.57), we find

$$\mathcal{V}_2(x) > 2\delta_3 \left[\delta_4 - \frac{C_0}{\delta_5} \right] \tag{21.63}$$

and

$$\mathcal{V}_1(x) > -\delta_4 \| f \|_\infty R^{-\rho+1} \kappa_0, \tag{21.64}$$

where $\kappa_0 \geq |\rho f - x \cdot \nabla f| \, \|x\|^{-1}$, for $\|x\| > R$. By adjusting δ_4, if necessary, we see that there exists $\epsilon_2 > 0$ such that

$$\langle u, j_2^2 [\mathcal{V}_1 + \mathcal{V}_2] u \rangle > \epsilon_2 \| j_2 u \|^2. \tag{21.65}$$

This equation, together with (21.62), establishes the nontrapping condition for potentials of the form $V(x) = f(x) \|x\|^{-\rho}$ for $\|x\| > R$. It follows that any potential that is a sum of such functions outside of some compact region will satisfy the nontrapping condition.

From this discussion, we see that there are nontrivial shape resonance potentials that satisfy conditions (V1)–(V4) of Chapter 20 (we simply have to take f to be the restriction of a function analytic in some truncated slab in $\mathbb{C}^n$). We will discuss the nontrapping condition for the Stark effect and for Stark ladder resonances in Chapter 23.

21.5 Notes

Nontrapping conditions play a significant role in controlling the resonances arising from the exterior of the potential. Nontrapping conditions also enter in the theory of resonances for the scattering of waves by obstacles (see Section 23.3 for a discussion of the Helmholtz resonator, the basic paper of Morawetz [Mo1], and the book [Mo2], and the book by Lax and Phillips [LP]). As in Theorem 21.4, nontrapping conditions in quantum mechanics enter into stability estimates. Nontrapping is closely related to the existence of *resonance-free domains*. In the shape resonance case, the nontrapping condition is used to prove that the exterior Hamiltonian, $H_1(\lambda, \theta)$, has no resonances in certain regions of the lower half of the complex plane. These ideas were studied in the papers of Briet, Combes, and Dulcos [BCD1], Combes, Dulcos, Klein, and Seiler [CDKS], Combes and Hislop [CH], DeBièvre and Hislop [DeBH], Helffer and Sjöstrand [HSj4] and Klein [Kl1]. They were further generalized by Gérard and Sjöstrand [GSj].

Nontrapping conditions also enter into *semiclassical resolvent estimates*. These estimates are a form of the limiting absorption principle for which the dependence of the boundary value of the resolvent on the semiclassical parameter is explicit. Let $H(h) \equiv -h^2 \Delta + V$ be a self-adjoint Schrödinger operator with a C^2 decaying potential. Suppose that the potential V satisfies a nontrapping condition at energy E with respect to a vector field f on the complement of the classically forbidden region. An example of such a semiclassical resolvent estimate is

$$\limsup_{\epsilon \to 0} \|(1 + \|x\|^2)^{-\frac{\alpha}{2}}(H(h) - E - i\epsilon)^{-1}(1 + \|x\|^2)^{-\frac{\alpha}{2}}\| \le Ch^{-1},$$

for any $\alpha > 1/2$ and h sufficiently small. Such estimates are used in the study of scattering in the semiclassical regime. We also mention the relation between the Mourre estimate (see [CFKS]), which in some sense generalizes the nontrapping condition, and semiclassical resolvent estimates. We refer the reader to some papers discussing these issues: Robert and Tamura [RT1], [RT2]; Yafaev [Y]; Jensen [J3]; Gérard and Martinez [GM1]; Gérard [G3]; Graf [Gr]; Hislop and Nakamura [HN]; and X. P. Wang [W1].

22

Theory of Quantum Resonances III: Resonance Width

22.1 Introduction and Geometric Preliminaries

The goal of this chapter is to prove exponentially small upper bounds on the imaginary part of resonances in the semiclassical regime. In Chapters 20 and 21, we proved the stability of a low-lying eigenvalue $\tilde{e}_i \in \sigma(K)$ of finite multiplicity $m_i \geq 1$, with respect to the perturbation $H(\lambda, \theta)$, $\text{Im}\,\theta > 0$, for all λ sufficiently large. We first showed in Theorem 20.5 that there exist m_i not necessarily distinct functions $\{e_{j(i)}(\lambda)\}_{j=1}^{m_i}$ such that

$$e_{j(i)}(\lambda) \in \sigma(H_0(\lambda)) \tag{22.1}$$

and

$$\lim_{\lambda \to \infty} e_{j(i)}(\lambda) = \tilde{e}_i. \tag{22.2}$$

Next, we considered an annular region $A(\tilde{e}_i)$, described in (20.41) and (20.46). For $a_1 > 0$ satisfying $0 < a_1 \leq \frac{1}{2}(\text{Im}\,\theta)b$, we defined

$$A(\tilde{e}_i) = \left\{ z \mid \frac{1}{2} a_1 < |\tilde{e}_i - z| < a_1 \right\}. \tag{22.3}$$

By (22.2), we can find $\lambda_0 > 0$ such that $\lambda > \lambda_0$ implies that $|e_{j(i)}(\lambda) - \tilde{e}_i| < \frac{1}{4}a_1$ and that $\sigma(H_0(\lambda)) \cap \{z \mid |z - \tilde{e}_i| < \frac{1}{2}a_1\} = \{e_{j(i)}(\lambda)\}$. We then proved in Theorem 20.4 that for λ sufficiently large, there exist eigenvalues $\{z_{i,k}(\lambda)\}$ of $H(\lambda, \theta)$ lying in $\{z \mid |z - \tilde{e}_i| < \frac{1}{2}a_1\} \cap \mathbb{C}^-$ of total algebraic multiplicity m_i. From (20.55), we conclude that

$$|\text{Re}\, z_{i,k}(\lambda) - \tilde{e}_i| \leq c_0 \lambda^{-\varepsilon}, \tag{22.4}$$

for some $\varepsilon > 0$ and c_0 depending on $\tilde{e}_i$, and

$$|\mathrm{Im}\ z_{i,k}(\lambda)| \leq c_0 \lambda^{-\varepsilon}. \tag{22.5}$$

Our aim is to improve this second result on Im $z_{i,k}(\lambda)$. For simplicity of notation, let us fix $\tilde{e}_i \in \sigma(K)$ and call it e_0 as in Chapter 21. Correspondingly, we write $e_{j(0)}(\lambda)$ and $z_{0,j}(\lambda)$. We consider the Agmon metric ρ_0, as in Definition 3.2, for $V(x; \lambda)$ and e_0:

$$\rho_0(x, y) = \inf_{\gamma \in \mathcal{P}_{x,y}} \left\{ \int_0^1 [V(\gamma(t); \lambda) - e_0]_+^{\frac{1}{2}} |\dot{\gamma}(t)| dt \right\}, \tag{22.6}$$

where $\gamma \in \mathcal{P}_{x,y}$ satisfies $\gamma \in AC[0, 1]$, $\gamma(0) = x$, and $\gamma(1) = y$. Recall that for two subsets $T, S \subset \mathbb{R}^n$, the ρ_0-distance between them is defined as

$$\rho_0(S, T) = \inf_{\substack{x \in S \\ y \in T}} \rho_0(x, y). \tag{22.7}$$

Our improvement of (22.5) is the following.

Theorem 22.1. *Let $S_0(\lambda)$ be the distance in the Agmon metric ρ_0 (22.6) between $S^-(e_0; \lambda)$ and $S^+(e_0; \lambda)$. Then for any $\varepsilon > 0$ and for all λ sufficiently large, $\exists\, c_\varepsilon > 0$ such that*

$$|z_{0,k}(\lambda) - e_{j_k(0)}(\lambda)| \leq c_\varepsilon e^{-2(1-\varepsilon) S_0(\lambda)} \tag{22.8}$$

for some j_k and, in particular,

$$|\mathrm{Im}\ z_{0,k}(\lambda)| \leq c_\varepsilon e^{-2(1-\varepsilon) S_0(\lambda)}. \tag{22.9}$$

Furthermore, we have

$$S_0(\lambda) = \lambda d_A^0(0, S^+(0)) + \mathcal{O}\left(\lambda^{-\frac{1}{2}}\right) \tag{22.10}$$

as $\lambda \to \infty$, where

$$d_A^0(x, y) = \inf_{\gamma \in \mathcal{P}_{xy}} \left\{ \int_0^1 [V(\gamma(t)) - V(0)]_+^{\frac{1}{2}} |\dot{\gamma}(t)| dt \right\} \tag{22.11}$$

is independent of λ.

Let us note that the corresponding real part (22.8) does not really improve (22.4) and that (22.9) is the main result. The exponentially small imaginary part is characteristic of resonances that have their origin in tunneling phenomena. By contrast, resonances arising from a perturbation of the form $H_0 + \lambda V$ will generally have a nonvanishing, imaginary, part second-order in λ. The width is then predicted by Fermi's golden rule (see [RS4]). As we have discussed, the exponentially small shift (22.9) cannot be predicted in any finite-order perturbation theory in λ.

To prove (22.8), we construct approximate eigenfunctions of $H(\lambda, \theta)$ and $z_{0,k}(\lambda)$ from the eigenfunctions of $H_0(\lambda)$ for the cluster of eigenvalues $\{e_{j(0)}(\lambda)\}$. This is a good approximation because the remainder term is localized in the CFR(e_0) for the potential V_0. Thus, we can use the exponential decay of the corresponding eigenfunctions of $H_0(\lambda)$ to control the remainder.

Let us recall some relevant geometry of V discussed in Chapters 20 and 21. For $e_0 \in \sigma(K)$, we define

$$\begin{aligned} e' &= e_0 + \lambda^{\delta}, \\ e &= e_0 + 2\lambda^{\delta}, \end{aligned}$$

for $\delta > 1/2$. The potential $V_0(x; \lambda)$, defined in (20.36), satisfies $V_0 = V$ on Int $S^+(e'; \lambda)$. We used a partition of unity $\{j_i\}_{i=0}^{1}$, constructed in Section 20.7, satisfying $\sum j_i^2 = 1$ and

$$\text{supp}(\nabla j_i) \subset \text{Int } S^+(e'; \lambda) \setminus \text{Int } S^+(e; \lambda), \tag{22.12}$$

so that

$$j_0 \mid \text{Int } S^+(e; \lambda) = 1, \tag{22.13}$$

$$j_1 \mid \text{Ext } S^+(e', \lambda) = 1. \tag{22.14}$$

Given our construction of the spectral deformation family of $H(\lambda)$, if $u \in C_0^{\infty}(\text{Int } S^+(e'; \lambda))$, then

$$H(\lambda, \theta)u = H_0(\lambda)u. \tag{22.15}$$

This observation is essential for the construction of approximate eigenfunctions.

22.2 Exponential Decay of Eigenfunctions of $H_0(\lambda)$

Let $\{e_{j(i)}(\lambda)\}_{j=1}^{m_i}$ be the set of eigenvalues of $H_0(\lambda)$ satisfying (22.2). For simplicity, we write $e_j(\lambda)$ for these eigenvalues, e_0 for the unperturbed eigenvalue, and m_0 for the total multiplicity. Let $\{u_j\}_{j=1}^{m_0}$ be an orthonormal set of corresponding eigenfunctions for $H_0(\lambda)$: $H_0(\lambda)u_j = e_j(\lambda)u_j$. Because of the choice of the partition $\{j_i\}_{i=0}^{1}$ in (22.12)–(22.14), and the convergence (22.2), we know that supp$|\nabla j_i| \subset$ CFR$(e_j(\lambda))$ for the potential $V(x; \lambda)$ for all large λ. We can then apply the methods of Chapter 3 on the decay of eigenfunctions for eigenvalues below the bottom of the essential spectrum. It is important to note the following difference. Here, we want the decay in λ of the eigenfunctions u_j restricted to a subset of the classically forbidden region rather than decay in x. This results in some simplifications. In particular, as the CFR(e_0) for V_λ is bounded in the Agmon metric, the technicalities of Lemma 3.7 are not needed. Recall, finally, from Proposition 3.3 that ρ_0 as given in (22.6) satisfies

$$|\nabla_x \rho_0(x, y)|^2 \leq (V(x;\lambda) - e_0)_+ \tag{22.16}$$

almost everywhere. We need the following variation of Theorem 3.4.

Theorem 22.2. *Let χ be supported in $I_e \equiv \text{Int } S^+(e';\lambda) \setminus \text{Int } S^+(e;\lambda)$, and recall that $S_0(\lambda)$ is defined in Theorem 22.1 as the Agmon distance from $S^-(e_0;\lambda)$ to $S^+(e_0;\lambda)$. Then for any $\varepsilon > 0 \; \exists \; \lambda_\varepsilon$ and constants $c_\varepsilon, c'_\varepsilon > 0$ such that for all $\lambda > \lambda_\varepsilon$,*

$$\|\chi u_j\| \leq c_\varepsilon e^{-(1-\varepsilon)S_0(\lambda)} \tag{22.17}$$

and

$$\|\chi \nabla u_j\| \leq c'_\varepsilon e^{-(1-\varepsilon)S_0(\lambda)}. \tag{22.18}$$

Proof. The first result (22.17) follows from the proof of Theorem 3.4 with some minor modifications. We outline these here and leave the details to Problem 22.1. Let $f(x) \equiv (1-\varepsilon)\rho_0(0, x)$. Note that this is a function of bounded support. From (22.16) and (20.36), we find that

$$[V_0(x;\lambda) - |\nabla f|^2 - e_j(\lambda)] \mid \text{Ext } S^-(e';\lambda) \geq c_0 \varepsilon \lambda^\delta, \tag{22.19}$$

for $\delta > 1/2$, as above, and some $c_0 > 0$. Here we used the fact that $|e_j(\lambda) - e_0| < c_1 \lambda^{-1/5}$, as follows from Theorem 20.5. Consequently, if $\phi \in H^1(\mathbb{R}^n)$, $\text{supp } \phi \subset \text{Ext } S^-(e', \lambda)$, the calculation of Lemma 3.6 yields

$$\text{Re}\langle e^f \phi, (H_0(\lambda) - e_j(\lambda))e^{-f}\phi\rangle \geq \|\nabla \phi\|^2 + c_0 \varepsilon \lambda^\delta \|\phi\|^2. \tag{22.20}$$

We next define a smooth cut-off function χ_0 by

$$\chi_0(x) = \begin{cases} 1, & x \in \text{Ext } S^-(e;\lambda), \\ 0, & x \in \text{Int } S^-(e';\lambda). \end{cases} \tag{22.21}$$

We see from (20.50) that $|\nabla \chi_0| = \mathcal{O}(\lambda^{\frac{1}{2}-\delta})$. Setting $\phi = e^f \chi_0 u_j$ in the left side of (22.20), we obtain an upper bound:

$$\begin{aligned} &\text{Re}\langle e^{2f} \chi_0 u_j, (H_0(\lambda) - e_j(\lambda))\chi_0 u_j\rangle \\ &\quad = \langle [|\nabla \chi_0|^2 + 2(\nabla \chi_0 \cdot \nabla f)\chi_0] e^{2f} u_j, u_j\rangle \\ &\quad \leq c_0 \|\nabla \chi_0\|_\infty (\|\nabla \chi_0\|_\infty + \|\xi \nabla f\|_\infty) e^{2(1-\varepsilon)\rho_0(0, S^-(e;\lambda))}, \end{aligned} \tag{22.22}$$

where $\xi = 1$ on $\text{supp}(\nabla \chi_0)$. Combining this with the lower bound (22.20), we obtain

$$\begin{aligned} &\|\nabla(e^f \chi_0 u_j)\|^2 + \varepsilon c_0 \lambda^\delta \|e^f \chi_0 u_j\|^2 \\ &\qquad \leq c_1 \lambda^{2\delta - \frac{1}{2}} e^{2(1-\varepsilon)\rho_0(0, S^-(e;\lambda))}. \end{aligned} \tag{22.23}$$

Let us define $\tilde{\rho}_0(x) \equiv \rho_0(0, x) - \rho_0(0, S^-(e, \lambda))$. Then (22.23) immediately implies that

$$\int e^{2(1-\varepsilon)\tilde{\rho}_0(x)} \chi_0^2(x) u_j^2(x) \leq c_0 \varepsilon^{-1} \lambda^{\delta - \frac{1}{2}}. \tag{22.24}$$

Furthermore, since we have

$$\|e^f \chi_0 \nabla u_j\| \le \|\nabla(e^f \chi_0 u_j)\| + \varepsilon e_0 \lambda^\delta \|e^f \tilde{\chi}_0 u_j\|,$$

where $\tilde{\chi}_0$ is supported on a slightly bigger set then χ_0, we obtain, from (22.23) and (22.24),

$$\int e^{2(1-\varepsilon)\tilde{\rho}_0(x)} \chi_0^2(x) |\nabla u_j(x)|^2 \le c\varepsilon^{-1} \lambda^{2\delta - \frac{1}{2}}. \tag{22.25}$$

Let χ be supported in I_e as in the statement of the theorem. From (22.24), we obtain

$$\begin{aligned} \|\chi u_j\|^2 &= \int \left[e^{2(1-\varepsilon)\tilde{\rho}_0(x)} \chi_0^2(x) u_j^2(x) \right] \left[\chi^2(x) e^{-2(1-\varepsilon)\tilde{\rho}_0(x)} \right] \\ &\le c\varepsilon^{-1} \lambda^{\delta - \frac{1}{2}} e^{-2(1-\varepsilon)\tilde{\rho}_0(S^+(e;\lambda))}. \end{aligned} \tag{22.26}$$

We must estimate the exponent in (22.26). First, we show that for all large λ,

$$\rho_0(S^\pm(e_0, \lambda), S^\pm(e, \lambda)) \le c\lambda^{(2\delta - 3)/2}. \tag{22.27}$$

Let $x \in S^-(e_0, \lambda)$ and $y \in S^-(e, \lambda)$. Since $V_\lambda \le e_0 + 2\lambda^\delta$ between these surfaces, it follows easily from (22.6) and the choice of $\gamma(t) = (1-t)x + ty$ that the distance in (22.27) is bounded above by $\lambda^{\delta/2}\|x - y\|$. Furthermore, from the definitions of the inner turning surfaces S^-, we have

$$\begin{aligned} V\left(\lambda^{-\frac{1}{2}} y\right) - V\left(\lambda^{-\frac{1}{2}} x\right) &= 2\lambda^{\delta-1} = \int_0^1 \frac{d}{dt} V(\lambda^{-\frac{1}{2}} \gamma(t)) dt \\ &= \int_0^1 \lambda^{-\frac{1}{2}} (y - x) \cdot \nabla V(\lambda^{-\frac{1}{2}} \gamma(t)) dt, \end{aligned}$$

from which we conclude that

$$\left[2\lambda^{(3-2\delta)/2} (y - x) \right] \cdot \int_0^1 \nabla V(\gamma(t)) dt = 1.$$

Since the integral is bounded above independent of λ, it follows that $\|y - x\| = \mathcal{O}(\lambda^{(2\delta-3)/2})$ as $\lambda \to \infty$. Next, we verify the estimate

$$c_1 \lambda^{\frac{(\delta+1)}{2}} \le \rho_0(S^-(e;\lambda), S^+(e;\lambda)) \le c_2 \lambda^{\frac{(\delta+1)}{2}}. \tag{22.28}$$

The lower bound follows from the fact that between the two surfaces $(V(x;\lambda) - e_0) \ge 2\lambda^\delta$ and from estimate (20.19). The upper bound follows because $V(x)$ is bounded. Finally, from the triangle inequality, we have

$$S_0(\lambda) \le \rho_0(S^-(e_0, \lambda), S^+(e, \lambda)) - \rho_0(S^+(e;\lambda), S^+(e_0;\lambda)).$$

Combining this with (22.27) and (22.28), and noting that $\rho_0(S^-(e;\lambda), S^+(e;\lambda)) < S_0(\lambda)$, we obtain

$$\begin{aligned} \tilde{\rho}_0(S^+(e, \lambda)) &\ge S_0(\lambda) - \rho_0(S^-(e_0;\lambda), S^-(e;\lambda)) \\ &\quad - \rho_0(S^+(e;\lambda), S^+(e_0;\lambda)) \\ &\ge S_0(\lambda)(1 - c\lambda^{(\delta-4)/2}), \end{aligned} \tag{22.29}$$

for some $c > 0$. Since we can take $1/2 < \delta < 1$, the result now follows from (22.26) and (22.29) by taking λ sufficiently large. □

Problem 22.1. Beginning with estimate (22.25), derive estimate (22.18) for ∇u_j.

Problem 22.2. Prove estimates (22.10) and (22.11) on the asymptotic behavior of $S_0(\lambda)$.

Problem 22.3. The goal of this problem is to obtain decay estimates on the *resonance eigenfunctions* localized in CFR(e'; V_λ). Let $\{\phi_j\}_{j=1}^{m_0}$ be an orthonormal basis of generalized eigenfunctions (see Section 6.2) for $H(\lambda, \theta)$ in Ran $P(\lambda, \theta)$. Let $\{u_k\}_{k=1}^{m_0}$ be an orthonormal basis for Ran $P_0(\lambda)$, as used above.

(1) Show that $P(\lambda, \theta)f = \sum_{j=1}^{m_0} \langle \overline{\phi}_j, f\rangle \phi_j$. (*Hint*: The condition $H(\lambda, \theta)^* = \overline{H}(\lambda, \theta)$ (complex conjugate) implies that $P(\lambda, \theta)^* = \overline{P}(\lambda, \theta)$.)

(2) Let $M_{ij} \equiv \langle \overline{\phi}_i, u_j\rangle$; show that the $m_0 \times m_0$ matrix M is invertible and that $\|M^{-1}\| < c_0$, for all large λ.

(3) Let j_0 be as in (22.12) and (22.13). Prove the following geometric resolvent formula (a variant of Lemma 19.16):

$$R(z)j_0 - j_0 R_0(z) = -R(z)W_0 R_0(z), \tag{22.30}$$

where $W_0 = [\Delta, j_0]$.

(4) Employing formula (22.30), derive the relation

$$P(\lambda, \theta)j_0 u_k = j_0 u_k + r_{\lambda,k} \tag{22.31}$$

by integrating over Γ_0 in $A(\tilde{e}_0)$. The remainder is given by

$$r_{\lambda,k} = (2\pi i)^{-1} \oint_{\Gamma_0} R(z)W_0 u_k (e_k(\lambda) - z)^{-1} dz.$$

Estimate $(1 - j_0)u_k$ and $r_{\lambda,k}$ using Theorem 22.2 (in particular, (22.24) and (22.25)), the stability theorem, Theorem 20.7, and (20.50).

(5) Make use of parts (1) and (2) and formula (22.31) to obtain the identity

$$\phi_i = \sum_{j=1}^{m_0} (M^T)^{-1}_{ij} (j_0 u_j + r_{\lambda,j}),$$

and conclude that for $\xi = 1$ on CFR(e; V_λ),

$$\int e^{2(1-\varepsilon)\tilde{\rho}_0(x)} \xi(x)^2 \mid \phi_i(x)|^2 \leq c_0 \lambda^{\frac{1}{2}},$$

where $\tilde{\rho}_0$ is defined after (22.23). Extend this result to $\partial_i \partial_j \phi_k$ using (22.15) and condition (V1).

22.3 The Proof of Estimates on Resonance Positions

There are at least two techniques available for estimating resonance positions. We will prove Theorem 22.1 by constructing an approximate basis for the resonance subspace Ran $P(\lambda, \theta)$. We outline in Problem 22.7 another approach, used in [HSj3] and [Si5], based on estimates of the resonance eigenfunctions on the surface $S^+(e;\lambda)$. We also refer to the paper of Howland [Ho2], which discusses the resonance width in some simple one-dimensional models.

Lemma 22.3. *Let* $\{u_j\}_{j=1}^{m_0}$ *be an orthonormal family of eigenfunctions of* $H_0(\lambda)$ *for eigenvalues* $\{e_j(\lambda)\}_{j=1}^{m_0}$ *as in Section* 22.2 *with* $\lim_{\lambda\to\infty} e_j(\lambda) = \tilde{e}_0$. *Let* $\{j_i\}_{i=0}^{1}$ *be the partition of unity constructed in* (22.12)–(22.14). *We define functions* $\psi_j \equiv j_0 u_j$, $j = 1, \ldots, m_0$. *Then for any* $1/2 < \delta < 1$ $\exists \lambda_0$ *such that* $\lambda > \lambda_0$ *implies*

$$H(\lambda, \theta)\psi_j = e_j(\lambda)\psi_j + r_j(\lambda), \tag{22.32}$$

where

$$\|r_j(\lambda)\| \leq c_0 e^{-(1-\varepsilon)S_0(\lambda)} \tag{22.33}$$

and

$$\langle \psi_i, \psi_j \rangle = \delta_{ij} + \mathcal{O}(e^{-2(1-\varepsilon)S_0(\lambda)}). \tag{22.34}$$

Proof. By the choice of j_0, (22.15) and (20.36), we have $j_0 u_j \in D(H(\lambda, \theta))$ and

$$H(\lambda, \theta)\psi_j = H_0(\lambda) j_0 u_j = e_j(\lambda)\psi_j + [-\Delta, j_0]u_j.$$

As $\|\nabla j_0\|_\infty = \mathcal{O}(\lambda^{\frac{1}{2}-\delta})$ and $\text{supp}|\nabla j_0| \subset I_e$, Theorem 22.2 implies that

$$\|[-\Delta, j_0]u_j\| \leq c_1 e^{-(1-\varepsilon)S_0(\lambda)},$$

which proves (22.32) and (22.33). As for (22.34), we have

$$\langle \psi_i, \psi_j \rangle = \langle j_0^2 u_i, u_j \rangle = \delta_{ij} + \langle (j_0^2 - 1)u_i, u_j \rangle.$$

Since $\text{supp}(j_0^2 - 1) \subset I_e$, Theorem 22.2 implies the result. □

This lemma shows that the set $\{\psi_j\}_{j=1}^{m_0}$ behaves as a set of eigenvectors for $H(\lambda, \theta)$ up to exponentially small corrections. We want to show that the vectors $\{P(\lambda, \theta)\psi_j\}$ remain linearly independent for all large λ and that they diagonalize $H(\lambda, \theta)$ on Ran $P(\lambda, \theta)$ up to exponentially small corrections. To this end, we need a slight modification of the *geometric resolvent formula* of Lemma 19.16. For $i = 0, 1$, let $R_i(z) = (H_i(\lambda) - z)^{-1}$, where $H_1(\lambda) = H_1(\lambda, \theta)$, as above. We assume that $z \in \left(\bigcap_{i=0}^{1} \rho(H_i(\lambda))\right) \cap \rho(H(\lambda, \theta))$. For each i, we write

$$R(z) j_i - j_i R_i(z) = R(z) W_i R_i(z), \tag{22.35}$$

where, as above,

$$W_i \equiv [\Delta, j_i].$$

Multiplying (22.35) on the right by j_i and summing over i, the condition $\sum_{i=0}^{1} j_i^2 = 1$ yields

$$R(z) = \sum_{i=0}^{1} j_i R_i(z) j_i + R(z)K(z), \tag{22.36}$$

where the geometric interaction term, $K(z)$, is given by

$$K(z) \equiv \sum_{i=0}^{1} W_i R_i(z) j_i. \tag{22.37}$$

We write (22.36) as

$$R(z)(1 - K(z)) = \sum_{i=0}^{1} j_i R_i(z) j_i. \tag{22.38}$$

(Note that this formula is quite general and holds for any finite partition such that $\sum j_i^2 = 1$.) This formula is similar to ones encountered in Fredholm theory. For $z \in A(\tilde{e}_0)$, as in (22.3), the estimates of Theorems 20.7 and 20.8, together with $\|\nabla j_i\|_\infty = \mathcal{O}(\lambda^{\frac{1}{2}-\delta})$, $\delta > 1/2$, imply that for all large λ,

$$\|K(z)\| < \frac{1}{2}, \tag{22.39}$$

so that $(1 - K(z))$ is boundedly invertible. As a consequence, we have from (22.38),

$$R(z) = \left(\sum_{i=0}^{1} j_i R_i(z) j_i \right) (1 - K(z))^{-1}, \tag{22.40}$$

for $z \in A(\tilde{e}_0)$. Inserting this expression for $R(z)$ into the right side of (22.36) yields the desired formula,

$$R(z) = \left(\sum_{i=0}^{1} j_i R_i(z) j_i \right) [1 + K(z)(1 - K(z))^{-1}], \tag{22.41}$$

which is valid on $A(\tilde{e}_0)$ for all large λ,

Let $\mathcal{E}_0(\lambda)$ be the subspace of $L^2(\mathbb{R}^n)$ spanned by $\{u_j\}_{j=1}^{m_0}$, and let $\mathcal{F}_0(\lambda)$ be the subspace generated by $\{P(\lambda, \theta)\psi_j\}_{j=1}^{m_0}$, with $\psi_j \equiv j_0 u_j$ and $P(\lambda, \theta)$ given in (20.42). The projector $P(\lambda, \theta) \neq 0$, by Theorem 20.3. We will show that $\mathcal{E}_0(\lambda)$ is a good approximation to the $H(\lambda, \theta)$-invariant subspace $\mathcal{F}_0(\lambda)$. Moreover, $H(\lambda, \theta)$ is approximately diagonal on $\mathcal{F}_0(\lambda)$. The key lemma, using the results of Lemma 22.3, is the following.

Lemma 22.4. *The $H(\lambda, \theta)$-invariant subspace $\mathcal{F}_0(\lambda)$ is m_0-dimensional. In the basis $\{P(\lambda, \theta)\psi_j\}$ for $\mathcal{F}_0(\lambda)$, we have for any $\varepsilon > 0$ and all λ sufficiently large,*

$$\begin{aligned} &\langle \psi_j, P(\lambda, \theta) H(\lambda, \theta) P(\lambda, \theta) \psi_k \rangle \\ &\quad = e_j(\lambda)\delta_{jk} + \mathcal{O}(e^{-2(1-\varepsilon)S_0(\lambda)}). \end{aligned} \tag{22.42}$$

Proof. It is not difficult to see that $\{P(\lambda,\theta)\psi_j\}$ is a set of m_0-linearly independent vectors using (22.34) and the fact that $P(\lambda,\theta)$ is well approximated by $P_0(\lambda)$, as in (20.53). We leave this as Problem 22.4. Let Γ_0 be a simple closed contour in $A(\tilde{e}_0)$ with clockwise orientation. From the definition of $P(\lambda,\theta)$ in (20.42), we rewrite the left side of (22.42) as

$$\begin{aligned}\langle\psi_j, P(\lambda,\theta)H(\lambda,\theta)P(\lambda,\theta)\psi_k\rangle \\ = (2\pi i)^{-1}\oint_{\Gamma_0}\langle\psi_j, zR(z)\psi_k\rangle dz. \qquad (22.43)\end{aligned}$$

We insert formula (22.41) into the right side of (22.43) and obtain two terms, I_1 and I_2. The first term is

$$\begin{aligned} I_1 &= (2\pi i)^{-1}\sum_{i=0}^{1}\oint_{\Gamma_0}\langle\psi_j, j_i R_i(z) j_i\psi_k\rangle z\,dz \\ &\equiv (2\pi i)^{-1}(I_{1,0}+I_{1,1}). \qquad (22.44)\end{aligned}$$

The second term involves $K(z)$ and will be evaluated shortly. The j_0-contribution to I_1, $I_{1,0}$, is written as

$$\begin{aligned} I_{1,0} &= \oint_{\Gamma_0}\langle j_0^2 u_j, R_0(z) j_0\psi_k\rangle z\,dz \\ &= \left(\oint_{\Gamma_0} z(e_j(\lambda)-z)^{-1}dz\right)\langle\psi_j,\psi_k\rangle \\ &\quad + \oint_{\Gamma_0}\langle(j_0^2-1)u_j, R_0(z)j_0\psi_k\rangle z\,dz. \qquad (22.45)\end{aligned}$$

By Lemma 22.3, the first integral on the right in (22.45) is

$$(2\pi i)e_j(\lambda)\delta_{jk} + \mathcal{O}(e^{-2(1-\varepsilon)S_0(\lambda)}). \qquad (22.46)$$

As for the second integral in (22.45), we use the fact that

$$R_0(z)j_0\psi_k = (e_k(\lambda)-z)^{-1}u_k + R_0(z)(j_0^2-1)u_k. \qquad (22.47)$$

The first term contributes $\mathcal{O}(e^{-2(1-\varepsilon)S_0(\lambda)})$. Since $\mathrm{dist}(\sigma(H_0(\lambda)),\Gamma_0)\ge\frac{1}{4}a_0$, $\|R_0(z)\|$ is uniformly bounded on Γ_0, and we see from Theorem 22.2 that the second term in (22.47) contributes the same order. Hence, $I_{1,0}$ is estimated as in (22.46). As for $I_{1,1}$, we have

$$I_{1,1} = \oint_{\Gamma_0}\langle j_1 j_0 u_j, R_1(z) j_1 j_0 u_k\rangle z\,dz.$$

We proved in Theorem 20.8 that $\|R_1(z)\|$ is uniformly bounded on Γ_0. Since $j_1 j_0$ is supported in $\mathrm{Int}\,S^+(e';\lambda)\setminus\mathrm{Int}\,S^+(e;\lambda)$, Theorem 22.2 is applicable and thus

$$I_{1,1} = \mathcal{O}(e^{-2(1-\varepsilon)S_0(\lambda)}). \qquad (22.48)$$

Consequently, $(2\pi i)I_1$ is estimated as in (22.46). Next, we turn to I_2, given by

$$I_2 = (2\pi i)^{-1} \sum_{i=0}^{1} \oint_{\Gamma_0} \langle \psi_j, j_i R_i(z) j_i (1 - K(z))^{-1} K(z) \psi_k \rangle z \, dz. \tag{22.49}$$

The important point is $K(z)$ has the right support properties. To estimate $K(z)\psi_k$, we treat the $i = 0$ and $i = 1$ contributions to $K(z)$, as in (22.37), separately.

$$\underline{\underline{i=0}} \quad W_0 R_0(z) j_0^2 u_k = (e_k(\lambda) - z)^{-1} W_0 u_k + W_0 R_0(z)(j_0^2 - 1) u_k.$$

This is easily seen to be $\mathcal{O}(e^{-(1-\varepsilon)S_0(\lambda)})$; $W_0 u_k$ and $(j_0^2 - 1)u_k$ are estimated in Theorem 22.2 and $W_0 R_0(z)$ is bounded on Γ_0 by $c_0 \lambda^{\frac{1}{2}-\delta}$ as in (20.51). The $i = 1$ term is easier:

$$\underline{\underline{i=1}} \quad W_1 R_1(z) j_1 j_0 u_k = \mathcal{O}(e^{-(1-\varepsilon)S_0(\lambda)}),$$

from Theorem 22.2 and (20.52). Hence we have

$$\|K(z)\psi_k\| = \mathcal{O}(e^{-(1-\varepsilon)S_0(\lambda)}). \tag{22.50}$$

Returning to (22.49), consider the $i = 0$ contribution to the sum. We write the integrand as

$$\begin{aligned} &\langle (j_0^2 - 1)u_j, R_0(z) j_0 (1 - K(z))^{-1} K(z) \psi_k \rangle \\ &\quad + (e_j(\lambda) - z)^{-1} \langle j_0 u_j, (1 - K(z))^{-1} K(z) \psi_k \rangle. \end{aligned} \tag{22.51}$$

The first term in (22.51) is $\mathcal{O}(e^{-2(1-\varepsilon)S_0(\lambda)})$ from Theorem 22.2 and (22.50). As for the second term in (22.51), we again use the fact that $K(z)$ has good support properties. Let $\tilde{\chi}$ be such that $\tilde{\chi} K(z) = K(z)$. Then, we can apply Theorem 22.2 to $\tilde{\chi} j_0 u_j$. Hence the $i = 0$ term in (22.49) is $\mathcal{O}(e^{-2(1-\varepsilon)S_0(\lambda)})$. As for $i = 1$ in (22.49), we write

$$\langle j_1 j_0 u_j, R_1(z) j_1 (1 - K(z))^{-1} K(z) \psi_k \rangle,$$

which is easily seen to be $\mathcal{O}(e^{-2(1-\varepsilon)S_0(\lambda)})$ by Theorems 20.8 and 22.2, (22.39), and (22.50). Collecting the estimates for I_2 and (22.46) for I_1, we find that

$$\langle \psi_j, P(\lambda,\theta) H(\lambda,\theta) P(\lambda,\theta) \psi_k \rangle = e_j(\lambda)\delta_{jk} + \mathcal{O}(e^{-2(1-\varepsilon)S_0(\lambda)}),$$

proving the lemma. □

Problem 22.4. Prove that the sets $\{\psi_k\}$ and $\{P(\lambda,\theta)\psi_k\}$ consist of m_0 linearly independent vectors for all λ sufficiently large.

The ideas concerning the "closeness" of two closed subspaces of a Hilbert space and approximate eigenfunctions, which are used in the proof of Theorem 22.5, are of importance in their own right. We explore this in the next two problems. This material comes from [HSj1].

Problem 22.5. Let $\mathcal{E}$ and $\mathcal{F}$ be two closed subspaces of a Hilbert space $\mathcal{H}$, and let $P_{\mathcal{E}}$ and $P_{\mathcal{F}}$ be the corresponding orthogonal projections. We define the distance from $\mathcal{E}$ to $\mathcal{F}$ by $\tilde{d}(\mathcal{E},\mathcal{F}) \equiv \|(1 - P_{\mathcal{F}})P_{\mathcal{E}}\|$.

(1) Prove that this function is *not symmetric*, that $\tilde{d}(\mathcal{E}, \mathcal{F}) = 0$ if and only if $\mathcal{E} \subset \mathcal{F}$, and that it satisfies a triangle inequality: For any other closed subspace $\mathcal{G} \subset \mathcal{H}$, we have

$$\tilde{d}(\mathcal{E}, \mathcal{G}) \leq \tilde{d}(\mathcal{E}, \mathcal{F}) + \tilde{d}(\mathcal{F}, \mathcal{G}).$$

(2) Prove that if $\tilde{d}(\mathcal{E}, \mathcal{F}) < 1$, then the map $P_{\mathcal{F}} \mid \mathcal{E} : \mathcal{E} \mapsto \mathcal{F}$ is injective.

(3) Prove that if $\tilde{d}(\mathcal{E}, \mathcal{F}) < 1$ and $\tilde{d}(\mathcal{F}, \mathcal{E}) < 1$, then $P_{\mathcal{F}} \mid \mathcal{E} : \mathcal{E} \mapsto \mathcal{F}$ and $P_{\mathcal{E}} \mid \mathcal{F} : \mathcal{F} \mapsto \mathcal{E}$ are bijective with bounded inverses.

Problem 22.6. We combine the previous problem with the existence of approximate eigenvectors to obtain a perturbation result about the discrete spectrum of a self-adjoint operator. Prove the following proposition (for a more general version, see [HSj1]).

Proposition 22.5. *Let A be a self-adjoint operator, and let $I \subset \mathbb{R}$ be a compact interval. Suppose that $\sigma(A) \cap I$ is discrete and that $dist(I, \sigma(A) \backslash (\sigma(A) \cap I)) \geq a > 0$. Suppose that there exist real numbers $\{\mu_1, \ldots, \mu_k\} \subset I$ and k linearly independent vectors ψ_i and vectors r_i such that*

$$A\psi_i = \mu_i \psi_i + r_i$$

and that

$$\|r_i\| < \epsilon,$$

for some $\epsilon > 0$ and for $i = 1, \ldots, k$. Let $\mathcal{E}$ be the subspace spanned the vectors ψ_i, and let $\mathcal{F}$ be the subspace spanned by the eigenvectors for A and the μ_i. Then, there is a constant C_0, depending only on the vectors ψ_i, such that

$$\tilde{d}(\mathcal{E}, \mathcal{F}) < C_0 k^{\frac{1}{2}} \epsilon a^{-1}.$$

If this distance is less than one, then A has at least k eigenvalues in I.

Theorem 22.6. *Let $\{z_{0,j}(\lambda)\}$ be the eigenvalues of $P(\lambda, \theta)H(\lambda, \theta)$. Then there exists a permutation π of $\{1, \ldots, m_0\}$ such that for any $0 < \varepsilon < 1$ $\exists$ $c_\varepsilon > 0$ and $\lambda_\varepsilon > 0$ such that $\lambda > \lambda_\varepsilon$ implies*

$$|z_{0,j}(\lambda) - e_{\pi(j)}(\lambda)| \leq c_\varepsilon e^{-2(1-\varepsilon)S_0(\lambda)},$$

for $j \in \{1, \ldots, m_0\}$, where $\{e_j(\lambda)\}$ are the eigenvalues of $H_0(\lambda)$ converging to $\tilde{e}_0 \in \sigma(K)$. In particular,

$$|\text{Im } z_{0,j}(\lambda)| \leq c_\varepsilon e^{-2(1-\varepsilon)S_0(\lambda)}.$$

Proof. Lemmas 22.3 and 22.4 establish that $\{P(\lambda, \theta)\psi_j\}$ is a basis for Ran $P(\lambda, \theta)$ and in that basis,

$$P(\lambda, \theta)H(\lambda, \theta) = (e_j(\lambda)\delta_{jk})P(\lambda, \theta) + \mathcal{O}\left(e^{-2(1-\varepsilon)\frac{S_0}{\lambda}}\right).$$

The result follows by taking the Jordan canonical form of the matrix $P(\lambda, \theta) H(\lambda, \theta)$. □

Problem 22.7. In this problem, we outline another method for obtaining estimates on Im $z_{0,j}(\lambda)$ in dimension $n \geq 3$. Let $\Omega_e \equiv \text{Int } S^+(e; \lambda)$. Because of our choice of the spectral deformation family, if $f \in C_0^\infty(\Omega_e)$,

$$H(\lambda, \theta) f = H(\lambda) f. \tag{22.52}$$

Suppose that ϕ_j satisfies

$$H(\lambda, \theta)\phi_j = z_{0,j}(\lambda)\phi_j. \tag{22.53}$$

Take the inner product of (22.53) with ϕ_j on $L^2(\Omega_e)$ and use (22.52) to derive

$$z_{0,j}(\lambda) \int_{\Omega_e} |\phi_j|^2 = \int_{\Omega_e} \overline{\phi}_j \cdot H(\lambda)\phi_j. \tag{22.54}$$

Integrate by parts on the right in (22.54) and derive the formula

$$\text{Im } z_{0,j}(\lambda) = -\left(\text{Im} \int_{\partial\Omega_e} \overline{\phi}_j \frac{\partial \phi_j}{\partial v}\right) \left(\int_{\Omega_e} |\phi_j|^2\right)^{-1}, \tag{22.55}$$

where $\frac{\partial}{\partial v}$ is the outward normal derivative for $\partial\Omega_e$. Let $\tilde{\rho}_0(x)$ be the modified Agmon distance function defined after (22.23). Using the Schwarz inequality and the results of Problem 22.3, conclude that

$$|\text{Im } z_{0,j}(\lambda)| \leq c_0 e^{-2\tilde{\rho}_0(S^+(e;\lambda))} \left\{\int_{\partial\Omega_e} e^{2\tilde{\rho}_0} |\phi_j|^2\right\}^{\frac{1}{2}} \left\{\int_{\partial\Omega_e} e^{2\tilde{\rho}_0} \left|\frac{\partial \phi_j}{\partial v}\right|^2\right\}^{\frac{1}{2}}, \tag{22.56}$$

for some constant $c_0 > 0$ independent of λ and ϕ_j. To conclude the estimate, we need a case of the Sobolev trace theorem (see [Ag2] for the general theorem and the proof): For $\Omega \subset \mathbb{R}^n$ sufficiently regular, the map $H^1(\Omega) \to L^2(\partial\Omega)$ is continuous. Conclude from this and (22.56) that for any regular region W_e containing Ω_e,

$$|\text{Im } z_{0,j}(\lambda)| \leq c_0 e^{-2\tilde{\rho}_0(S^+(e;\lambda))} \|e^{\tilde{\rho}_0}\phi_j\|_{H^1(W_e)} \|e^{\tilde{\rho}_0}\nabla\phi_j\|_{H^1(W_e)}.$$

Make a judicious choice of W_e and, using the results of Problem 22.3 together with the bounds on the Agmon metric in the proof of Theorem 22.2, complete the proof of (22.9).

23

Other Topics in the Theory of Quantum Resonances

The purpose of this last chapter is to survey various directions in which the theory developed in Chapters 16–22 has been extended. It will necessarily be less detailed, and the reader is encouraged to work out the proofs or consult the literature. As earlier, we will discuss resonances in two-body situations only. There is a growing literature on resonances for N-body Schrödinger operators in external fields. In the first two sections, we will discuss two-body Stark and Zeeman problems. The third section is an introduction to resonances occurring in the classical scattering of waves by obstacles. Finally, we conclude in Section 23.4 by listing other directions in the theory of resonances. We have not discussed the approach to quantum resonances developed by Helffer and Sjöstrand [HSj4] and their collaborators. This work uses techniques outside of those developed here. The results obtained by these methods are, in general, much sharper than those given in this book. For example, they obtain asymptotic expansions for the resonance widths rather than simply upper bounds in certain cases. We mention some of these results in Section 23.4. The interested reader will find reference to works using these techniques in the References section of this book.

23.1 Stark and Stark Ladder Resonances

The shift of the energy levels of a hydrogen atom due to the perturbation by an external, constant electric field was computed by Schrödinger [Sch] in his second paper on quantum mechanics. Soon afterwards, the topic was treated again by Oppenheimer [Op]. He showed that the energy levels of the hydrogen atom disappear when the field is turned on. This calls into question the meaning of the calculation by Schrödinger since it indicated that the perturbation expansion does not converge. However, the calculations do give estimates in reasonable agreement with experiments. Although Oppenheimer did not realize this, the Rayleigh–Schrödinger series is an example of a strong asymptotic series, discussed in Chapter 19. Oppenheimer interpreted the states as metastable states. He found that the width Γ of these states is approximately

$$\Gamma \sim e^{-\frac{4}{3}|E_n|^{\frac{3}{2}}\|F\|^{-1}}, \tag{23.1}$$

where $\|F\| > 0$ is the strength of the electric field $F \in \mathbb{R}^n$, and E_n is the nth hydrogen atom energy level. Much later, Titchmarsh [T2] developed the mathematical theory of the Hamiltonian,

$$H(F) = -\Delta + F \cdot x - \|x\|^{-1} \tag{23.2}$$

on $L^2(\mathbb{R}^3)$. He proved that $H(F)$ is self-adjoint and that the spectrum is $\mathbb{R}$ with no eigenvalues. Furthermore, Titchmarsh proved that the resolvent of $H(F)$ admits a meromorphic continuation across $\mathbb{R}$. He proved the existence of poles near the eigenvalues of $H(0)$ whose imaginary parts were bounded above by an expression similar to the one found by Oppenheimer, at least for $\|F\|$ sufficiently small. As we have discussed in Chapter 16, this is an example of quantum resonances in the semiclassical regime. Titchmarsh was able to prove these results because the Schrödinger operator (23.2) in three dimensions is separable in a certain coordinate system.

In this section, we will study Stark Hamiltonians,

$$H(F) = -\Delta + F \cdot x + V \tag{23.3}$$

on $L^2(\mathbb{R}^n)$, for which $H(0)$ has an isolated eigenvalue below the bottom of the essential spectrum. We will show that $H(F)$ has a nearby resonance, for $\|F\|$ small, whose width Γ is bounded above by a formula similar to (23.1),

$$\Gamma \leq C_0 e^{-\alpha\|F\|^{-1}},$$

where α is expressible in terms of the unperturbed eigenvalue (see Theorem 23.4). We call this bound the *Oppenheimer formula*. We will also study a related one-dimensional problem,

$$H(h, F) = -h^2 \frac{d^2}{dx^2} + Fx + v_p \tag{23.4}$$

on $L^2(\mathbb{R})$, where v_p is a periodic potential. Although $H(0)$ has no eigenvalues, the operator $H(F)$ will be shown to have resonances since the electric field forces the electron in one dimension to tunnel through finitely many potential barriers before escaping to infinity. The unperturbed eigenvalues are states localized to these finitely many potential wells. The resulting resonances are called *Stark ladder resonances*.

We first discuss the Stark Hamiltonian (23.3). We will assume the potential V satisfies the following conditions:

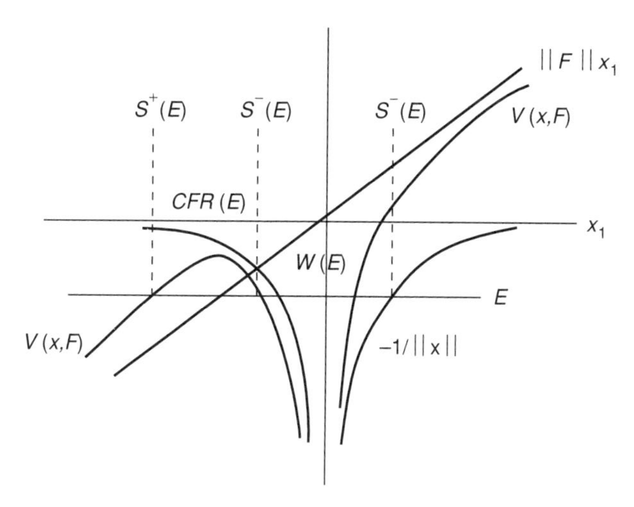

FIGURE 23.1. A cross-section of the potential $V(x, F) = -1/\|x\| + \|F\|x_1$.

(V1) V is relatively Δ-compact;

(V2) outside $B_R(0)$, for some $R \gg 0$, $V \in C^1(\mathbb{R}^n)$ and satisfies

$$|\partial^\alpha V(x)| \leq c_\alpha(1 + \|x\|)^{-\mu-|\alpha|},$$

for some $\mu > 0$ and $|\alpha| = 0, 1$;

(V3) the unperturbed Hamiltonian $H(0) = -\Delta + V$ has an isolated eigenvalue $E_0 < 0 = \inf \sigma_{\text{ess}}(H(0))$.

Under these conditions, we have the following standard result. The essential self-adjointness follows from the Faris–Lavine theorem; see [RS2].

Proposition 23.1. *Assume V satisfies* (V1)–(V3). *For any $F \in \mathbb{R}^n$, the Stark Hamiltonian $H(F) = -\Delta + F \cdot x + V$ is essentially self-adjoint on $C_0^\infty(\mathbb{R}^n)$. For $F \neq 0$, $\sigma(H(F)) = \mathbb{R}$ and there are no eigenvalues.*

Let us now consider the stability of the eigenvalue E_0 under the perturbation $H(F)$. We prove this using the machinery of Chapters 18, 20, and 21. For the operator K, we take $K \equiv H(0)$. The scaled potential $V(x; \lambda)$ is simply $V(x, F) = V + F \cdot x$. Note that $\|F\|^{-1}$ plays the role of the generalized semiclassical parameter λ. We define CFR(E_0, F) as in (20.3). Unlike the shape resonance case, the growth of V in the $\hat{F}$-direction results in an unbounded, classically forbidden region. Roughly speaking, the particle can tunnel through the potential barrier only in the direction of the electric field. As in the shape resonance case, CFR$(E_0; F)$ is bounded by two disjoint surfaces $S^-(E_0; F)$ and $S^+(E_0; F)$, but now $S^+(E_0; F)$ is unbounded. These features are sketched in Figure 23.1.

Spectral deformation of Stark Hamiltonians is based on the idea of translation analyticity. The corresponding vector field was introduced in Example 17.1 and studied in Problem 18.9. Let ϕ_θ be the diffeomorphism of $\mathbb{R}^n$ given by

$$\tilde{\phi}_\theta(x) = x + \theta \hat{F}, \quad \hat{F} \equiv F\|F\|^{-1}. \tag{23.5}$$

Then if U_θ is the corresponding unitary map as defined in (16.19), we easily find that

$$U_\theta(-\Delta + F \cdot x)U_\theta^{-1} = (-\Delta + F \cdot x) + \theta\|F\|. \tag{23.6}$$

If V is the restriction to $\mathbb{R}^n$ of a function analytic in a strip $\{z \in \mathbb{C}^n \mid |\text{Im } z| < \theta_0\}$, then we can define the family

$$H(F, \theta) \equiv U_\theta H(F) U_\theta^{-1} = -\Delta + F \cdot x + V_\theta + \theta\|F\|, \tag{23.7}$$

where

$$V_\theta(x) \equiv V(x + \theta \hat{F}). \tag{23.8}$$

This generates a suitable spectral deformation family in the sense of Section 16.2. As opposed to the dilation analytic case discussed in Example 16.3, $\sigma_{\text{ess}}(H(F, \theta))$ is the line $\mathbb{R} + i(\text{Im } \theta)F$. The resonances lie in the strip $\text{Im } \theta < \text{Im } z < 0$, for $\text{Im } \theta < 0$. This theory is not completely satisfactory, however, because the Coulomb potential is not translation analytic, and so the Stark Hamiltonian (23.2) is not translation analytic. This problem was originally solved by Herbst [He3], who used dilation analyticity. Here we note that it suffices to implement the spectral deformation family in the half-space bounded by $S^+(E_0; F)$ where $F \cdot x < 0$. In this half-space, the force from the electric field will accelerate the particle to infinity. We expect a nontrapping condition to hold there.

To sketch this idea, let $E' = E_0 + \|F\|$ and $E = E_0 + 2\|F\|$, respectively analogous to the energies e' and e of Chapter 20. Let ϕ_E be a function given by

$$\phi_E(u) = \begin{cases} 1 - e^{-(E-u)}, & u < E, \\ 0, & u > E, \end{cases} \tag{23.9}$$

so that $\phi_E \to 1$ as $u \to -\infty$. We define a vector field $v_{E,F}(x)$ by

$$v_{E,F}(x) \equiv -\hat{F}\phi_E(V(x; F)). \tag{23.10}$$

This vector field points into the half-space $F \cdot x < 0$ and approaches $-\hat{F}$ as $x \cdot \hat{F} \to -\infty$. It is not hard to check that $V(x, F)$ is nontrapping at energy E with this vector field on the exterior of $S^+(E'; F)$. The first term in (21.7) is

$$-v_{E,F} \cdot \nabla V(x; F) = (-\hat{F} \cdot \nabla V(x) + \|F\|)\phi_E(V(x; F)). \tag{23.11}$$

Now, because $\|x\| = \mathcal{O}(\|F\|^{-1})$, condition (V2) implies that $|\hat{F} \cdot \nabla V(x)| = \mathcal{O}(\|F\|^{\mu+1})$ on Ext $S^+(E'; F)$, for $\|F\|$ sufficiently small. Since $\phi_E' < 0$ and vanishes as $x \cdot F \to -\infty$, and $[E - V(x; F)] > \|F\|$ in Ext $S^+(E'; F)$, one finds for $u \in H^1(\text{Ext } S^+(E'; F))$,

$$\langle u, S_E(v_{E,F}; V(x; F))u\rangle \geq c_0\|F\| \, \|u\|^2, \tag{23.12}$$

for $\|F\|$ small enough. With estimate (23.12), it is not difficult to prove an analogue of Theorem 21.4 in this case. The β on the right side in (21.22) will be replaced by $\|F\|\beta$.

We use the vector field $v_{\tilde{E},F}$ as in (23.10) to construct a spectral deformation family for $H(F)$. Instead of $\tilde{\phi}_\theta$ in (23.5), we define a family of diffeomorphisms ϕ_θ by

$$\phi_\theta(x) = x + \theta v_{E,F}(x). \tag{23.13}$$

We impose the following condition on V:

(V4) V is the restriction to $\mathbb{R}^n$ of a function V analytic in a truncated strip

$$\mathcal{S}_{\theta_0,R} \equiv \{z \in \mathbb{C}^n \mid -\text{Re } z > R, |\text{Im } z_i| < \theta_0, \ i = 1, \ldots, n\},$$

for some $\theta_0 > 0$ and $R > 0$.

Following the construction of Chapter 18, we obtain an analytic type-A family $H(F,\theta)$, $|\text{Im }\theta| < \theta_0$, given by

$$H(F,\theta) = p_\theta^2 + F \cdot x + V \circ \phi_\theta + \theta F \cdot v_{E,F}(x). \tag{23.14}$$

One must establish the analogues of Proposition 18.2 and Theorem 18.6 on the essential spectrum. Utilizing the fact that $v_{E,F} \to 1$ as $x \cdot F \to -\infty$, we obtain the following proposition.

Proposition 23.2. $\sigma_{\text{ess}}(H(F,\theta)) = \{\lambda + i(\text{Im }\theta)\|F\| \mid \lambda \in \mathbb{R}\}$ *for* $|\text{Im }\theta| < \theta_0$.

Having established the main technical tools, we can now prove the stability of $E_0 \in \sigma_d(H(0))$ following the discussion of Chapter 20. We note that (20.50) is replaced here by $\|\nabla j_i\| = \mathcal{O}(\|F\|)$, and the annulus $A(e_0)$ in (20.41) is defined here by

$$A_F(E_0) \equiv \{z \mid a_0\|F\| < |z - E_0| < \|F\|a_1\}. \tag{23.15}$$

Theorem 23.3. *Assume that V satisfies* (V1)–(V4) *and that* $E_0 < \inf \sigma_{\text{ess}}(H(0))$ *is an isolated eigenvalue. Then E_0 is stable under the perturbation $H(F)$ given in* (23.3) *for all $\|F\|$ sufficiently small.*

Let $z_k(F)$ be the resonances of $H(F)$ close to E_0. It remains to establish the Oppenheimer formula for the resonance width, $\Gamma = |\text{ Im} z_k(F)|$. The Agmon metric is given by

$$ds_A^2 = (V(x) + F \cdot x - E_0)_+ dx^2.$$

Let $\rho_{0,F}(x,y)$ be the corresponding distance function (see (22.6)). We define $S_0(F)$ in analogy with $S_0(\lambda)$ in Theorem 22.1 by

$$S_0(F) = \rho_{0,F}(S^-(E_0;F), S^+(E_0,F)).$$

Theorem 23.4. *For any $\varepsilon > 0$ and for all $\|F\|$ sufficiently small, $\exists c_\varepsilon > 0$ such that*

$$|z_k(F) - E_0| \leq c_\varepsilon e^{-2(1-\varepsilon)S_0(F)},$$

where $\{z_k(F)\}$ *are the resonances near* E_0. *Furthermore, we have*

$$S_0(F) = \frac{2}{3}(-E_0)^{\frac{2}{3}} \|F\|^{-1} + \mathcal{O}(\|F\|^{-1+\sigma})$$

where $\sigma = \min(1, \mu)$, μ *as in* (V2).

We now discuss the one-dimensional Stark ladder problem (23.4). This section follows the paper of Combes and Hislop [CH]. Related results appear in the papers of Agler and Froese [AF], Bentosela [Ben], Bentosela, Carmona, Duclos, Simon, Soulliard, and Weder [BCDSSW], Bentosela and Grecchi [BG], Buslaev and Dmitrieva [BD], Herbst and Howland [HH], Jensen [J4, J5], and Nenciu and Nenciu [NN1, NN2]. Unlike the Stark problem we studied above, $\sigma(H(0))$ has no eigenvalues. The spectrum of a one-dimensional periodic Schrödinger operator is well known and consists of bands of essential spectrum separated by gaps. The proof of the following proposition can be found, for example, in [RS4].

Proposition 20.5. *Let* v_p *be a periodic potential on* $\mathbb{R}$ *with period* τ. *Let* $\{\alpha_i\}_{i=1}^{\infty}$, *respectively* $\{\beta_i\}_{i=1}^{\infty}$, *be the eigenvalues of* $-d^2/dx^2 + v_p$ *on* $L^2[0, \tau]$ *with periodic, respectively anti-periodic, boundary conditions. Then* $\sigma(H(0)) = \bigcup_{i=1}^{\infty}[\alpha_i, \beta_i]$.

As an example, let $v_p(x) = \cos x$. This potential has an infinite number of open gaps, that is, for infinitely many i, the bands $[\alpha_i, \beta_i]$ do not overlap. The nature of the spectrum can be understood from a semiclassical viewpoint as follows from [BCD3], [JLMS1], [Sim7], and [Ou]. Let $H(h) = -h^2(d^2/dx^2) + \cos x$, where h is a small parameter. We consider an infinite direct sum of identical operators h_n on $L^2[2\pi n, 2\pi(n+1)]$, with Dirichlet boundary conditions at the endpoints. The operator h_n has the same symbol as $H(h)$. Each Hamiltonian h_n has identical discrete spectrum $\{e_i\}_{i=1}^{\infty}$. Each eigenvalue has multiplicity 1. The operator $\oplus_{n=-\infty}^{\infty} h_n$ on $L^2(\mathbb{R})$ has the same set of eigenvalues, but each has infinite multiplicity. For the eigenvalues below 1, the wave functions decay exponentially into the classically forbidden region between the wells. It can be shown that for h sufficiently small, the effect of the perturbation by the Dirichlet boundary conditions is to split the infinite degeneracy of the low-lying eigenvalues and form a finite-width cluster of spectrum about the unperturbed eigenvalue. The width of the band has been shown by Simon [Sim7] to be exponentially small in h. Hence, the bands can be viewed as a multiwell tunneling phenomenon. Since there is an infinite number of identical wells and barriers, the wave functions spread out and become "delocalized." That is, a wave packet corresponding to energies in a band has identical probability density in each well. This implies that any such wave function cannot be square integrable, so there are no eigenvalues.

We now consider the addition of an electric field. The resulting potential $\cos x + Fx$, for $F > 0$, is sketched in Figure 23.2. The new feature is that an electron state will be localized in only a finite number of wells provided $0 < F < \|v'\|_\infty$. As $x \to +\infty$, the wave function enters the forbidden region and decays. However, in the opposite direction, after the wave function tunnels through finitely many barriers, it can escape to $-\infty$. This is the origin of the resonances. If $e_i \in h_0$ as

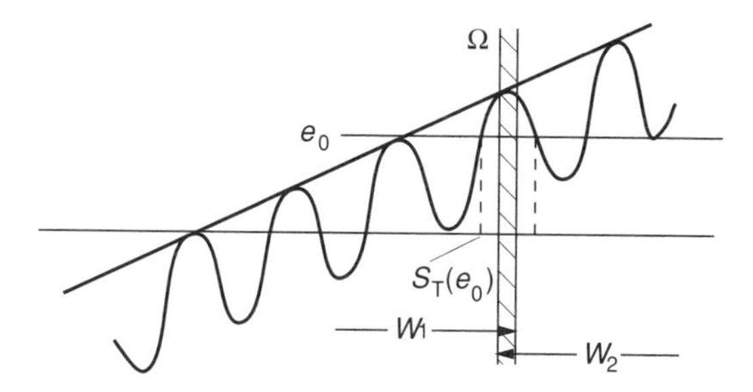

FIGURE 23.2. A Stark ladder potential $V_F(x) = \cos x + Fx$, for $F > 0$.

above, there will be a nearby resonance z_i of $H(F)$. Since $e_i + kF\tau$, $k \in \mathbb{N}$, are identical eigenvalues with respect to the potential wells, we obtain a sequence or *ladder of resonances* $z_i + kF\tau$, $k \in \mathbb{N}$.

To formulate this theory precisely, let us assume

(V1) The potential v_p, a real-valued, nonconstant, periodic function with period τ, is the restriction to $\mathbb{R}$ of a function $v(z)$ analytic in a strip

$$S_\eta \equiv \{z \in \mathbb{C} \mid |\text{Im } z| < \eta\},$$

for some $\eta > 0$.

We write V_F for the Stark ladder potential $V_F \equiv Fx + v_p$. From now on, we assume that $F \neq 0$. Let s be any number so that $v_p(s) = \max v$. The restriction $V_0 \equiv V_F \mid [s, s+\tau)$ is a potential with exactly one potential well. We extend V_0 to a potential $\tilde{V}_0$ on $\mathbb{R}$ so that $\tilde{V}_0 \in C(\mathbb{R})$ and $\tilde{V}_0(x) = \mathcal{O}(|x|)$ as $|x| \to \infty$. Let $\tilde{h}_0 \equiv -D_x^2 + \tilde{V}_0$, where $D_x \equiv d/dx$. It follows from Theorem 10.7 that $\sigma(\tilde{h}_0)$ is purely discrete. We write $\sigma(\tilde{h}_0) = \{\tilde{e}_i\}_{i=1}^\infty$. The lowest eigenvalues are insensitive to the extension of V_0 outside $[s, s+\tau)$ for small $|F|$, by the discussion of Chapter 11. We define a classical turning point $S_T(\tilde{e}_0)$ for a particle with energy $\tilde{e}_0 \in \sigma(\tilde{h}_0)$ by

$$S_T(\tilde{e}_0) = \min\{x \mid Fx + v_p(x) = \tilde{e}_0\}.$$

For any $\varepsilon_1 > 0$, we can find a small interval $[S_T(\tilde{e}_0)+\eta_1, S_T(\tilde{e}_0)+\eta_2] \equiv I_{\varepsilon_1}$ such that $(V - \tilde{e}_0) \mid I_{\varepsilon_1} > \varepsilon_1$. We partition $\mathbb{R}$ into two overlapping sets $W_1 = (-\infty, S_T + \eta_2]$ and $W_2 = [S_T + \eta_1, \infty)$, and set $\Omega = W_1 \cap W_2$. Note that $\Omega \subset \text{CFR}(\tilde{e}_0)$. We define $H_1(h, F)$ to be $H(h, F) \mid W_1$ with Dirichlet boundary conditions at $S_T + \eta_2$. Let $\tilde{V}_2$ denote an extension of $V_F \mid W_2$ to $\mathbb{R}$ so that $\tilde{V}_2(x) = \mathcal{O}(|x|)$ as $x \to -\infty$. Let $H_2(h, F)$ be the self-adjoint operator on $\mathbb{R}$ given by $-h^2 D_x^2 + \tilde{V}_2$. The spectrum of $H_1(h, F)$ is $\mathbb{R}$ for $F > 0$, and the spectrum of $H_2(h, F)$ is purely discrete for $F > 0$.

Using the multiwell tunneling estimates of [BCD3], one can prove the existence of an eigenvalue $e_0 \in \sigma(H_2(h, F))$ near $\tilde{e}_0$ separated at a distance of $\mathcal{O}(h^{2+\varepsilon})$ from the other eigenvalues of $H_2(h, F)$. It is convenient to take as an approximate Hamiltonian

$$H_0(h, F) = H_1(h, F) \oplus H_2(h, F), \tag{23.16}$$

acting on $\mathcal{H}_0 = L^2(W_1) \oplus L^2(\mathbb{R})$. The spectrum of the operator $H_0(h, F)$ is $\mathbb{R}$, and it has an *embedded eigenvalue* e_0.

We now apply the tools of Chapters 16, 18, and 20 to $H_0(h, F)$ and $H(h, F)$. Note that unlike the shape resonance problem, the unperturbed operator $H_0(h, F)$ has an embedded eigenvalue. However, this operator acts on the Hilbert space $\mathcal{H}_0 = L^2(W_1) \oplus L^2(\mathbb{R})$, whereas the perturbed operator $H(h, F)$ acts on $\mathcal{H} = L^2(\mathbb{R})$. Since these two spaces differ in the classically forbidden region, we expect that we can use a geometric resolvent formula and tunneling estimates to relate the two.

We first consider the nontrapping property of the potential V_F. Recall the generalized virial $S_E(x; V_F, f)$ given in (22.9) relative to a vector field f. In the present situation, we have the following proposition.

Proposition 23.6. *Assume v_p satisfies* (V1) *and that $F > 0$. There exists an interval $I_0 \ni e_0$ such that for any $E \in I_0$, $\exists$ bounded vector field $f_E \in C^\infty(\mathbb{R})$ with f_E' bounded,* $\text{supp}\ f_E \subset (-\infty, S_T(e_0) + \delta_1]$, *any $\delta_1 < \eta_1$, and constants $\delta_0 > 0$, $\varepsilon_0 > 0$ such that for any $E' \in I_0$, and $x \in (-\infty, S_T(e_0) + \delta_0]$,*

$$S_{E'}(x, V, f_E) \geq \varepsilon_0 > 0. \tag{23.17}$$

The idea of the proof is to construct f_E as the sum of two vector fields. One vector field, f_1, controls the virial as $x \to -\infty$. It is constructed using formula (21.10) and has the property that it approaches a constant as $x \to -\infty$. Hence, as in the discussion of the Stark resonances, we can compute the essential spectrum. Let $I_0 = [I_0^-, I_0^+]$. For some $x_0 < S_T(I_0^-)$, define for $x < x_0$

$$f_1(x) = \frac{1}{2}[E - V(x)]^{-\frac{1}{2}} \int_{S_T(I_0^-)}^{x} a(s)[E - V(s)]^{-\frac{1}{2}} ds, \tag{23.18}$$

for some real bounded function $a > 0$. The other vector field, f_2, controls the virial near the turning point $S_T(I_0^-)$. We set

$$f_2(x) = x.$$

We now recall from Problem 21.3 that f_1 is constructed so that $S_E(x; V_F, f_1) \geq a(x)$ on the interval $(-\infty, x_0]$. We choose a to approach a positive constant at $-\infty$. We also easily compute that

$$S_E(x, V, f_2) = -2(V(x) - E) - xV'(x), \tag{23.19}$$

which is positive near $S_T(e_0)$. Hence, it remains to choose cut-off functions g_1 and g_2 so that the vector field

$$f = \sum_{i=1}^{2} g_i f_i$$

satisfies (23.17). Substituting this f into the left side of (23.17), one sees from (23.18) and (23.19) how g_1 and g_2 must be chosen so that the conditions of the proposition are satisfied.

The spectral deformation method of Chapters 17 and 18 can now be implemented for $H(h, F)$ and $H_1(h, F)$ using the vector field f. The function $a(s)$ can be chosen to be $\alpha V'(x), \alpha > 0$, locally near the turning point where it is positive, and to satisfy $a(s) \to a_0 > 0$ as $x \to -\infty$. Hence, f is an approximation to the generator of translations as discussed in the Stark effect (see (23.5)). We write $H_\theta(h, F)$ and $H_{1,\theta}(h, F)$ for the spectral deformation families constructed with the vector field f for energy e_0. One checks that

$$\sigma_{\text{ess}}(H_\theta(h, F)) = \{\lambda - i a_0 \operatorname{Im} \theta \mid \lambda \in \mathbb{R}\},$$

and similarly for $H_{1,\theta}(h, F)$.

The stability estimate, resulting from the nontrapping condition of Proposition 23.6, for $H_{1,\theta}(h, F)$ can be established along the same lines as in Chapter 21. As above, we fix $0 < F < \|v_p'\|_\infty$ and take h sufficiently small. We denote by J_θ the derivative of the diffeomorphism $x \mapsto x + \theta f(x)$.

Proposition 23.7. *Assume that v_p satisfies* (V1). *Let f be the vector field of Proposition 23.6 for e_0. For $0 < \theta_0$, let $\mathcal{O}_\beta \subset \mathbb{C}$ be the region*

$$\mathcal{O}_\beta = \{z \mid \left(\frac{\varepsilon_0 \beta}{2} - \operatorname{Im} J_{i\beta}^2(E - z)\right) > 0 \,\forall E \in I_0 \text{ and } \forall x \in (-\infty, S_T(e_0) + \delta_0]\},$$

where ε_0, δ_0, and I_0 are as in Proposition 23.6. Then for any $z \in \mathcal{O}_\beta \cap \{\lambda + iy \mid 0 > y > -\theta_0, \lambda \in \mathbb{R}\}$, one has for $F > 0$ and h small enough,

(i) $z \in \rho(H_{1,i\beta}(h, F))$;

(ii) $\|(H_{1,i\beta}(h, F) - z)^{-1}\| \leq c_1 \left(\frac{\varepsilon_0 \beta}{2} - c_2 h\right)^{-1}$, *for some* $c_1, c_2 > 0$.

Given this stability estimate, we can proceed to the geometric perturbation theory. Since $\mathcal{H}_0$ differs from $\mathcal{H} = L^2(\mathbb{R})$, we must introduce an identification map $J: \mathcal{H}_0 \to \mathcal{H}$. Let $\{J_i\}_{i=1}^2$ be an almost everywhere differentiable partition of unity with supp $J_i \subset W_i$ and supp $J_i' \subset \Omega$. Define J by

$$J(u_1 \oplus u_2) = J_1 u_1 + J_2 u_2.$$

Let $\{\tilde{J}_i\}_{i=1}^2$ be another pair of such functions with $\tilde{J}_i J_i = J_i$. Then $\tilde{J}$ is defined analogously to J and $\tilde{J}^*: \mathcal{H} \to \mathcal{H}_0$ satisfies $J\tilde{J}^* = 1_{\mathcal{H}}$. Writing $R(z) \equiv (H_\theta(h, F) - z)^{-1}$ and $R_0(z) \equiv (H_{0,\theta}(h, F) - z)^{-1}$, we have the geometric resolvent equation

$$R(z)J = J R_0(z) + R(z)\tilde{J} M R_0(z), \tag{23.20}$$

where M is defined as

$$M(u_1 \oplus u_2) = h^2(D_x J_1' + J_1' D_x)u_1 \oplus h^2(D_x J_2' + J_2' D_x)u_2.$$

The analysis of (23.20) in an h-dependent neighborhood of e_0 in $\mathbb{C}$ follows the general argument of Chapter 20. However, the estimate on $(H_2(h, F) - z)^{-1}$ following from the analysis of its eigenvalues is not good enough. It is necessary to use stronger bounds on the resolvent localized in Ω. This requires estimates on $MR_0(z)$, which follow from the detailed analysis of multiwell tunneling in [BCD3]. The net result of this is the following theorem.

Theorem 23.8. *For F satisfying $0 < F < \|v_p'\|_\infty$, fixed, and for any $\varepsilon > 0$, $\exists\, h_F > 0$ such that for all $h < h_F$, $H(h, F)$ has an infinite ladder of spectral resonances $z_0(h) + kF\tau$, $k \in \mathbb{N}$, where*

$$|z_0(h) - e_0| < ch^{2+\varepsilon},$$

for some $c > 0$.

The proof of an Oppenheimer formula like (23.1) for the resonance width is possible but more delicate since the trapping region at e_0 consists of several wells separated by barriers. We refer to [CH] for the details.

23.2 Resonances and the Zeeman Effect

The theory of an atom in a constant, external magnetic field B is closely connected with the theory of resonances. Resonances appear in the theory of the asymptotic expansion of the ground state energy $E_0(B)$ as $B \to 0$. The theory of the asymptotic expansion of the ground state energy has a rather colorful history, and we refer to the review article of Simon for this and an overview of asymptotic perturbation theory [Sim12]. Resonance theory is employed as a computational tool for estimating the coefficients in the asymptotic expansion. For a treatment of the Zeeman effect in the manner of Chapters 20–22, see Briet [Br].

Let us recall from Section 16.4 that the Hamiltonian $H(B)$ for a particle under the influence of a potential V and a constant magnetic field $(0, 0, B)$ is

$$H(B) = -\Delta + \frac{B^2}{4}(x^2 + y^2) - BL_z + V, \tag{23.21}$$

where we set $\alpha = 1$ and write L_z for azimuthal angular momentum $L_z = -i(x\partial_y - y\partial_x)$. Let us suppose V satisfies these conditions:

(V1) V is relatively $-\Delta$-compact and vanishes at infinity;

(V2) the ground state of $H_0 \equiv -\Delta + V$ is a nondegenerate, isolated eigenvalue $E_0 < 0 = \inf \sigma_{\text{ess}}(H_0)$.

We can consider, for example, a Coulomb potential $V(x) = -\|x\|^{-1}$, in which case $E_0 = -\frac{1}{4}$. For simplicity, let us consider the Hamiltonian that is obtained from (23.21) by restricting to the zero azimuthal angular momentum subspace and by setting $B^2/4 \equiv \lambda$:

$$H(\lambda) \equiv -\Delta + \lambda(x^2 + y^2) + V. \tag{23.22}$$

In [AHS], the following analyticity and stability results were established for $H(\lambda)$.

Proposition 23.9.

(1) *The Hamiltonian $H(\lambda)$ on the domain $H^2(\mathbb{R}^3) \cap D(x^2 + y^2)$ is an analytic type-A family of operators for $|\arg \lambda| < \pi - \eta$, $\eta > 0$.*

(2) *The eigenvalue E_0 is stable under the perturbation $H(\lambda)$. There exists an eigenvalue $E_0(\lambda)$ near E_0 for λ in the domain $\Omega_0 \equiv \{0 < |\lambda| < \lambda_0, |\arg \lambda| < \pi\}$ and*

$$|E_0(\lambda) - E_0| = o(|\lambda|) \text{ as } |\lambda| \to 0.$$

(3) *$E_0(\lambda)$ admits an analytic condition through the cut $\overline{\mathbb{R}^-}$. For each $\eta > 0$, $\exists \lambda(\eta) > 0$ such that $E_0(\lambda)$ extends analytically to*

$$\Omega_\eta \equiv \{0 < |\lambda| < \lambda(\eta), \ |\arg \lambda| < 2\pi - \eta\}.$$

Notice that $H(\lambda)$ is not an analytic family at $\lambda = 0$. This means that the analytic perturbation theory of Chapter 15 is not applicable and one must instead use the results of Chapter 19. The Rayleigh–Schrödinger series of Section 15.5 can be computed term by term for $E_0(\lambda)$. Although the series cannot converge at $\lambda = 0$, we may still ask what information is contained in this series, especially since the first- and second-order terms give good predictions of the shift of the energy levels. The series is an *asymptotic series* in λ as $\lambda \to 0$ within Ω_0. That is, for each $N \in \mathbb{N}$, as $\lambda \to 0$,

$$\left| E_0(\lambda) - E_0 - \sum_{j=1}^{N-1} a_j \lambda^j \right| = \mathcal{O}(|\lambda|^N). \tag{23.23}$$

We write

$$E_0(\lambda) \sim E_0 + \sum_{j=1}^{\infty} a_j \lambda^j \tag{23.24}$$

when (20.23) holds. Concerning the series (23.24), we have the following result of [AHS] on the coefficients λ_j.

Proposition 23.10. *The analytic continuation of $E_0(\lambda)$ into Ω_η admits an asymptotic expansion of the form* (23.24)*, where $\{a_j\}$ are given by the Rayleigh–Schrödinger coefficients* (15.8)*. These coefficients satisfy the bound*

$$|a_j| \leq e^{j+1}(2j)! \tag{23.25}$$

as $j \to \infty$. They are given by

$$a_j = \frac{(-1)^j}{\pi} \int_0^{\lambda_0} \frac{\text{Im } E_0(-\lambda - i0)}{\lambda^{j+1}} d\lambda + \mathcal{O}(\lambda_0^{-j-1}), \tag{23.26}$$

for some $\lambda_0 > 0$.

Formula (23.26) relates the Rayleigh–Schrödinger coefficient a_j to the imaginary part of the boundary value of the analytically continued ground state energy $E_0(-\lambda - i0)$. (Recall from (23.21) and (23.22) that $\lambda = B^2/4 > 0$.) The main idea is that we can relate this imaginary part to the imaginary part of a *resonance* of $H(-\lambda)$, for $\lambda > 0$, with real part near E_0. Note that this corresponds to a purely imaginary magnetic field. Let us see why $H(-\lambda)$ should have a resonance near E_0 in the Coulomb case. The Hamiltonian $H(-\lambda)$, $\lambda > 0$, is

$$H(-\lambda) = -\Delta - \lambda(x^2 + y^2) - \frac{1}{r}. \tag{23.27}$$

Let (ρ, ϕ, z) be cylindrical coordinates on R^3. The potential at $z = 0$, $V(\rho, 0) = -\lambda\rho^2 - \rho^{-1}$, is sketched in Figure 16.4. It has the familiar shape resonance form. As in the Stark problem, the CFR(E_0, V) is unbounded. Labeling the inner boundary by $S^-(E_0; \lambda)$ and the outer boundary by $S^+(E_0; \lambda)$, we see that $S^-(E_0; \lambda) \to \{\rho = 0\}$ and $S^+(E_0; \lambda) \to \{\rho = (-E_0/\lambda))^{1/2}\}$ as $z \to \infty$.

Helffer and Sjöstrand [HSj4] have proved the existence of a resonance $z_0(\lambda)$ for $H(-\lambda)$, with a Coulomb potential as in (23.27), near E_0, with the following properties.

Proposition 23.11. *For all* $\lambda > 0$ *sufficiently small,* $\exists$ *a resonance* $z_0(\lambda)$ *of* $H(-\lambda)$ *such that*

$$\text{Im } z_0(\lambda) = -(4\lambda)^{-\frac{3}{4}} \left(1 + \mathcal{O}\left(\lambda^{\frac{1}{2}}\right)\right) e^{-\frac{\pi}{8\lambda^{1/2}}}, \tag{23.28}$$

and

$$|\text{Re } z_0(\lambda) - E_0| \to 0 \ \ as \ \lambda \to 0.$$

The proof of this theorem relies on a change of coordinates to parabolic coordinates. This transforms the problem to an anharmonic oscillator with a potential of the form $V = x^2 - \alpha x^4$ for $\alpha > 0$. The form of this potential is similar to a shape resonance potential, and the semiclassical regime corresponds to large $\alpha > 0$.

Note that (23.28) is precisely the width as predicted from the Agmon metric at energy $E_0 = -1/4$ in the small λ limit. It remains to show that Im $E_0(-\lambda - i0) =$ Im $z_0(\lambda)$. Let us write $\tilde{H}(\lambda) \equiv H(-\lambda)$, $\lambda > 0$. We then have that $\tilde{H}(\lambda)$ is an analytic family in λ in a region $\{\lambda \in \mathbb{C} \setminus 0 \mid |\arg \lambda| > 0\}$, in analogy with Proposition 23.9. One can show that matrix elements of $(\tilde{H}(\lambda) - z)^{-1}$ and $(H(\lambda) - z)^{-1}$ coincide for $\lambda \in \mathbb{C} \setminus 0$, $0 < |\arg \lambda| < \pi$. Hence, the matrix elements of $(\tilde{H}(\lambda) - z)^{-1}$ provide continuations for those of $(H(\lambda) - z)^{-1}$ for λ with $|\arg \lambda| = \pi$. Upon examining the poles, we obtain that $z_0(\lambda)$, $\arg \lambda = -\pi$, is a continuation of $E_0(-\lambda - i\varepsilon)$ as $\varepsilon \to 0$. A consequence of estimate (23.28) on Im $E_0(-\lambda - i0)$ is the result for the Rayleigh–Schrödinger coefficient

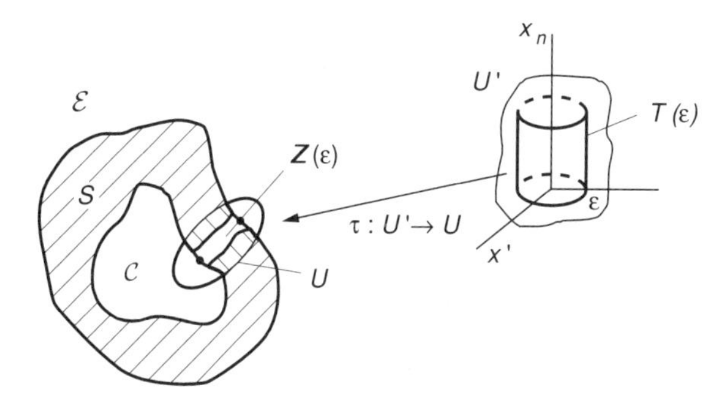

FIGURE 23.3. The Helmholtz resonator.

$$a_j = (-1)^{j+1}\left(\frac{1}{2}\right)\left(\frac{4}{\pi}\right)^{\frac{5}{2}}\left(\frac{16}{\pi^2}\right)^j\left(2j+\frac{1}{2}\right)!(1+\mathcal{O}(j^{-1})), \tag{23.29}$$

which strengthens (23.25).

23.3 Resonances of the Helmholtz Resonator

Up to this point, we have studied resonances formed by the phenomenon of quantum tunneling through a potential barrier. There is another resonance phenomenon that occurs in the scattering of waves by obstacles. We consider the wave equation

$$\partial_t^2 u = -\Delta u \tag{23.30}$$

on an unbounded domain $\mathbb{R}^n \setminus \tilde{\Omega}$, where $\tilde{\Omega}$ is a bounded obstacle. The Laplacian in (23.30) is defined with self-adjoint boundary conditions on the boundary of $\tilde{\Omega}$, which we denote by $\partial\tilde{\Omega}$. We restrict ourselves to Dirichlet boundary conditions (DBC) on $\partial\tilde{\Omega}$, for simplicity. We are interested in obstacles $\tilde{\Omega}$ called *resonators*. These obstacles have cavities that trap waves, traveling in from infinity, for some time, before they scatter and propagate back to infinity. A family of resonators can be constructed as follows. Let $S \subset \mathbb{R}^n$ be diffeomorphic to a spherical shell, that is, a region of the form $B_{R_2}(0)\backslash B_{R_1}(0)$, for some $R_2 > R_1 > 0$. For any $\epsilon > 0$, let $Z(\varepsilon) \subset S$ be diffeomorphic to a tube $T(\varepsilon) = \{(x', x_n) \in \mathbb{R}^{n-1} \times \mathbb{R} \mid \|x'\| < \varepsilon, 0 < x_n < 1\}$. A *Helmholtz resonator* is an obstacle of the form $\tilde{\Omega} = S \setminus Z(\varepsilon)$. The shell S partitions $\mathbb{R}^n \setminus S$ into two disjoint sets, a bounded cavity $\mathcal{C}$ and an unbounded exterior $\mathcal{E}$. A *Helmholtz resonator* then consists of a cavity $\mathcal{C}$ connected to the exterior $\mathcal{E}$ by a thin tube $Z(\varepsilon)$ of diameter ε; see Figure 23.3. The discussion in this section follows [HM]. For related results, we refer to the papers of Arsen'ev [Ar], Beale [Be], Fernandez [Fe], Gadyl'shin [Gad1, Gad2], Hislop [H], and references therein.

The scattering of waves by obstacles has been extensively studied. We refer the reader to the monograph by Lax and Phillips [LP] for an overview and a detailed discussion. Much of the recent progress concerning the scattering of waves by obstacles, including the structure of the scattering matrix, can be found in the book of Petkov and Stoyanov [PS]. We are interested in the *scattering resonances* of $-\Delta$ on $\mathbb{R}^n \setminus \tilde{\Omega}$, which we denote by $SR(-\Delta)$. These resonances correspond to solutions of (23.30) of the form $u(x,t) = e^{-i\lambda t} w(x)$, where $\lambda \in \mathbb{C}$ and w satisfies certain outgoing boundary conditions at infinity, in addition to the Dirichlet boundary condition $w \mid \partial\tilde{\Omega} = 0$. It is known that the set of such λ is discrete and corresponds to poles of the meromorphic continuation of the S-matrix for the problem (see the books mentioned above and the references therein). The following equivalent definition of the values λ^2 is our starting point.

Definition 23.12. *The scattering resonances of the Dirichlet Laplacian $-\Delta$ on $\mathbb{R}^n \setminus \tilde{\Omega}$, $\tilde{\Omega}$ compact, are the (square root of the) poles of the meromorphic continuation of the resolvent $R(z) \equiv (-\Delta - z)^{-1}$ from* $\operatorname{Im} z > 0$ *(where it is analytic).*

We mention that in the present case of obstacle scattering, the integral kernel $G(x, y; z)$ of $R(z)$ (Green's function) has, for $x \neq y$, a meromorphic continuation into $\operatorname{Re} z < 0$ with poles occurring at the scattering resonances of $-\Delta$.

Let us see how $SR(-\Delta)$ for a Helmholtz resonator fits into the semiclassical theory of resonances. Note that in this case, the operator $-\Delta$ does not directly depend on a parameter but rather depends on ε through the boundary conditions since the resonator depends on ϵ. Let us introduce the open, unbounded domain $\Omega(\varepsilon)$ defined by

$$\Omega(\varepsilon) \equiv \operatorname{Int}\overline{(\mathcal{C} \cup \mathcal{E} \cup Z(\varepsilon))}, \tag{23.31}$$

where $\operatorname{Int} \mathcal{O}$ denotes the interior of $\mathcal{O}$. Let H_ε be the Dirichlet Laplacian on $\Omega(\varepsilon)$. In the singular limit $\varepsilon \to 0$, the region $\Omega(\varepsilon)$ decomposes into two disjoint regions $\mathcal{C} \cup \mathcal{E}$. Correspondingly, H_ε approaches (in a sense to be made precise) the direct sum operator $H_0 = (-\Delta_{\mathcal{C}}) \oplus (-\Delta_{\mathcal{E}})$ on $L^2(\mathcal{C}) \oplus L^2(\mathcal{E})$. It is not too hard to show that $\sigma_{\text{ess}}(-\Delta_{\mathcal{E}}) = [0, \infty)$ using Weyl's theorem and the compactness of $\tilde{\Omega}(\varepsilon)$. Similarly, the compactness of $\mathcal{C}$ and DBC imply that $\sigma(-\Delta_{\mathcal{C}})$ is purely discrete. Hence, H_0 has discrete eigenvalues embedded in the essential spectrum. On the other hand, $\sigma(H_\varepsilon) = [0, \infty)$ and has no eigenvalues. Consequently, we expect that the perturbation H_ε of H_0 causes the eigenvalues of H_0 to move off the real axis and become resonances of H_ε. These resonances will be close to $\sigma(-\Delta_{\mathcal{C}})$. We see that $-\Delta_{\mathcal{C}}$ plays the role of K in the shape resonance problem. The parameter ε, the tube diameter, is the semiclassical parameter for this resonance problem. The cavity eigenvalues $\{\lambda_i\}$ physically represent the energies at which waves scattering off $\tilde{\Omega}(\varepsilon)$ will become trapped in $\mathcal{C}$ for some time. As the diameter ε of the tube narrows, the lifetimes of these states become longer.

Why does the tube act as a classically forbidden region as $\epsilon \to 0$? This is a consequence of the Dirichlet boundary conditions and the Poincaré inequality. For a bounded region $\Omega \subset \mathbb{R}^n$, there exists a constant $c_n > 0$ such that, $\forall u \in C_0^1(\Omega)$,

$$\|u\|_{L^2(\Omega)} \leq c_n |\Omega|^{\frac{1}{n}} \|\nabla u\|_{L^2(\Omega)}, \tag{23.32}$$

where $|\Omega|$ is the Lebesgue measure (the volume) of Ω. Inequality (23.32) can be extended to functions with one distributional derivative in $L^2(\Omega)$ and satisfying

$u \mid \partial\Omega = 0$. We will show that the Poincaré inequality together with the method of Agmon discussed in Chapter 3 imply that approximate and resonance eigenfunctions decay exponentially along the tube $Z(\epsilon)$.

The main stability result for $-\Delta_{\mathcal{C}}$ consists of two parts, as in the shape resonance case. We first consider the perturbation of $-\Delta_{\mathcal{C}}$ by adding the tube $Z(\epsilon)$ to the cavity $\mathcal{C}$. For this, we introduce an approximate operator, analogous to $H_0(\lambda)$. This operator, $H_0(\varepsilon)$, is the Dirichlet Laplacian on the bounded region $\mathcal{C}(\varepsilon) \equiv \text{Int}(\overline{\mathcal{C} \cup Z(\varepsilon)})$. We will show that an analogue of Theorems 20.5 and 22.2 hold for $H_0(\varepsilon)$. That is, the eigenvalues of $H_0(\varepsilon)$ are close to those of $-\Delta_{\mathcal{C}}$, and the corresponding eigenfunctions decay exponentially along the tube $Z(\varepsilon)$. We then consider the perturbation of an eigenvalue cluster of $H_0(\varepsilon)$ by H_ε. Since H_ε has only essential spectrum, we must first deform it off the positive real axis by the methods of spectral deformation. The main stability theorem is the following.

Theorem 23.13. *For each $\lambda_0 \in \sigma(-\Delta_{\mathcal{C}})$ of multiplicity N_0 $\exists$ complex neighborhood U of λ_0 such that for all ε sufficiently small, H_ε has exactly N_0 resonances $\rho_1(\varepsilon), \ldots, \rho_{N_0}(\varepsilon)$ (counting multiplicity) in U and $H_0(\varepsilon)$ has exactly N_0 eigenvalues $\lambda_1(\varepsilon), \ldots, \lambda_{N_0}(\varepsilon)$ in U tending to λ_0 as $\varepsilon \to 0$. Furthermore, for all $\delta > 0$ $\exists$ $c_\delta > 0$ such that for all $\varepsilon > 0$ sufficiently small,*

$$|\rho_j(\varepsilon) - \lambda_{k(j)}(\varepsilon)| \le c_\delta e^{-2(1-\delta)S(\varepsilon,\delta)/\varepsilon}, \tag{23.33}$$

$\forall j = 1, \ldots, N_0$, where $k(j)$ is a permutation of $\{1, \ldots, N_0\}$. The exponent $S(\varepsilon, \delta)$ is given by

$$S(\varepsilon, \delta) \equiv \max\{d_\varepsilon(x, y) \mid x, y \in \overline{Z(\varepsilon)},\ d(x, \mathcal{E} \cup \mathcal{C}) \ge \delta, d(y, \mathcal{E} \cup \mathcal{C}) \ge \delta\}$$

and

$$d_\varepsilon(x, y) = \inf\{\ell(\gamma) \mid \gamma\colon [0, 1] \to \Omega(\varepsilon),\ \gamma(0) = x \text{ and } \gamma(1) = y\},$$

with $\ell(\gamma)$ the Euclidean length of γ.

This theorem characterizes the resonances of H_ε close to the eigenvalues of the cavity. There may be other resonances due to the exterior boundary of $\tilde{\Omega}(\varepsilon)$, but we won't discuss them here. Since for all $\varepsilon > 0$ we have

$$S(\varepsilon, \delta) \ge d(\mathcal{C}, \mathcal{E}) - 2\delta > 0,$$

we see that the Euclidean distance along the tube $Z(\varepsilon)$ controls the width of the resonances. This is consistent with the earlier comments indicating that the tube behaves as a classically forbidden region. Thus, there is an effective tunneling through the tube. The origin of this tunneling is the Poincaré inequality, (23.32). Applying this inequality to the tube $Z(\epsilon)$, and noting that the right side of (23.32) is just the square root of the energy, we see that an eigenfunction of fixed energy cannot be supported in the tube for all small ϵ. An eigenfunction corresponding to a fixed eigenvalue is squeezed out of the tube as it narrows.

We now sketch some aspects of the proof of this theorem. We have introduced the analogues of K and $H_0(\lambda)$. As for the exterior region, let $\mathcal{E}(\varepsilon) \equiv \text{Int}(\overline{\mathcal{E} \cup Z(\varepsilon)})$, and denote by $H_1(\varepsilon)$ the Dirichlet Laplacian on $\mathcal{E}(\varepsilon)$. This operator is the analogue of $H_1(\lambda)$. Note that both $H_0(\varepsilon)$ and $H_1(\varepsilon)$ are defined on the forbidden region $Z(\varepsilon)$ (as in the Stark ladder problem where the two approximate operators agreed on a subset of the forbidden region). A spectral deformation family for H and $H_1(\varepsilon)$ is easily constructed on the exterior of $B_R(0)$, where R is chosen large enough so that $\tilde{\Omega}(\varepsilon) \subset B_R(0)$. We choose a vector field v on $\mathbb{R}^n$ so that

(1) $v = 0$ on $B_R(0)$;

(2) $v(x) = x - R\hat{x} - \frac{1}{2}\hat{x}$ on $\mathbb{R}^n \setminus B_{R+1}(0)$, where $\hat{x} \equiv x\|x\|^{-1}$;

(3) $\|v(x) - v(y)\| \leq \|x - y\|$, $x, y \in \mathbb{R}^n$.

Such a vector field approaches the generator of dilations as $\|x\| \to \infty$. All the results of Chapter 17 apply in this case. We denote the spectrally deformed families by $H_\varepsilon(\mu)$ and $H_1(\varepsilon; \mu)$, $\mu \in \{z \in \mathbb{C} \mid \text{Im } z \in i(-1, 1), |\text{Re } z| \text{ small}\}$. The essential spectrum of both families is $\{z \in \mathbb{C} \mid \arg z = -2\arg(1 + \mu)\}$.

The first step of the stability proof is to demonstrate that $\lambda_0 \in \sigma(-\Delta_{\mathcal{C}})$ is stable under the perturbation $H_0(\varepsilon)$.

Proposition 23.14. *Let $\lambda_0 \in \sigma(-\Delta_{\mathcal{C}})$ with multiplicity N_0. Then, for $n \geq 3$, $\exists\, \varepsilon_0$, $c > 0$ such that $\forall \varepsilon < \varepsilon_0$, $H_0(\varepsilon)$ has N_0 eigenvalues $\lambda_1(\varepsilon), \ldots, \lambda_{N_0}(\varepsilon)$ (counting multiplicity) satisfying for all $j = 1, \ldots, N_0$,*

$$|\lambda_0 - \lambda_j(\varepsilon)| \leq c\varepsilon^{\frac{n-2}{2}}. \tag{23.34}$$

For $n = 2$, the estimate is $\mathcal{O}(\varepsilon^\beta)$, for any $0 < \beta < 1/2$.

The proof of this proposition relies on a comparison of $H_0(\varepsilon)$ and the direct sum operator $-\Delta_{\mathcal{C}} \oplus h_\varepsilon \equiv \tilde{H}_0(\varepsilon)$, where h_ε is the Dirichlet Laplacian on $Z(\varepsilon)$. The effect of h_ε is small because the Poincaré inequality implies that $\inf \sigma(h_\varepsilon) \geq c\varepsilon^{-2}$. Hence, $(h_\varepsilon - z)^{-1}$ is analytic in any neighborhood of λ_0 for all ε sufficiently small. The resolvents $R_0(z)$ and $\tilde{R}_0(z)$ of $H_0(\varepsilon)$ and $\tilde{H}_0(\varepsilon)$, respectively, can be compared using Green's theorem since the two operators differ only by the addition of a Dirichlet boundary condition on the surface D_ε joining $Z(\varepsilon)$ and $\mathcal{C}$. For any $u, v \in L^2(\mathcal{C}(\varepsilon))$, we obtain

$$\langle u, (R_0(z) - \tilde{R}_0(z))v\rangle = \int_{D_\varepsilon} (\overline{TR_0(z)^* u})(B\tilde{R}_0(z)v), \tag{23.35}$$

where T is the restriction or trace map for D_ε,

$$B\tilde{R}_0(z)v = (n \cdot \nabla)_{\mathcal{C}}(-\Delta_{\mathcal{C}} - z)^{-1}v \oplus (n \cdot \nabla)_Z(h_\varepsilon - z)^{-1}v,$$

$(n \cdot \nabla)_{\mathcal{C}}$ is the outward normal for $\mathcal{C}$, and $(n \cdot \nabla)_Z$ is the outward normal for $Z(\varepsilon)$. The trace map is controlled by the Sobolev trace theorem as mentioned in Problem

22.7. Iterating (23.35), we again obtain an analogue of the geometric resolvent formula for which the interaction term is localized on the small disk D_ε. The proposition and bound (23.34) follow from a careful estimate of these operators.

The second step of the stability proof consists in proving the stability of the family of eigenvalues $\{\lambda_j(\varepsilon)\}_{j=1}^{N_0}$ of $H_0(\varepsilon)$ near λ_0 under the perturbation H_ε. We use a different form of geometric perturbation theory. Let $\mathcal{H}_0 \equiv L^2(\mathcal{C}(\varepsilon)) \oplus L^2(\mathcal{E}(\varepsilon))$ and $H_0(\varepsilon,\mu) \equiv H_0(\varepsilon) \oplus H_1(\varepsilon,\mu)$. We must compare $H_0(\varepsilon,\mu)$ acting on $\mathcal{H}_0$ with $H_\varepsilon(\mu)$ acting on $\mathcal{H} \equiv L^2(\Omega(\varepsilon))$. This is a two-Hilbert-space perturbation theory similar to the one encountered in the discussion of the Stark ladder resonances; see (23.20). Let $\{J_i\}_{i=1}^2$ be a partition of unity of $\mathbb{R}^n$ satisfying $\sum_{i=1}^2 J_i^2 = 1$ and

$$J_1 \mid \{x \mid d(x,\mathcal{E}) > 2\delta\} = 1$$

and

$$J_2 \mid \{x \mid d(x,\mathcal{E}) \le \delta\} = 1,$$

and so supp $\nabla J_1 \subset \{x \mid \delta \le d(x,\mathcal{E}) \le 2\delta\}$ and $|\nabla J_i|$ is independent of ε. We define a map $J\colon \mathcal{H} \to \mathcal{H}_0$ by

$$Ju = J_1 u \oplus J_2 u,$$

and so $J^*J = 1_{\mathcal{H}}$. Let $R(z) = (H_\varepsilon(\mu) - z)^{-1}$ and $R_0(z) = (H_0(\varepsilon,\mu) - z)^{-1}$. The geometric resolvent equation we need is

$$R(z) = J^* R_0(z) J + R(z) M R_0(z) J \tag{23.36}$$

on $\mathcal{H}$, where the interaction $M\colon \mathcal{H}_0 \to \mathcal{H}$ is given by

$$M(u_1 \oplus u_2) = [-\Delta, J_1]u_1 + [-\Delta, J_2]u_2.$$

(This should be compared with (23.20).) Note that M is supported in the tube $Z(\varepsilon)$. In analogy with (22.36)–(22.38), we define

$$K(z) \equiv J M R_0(z)\colon \mathcal{H}_0 \to \mathcal{H}_0.$$

The main technical lemma is the following. Let β be the exponent of ε appearing in Proposition 23.14 and c be the constant. Let Γ_ε be a contour around λ_0 of radius $2c\varepsilon^\beta$.

Lemma 23.15. *For any $\delta > 0$, $\exists\, c_\delta > 0$, $\varepsilon_0 > 0$ such that for $\varepsilon < \varepsilon_0$ and uniformly on Γ_ε,*

$$\|K(z)\| \le c_\delta \varepsilon^{2-\delta}. \tag{23.37}$$

Let χ be a function localized near supp$|\nabla J_i|$. The proof of this lemma follows from estimates on the localized resolvents $\chi(H_0(\varepsilon) - z)^{-1}$, analogous to Corollary 20.7, and $\chi(H_1(\varepsilon,\mu) - z)^{-1}$, analogous to Theorem 20.8. The former estimate follows from the Poincaré inequality. The latter estimate is much easier than in the shape resonance case. It is not hard to check that $H_1(\varepsilon,\mu)$ has no spectrum on and inside Γ_ε.

Given this lemma and (23.36), the proof of stability is easy. We solve (23.36) for $R(z)$ on Γ_ε to obtain (recall (23.37))

$$R(z) = J^* R_0(z) J + J^* R_0(z)(1 - K(z))^{-1} J M R_0(z) J. \tag{23.38}$$

This formula is integrated along Γ_ε. Denoting by $P(\varepsilon, \mu)$ and $P_0(\varepsilon)$ the resulting projections, we conclude that

$$\| P(\varepsilon, \mu) - J^* P_0(\varepsilon) J \| \le c_\delta \varepsilon^{2-\delta},$$

for all ε sufficiently small. We pass from $J^* P_0(\varepsilon) J$ to $P_0(\varepsilon)$ as in (20.54) using the decay of the eigenfunctions of $H_0(\varepsilon)$ in $Z(\varepsilon)$, which we now describe.

Proposition 23.16. *Let $\lambda(\varepsilon) \in \sigma(H_0(\varepsilon))$ satisfy $\lambda(\varepsilon) \to \lambda_0 \in \sigma(-\Delta_{\mathcal{C}})$ as $\varepsilon \to 0$, and let u_ε be a corresponding eigenfunction with $\|u_\varepsilon\| = 1$. Then for any $\delta > 0$ $\exists\, c_\delta > 0$ such that for all ε small enough*

$$\| e^{(1-\delta) d_\varepsilon(\cdot, \mathcal{C})/\varepsilon} u_\varepsilon \|_{L^2(Z(\varepsilon))} \le c_\delta, \tag{23.39}$$

and similarly for ∇u_ε (d_ε is defined in Theorem 23.13).

This proposition can be used in the arguments of Section 22.3 to obtain the upper bound (23.33) on the resonance width. The proof of Proposition 23.16 relies on the Poincaré inequality (23.32) and Agmon's weighted inequalities, as in Section 22.3. The form of the Poincaré inequality applicable here is

$$\int_{Z(\varepsilon)} |\phi|^2 \le \varepsilon^2 (1 + c\varepsilon) \int_{Z(\varepsilon)} |\nabla \phi|^2, \tag{23.40}$$

for appropriate ϕ on $\mathcal{C}(\varepsilon)$. Substituting $f\phi$ for ϕ in (23.40), where f is a real weight function, we obtain

$$\varepsilon^{-2}(1 + c\varepsilon)^{-1} \int_{Z(\varepsilon)} |f\phi|^2 \le \operatorname{Re} \int_{\mathcal{C}(\varepsilon)} f^2 \overline{\phi} (H_0(\varepsilon)\phi) + \int_{\mathcal{C}(\varepsilon)} |\nabla f|^2 |\phi|^2. \tag{23.41}$$

Finally, we choose $\phi = u_\varepsilon$ and $f = e^{\psi/\varepsilon}$, where $\psi(x) = (1 - \delta) d_\varepsilon(\mathcal{C}, x)$. Since $|\nabla \psi|^2 \le (1 - \delta)^2$, it is easy to check that (23.39) follows from (23.41).

This concludes the sketch of the proof of Theorem 23.13.

23.4 Comments on More General Potentials, Exponential Decay, and Lower Bounds

A. More General Potentials

It is not difficult to extend the theory of resonances presented here to potentials having local singularities [HS2]. It is much more difficult to relax the analyticity assumption. This is possible if V is exponentially decaying. In this case, the meromorphic continuation of the resolvent as a bounded operator between weighted Hilbert spaces can be proved rather directly. Resonances can be defined as poles of that continuation. It is also known that these poles coincide with the poles of the continuation of the S-matrix as a bounded operator on $L^2(S^{n-1})$. When the potential is expressible as a sum of an exponentially decaying potential and a short-range potential analytic in the sense of Section 16.2, Balslev and Skibsted [BS1] have proved the existence of the meromorphic continuation of the resolvent and the S-matrix. As in the exponentially decaying potential case, the resolvent extends as a bounded operator between exponentially weighted Hilbert spaces. They proved that the poles of the two continuations coincide and are of the same order. This does not seem to provide, however, a technique for proving the existence of resonances.

Nakamura [N4], [N5] studied a similar class of potentials within the framework of spectral deformations in momentum, rather than coordinate, space. This method (in its global form) has been introduced earlier by Cycon [Cy] and by Sigal [Si2]. Let $\nu: \mathbb{R}^n \to \mathbb{R}^n$ be a smooth vector field that is linear at infinity. If $\xi \in \mathbb{R}^n$ denotes the Fourier transform variable, consider the map $\phi_\theta: \mathbb{R}^n \to \mathbb{R}^n$ generated by ν:

$$\phi_\theta(\xi) = \xi + \theta\nu(\xi). \tag{23.42}$$

This induces a unitary map on $L^2(\mathbb{R}^n)$, for $|\theta|$ real and small enough, defined by

$$(U_\theta f)(x) = F^{-1}\left(J_\theta^{\frac{1}{2}}(Ff)\circ\phi_\theta\right)(x),$$

where F is the Fourier transform (see Appendix 4). The Fourier transform of the kinetic energy transforms nicely under the map U_θ induced by such a map ϕ_θ as in (23.42). One readily checks that the transformation is

$$\|\xi\|^2 = F(-\Delta)F^{-1} \to \|\xi + \theta\nu(\xi)\|^2 = FU_\theta(-\Delta)U_\theta^{-1}F^{-1}.$$

The transformed potential becomes more complicated because it is a convolution operator in momentum space. With this family of deformations, Nakamura extends the Aguilar–Balslev–Combes–Simon theory to potentials that are a sum of a smooth, exponentially decaying potential and an analytic potential with decaying derivatives. Furthermore, for such potentials that satisfy a nontrapping condition, outside a compact region Nakamura proves the existence of resonances in the semiclassical regime. This work extends the results of Chapter 20 to shape resonance potentials of compact support in coordinate space. The main techniques used by Nakamura is the calculus of h-pseudodifferential operators. In this approach, the resolvent estimates at nontrapping energies are derived from the Fefferman–Phong inequalities.

B. Exponential Decay Law

In physics literature (see, for example, [LL]), resonances are described as quasistationary states. These are states $\psi(x, t)$ that appear to be well localized for a period of time τ called the *lifetime*, but which asymptotically decay in time in the sense that for any ball of radius R,

$$\left[\int_{\|x\|<R} |\psi(x,t)|^2\right]\left[\int_{\|x\|<R} |\psi(x,0)|^2\right]^{-1} \leq ce^{-\Gamma t}, \tag{23.43}$$

as $t \to \infty$. The constant Γ is the resonance width and is related to τ by $\Gamma = \tau^{-1}$. The resonance energy is written as $E - \frac{1}{2}i\Gamma$. It is known that for a semibounded Hamiltonian (i.e., one for which $\exists M < \infty$ such that $-M < H$) an estimate like (23.43) cannot hold for all times. This follows from a Paley–Wiener theorem. Herbst [He4] observed, however, that this constraint does not apply to resonances for Stark Hamiltonians since these operators are not semibounded, and he proved the exponential decay of resonance states for such Schrödinger operators. In the semibounded case, one expects that an estimate of this type holds over several lifetimes (this is what is observed experimentally). Estimates on the time evolution of resonance states for exponentially decaying potentials in $\mathbb{R}^3$ were given by Skibsted [Sk1], [Sk3], and for more general problems admissible to spectral deformation methods, by Hunziker [Hu3]. The results of Hunziker apply, for example, to the shape resonance model discussed in Chapter 19. The time evolution of metastable states within a dilation analytic framework was also studied by Waxler [Wx]. Many of these results were unified and generalized by Gérard and Sigal [GS].

Let V be a real, exponentially decaying potential such that $e^{2s\|x\|}V \in L^p(\mathbb{R}^3)$, for some $s > 0$ and $p > 3/2$. Define operators $A = |V|^{1/2}$ and $B = |V|^{-1/2}V$ so that $V = AB$. The operator $A(-\Delta - z)^{-1}B$ is Hilbert–Schmidt for $\operatorname{Im} z > 0$. It has an analytic continuation from $\operatorname{Im} z > 0$ to $-s < \operatorname{Im} z^{1/2} \leq 0$ given by $A\tilde{R}_0(z)B$ where $\tilde{R}_0(z)$ is the analytic continuation of $(-\Delta - z)^{-1}$ as an operator between exponentially weighted Hilbert spaces. In this framework, $H = -\Delta + V$ on $L^2(\mathbb{R}^3)$ has a resonance at $z_0 \in \mathbb{C}_s^- \equiv \{z \mid -s < \operatorname{Im} z^{1/2} < 0\}$ if there exists $\eta \in L^2(\mathbb{R}^3)$ so that

$$A\tilde{R}_0(z)B\eta = -\eta.$$

This is an extension of the Birman–Schwinger principle to resonances. A resonance function for z_0 is defined by $f_{z_0} \equiv \tilde{R}_0(z_0)B\eta$. Such a function grows exponentially at infinity and so is not in $L^2(\mathbb{R}^3)$. Let χ_R be a characteristic function on $B_R(0)$. We are interested in the decay of the resonance function from $B_R(0)$ as $t \to \infty$. Skibsted proves that $\exists$ a function $\varepsilon(t)$ such that for some positive integer N, $|\varepsilon(t)| \ll 1$ for $0 < t\Gamma < N$, and

$$\|\chi_R e^{-itH}\chi_R f_{z_0}\|^2 \|\chi_R f_{z_0}\|^{-2} = e^{-\Gamma t}(1 + \varepsilon(t)).$$

This shows the exponential decay of the resonant state over several lifetimes (compare with (23.43)).

C. Lower Bounds on Resonance Widths

The main result of Chapter 22 is an upper bound on the resonance width. One might ask if it is possible to find a lower bound of the same form. This is possible using the techniques developed by Helffer and Sjöstrand [HSj4] in certain situations. This problem is similar to the one of the splitting of the lowest two eigenvalues for the double-well potential discussed in Chapter 12. There it was shown that there exist exponentially small upper and lower bounds for the splitting. Let us suppose that V is a potential of shape resonance type as discussed in Chapter 20. In addition, we must suppose that V is the restriction of an analytic function to $\mathbb{R}^n$ (at least outside the barrier). Recall that $d_A^0(0, S^+(0))$ is the distance in the Agmon metric for $(V(x) - V(0))$ at energy zero between 0 and the turning surface $S^+(0)$ (defined for $V(x) - V(0)$). Let $z_0(\lambda)$ be a resonance associated with the ground state energy e_0 of K as in Theorem 20.3. Then for all λ sufficiently large, $\exists c > 0$ such that

$$-\text{Im } z_0(\lambda) \geq c\lambda^{-\frac{1}{2}} e^{-2\lambda d_A^0(0, S^+(0))}.$$

Other related results, including asymptotic expansions, can be found in [HSj4].

Lower bounds for resonances for one-dimensional Schrödinger operators were obtained by Harrell [H2] using the methods of differential equations. Fernandez and Lavine [FL] obtained lower bounds for resonances widths for Schrödinger operators with potentials having compact support and for the Dirichlet Laplacian in the exterior of a star-shaped region in three dimensions. Lower bounds for resonances of one-dimensional Stark Hamiltonians with negative potentials of compact support were computed by Ahia [Ah].

Appendix 1. Introduction to Banach Spaces

A1.1 Linear Vector Spaces and Norms

Definition A1.1 *A linear vector space (LVS) X over a field F is a set with two binary operations: addition (a map of $X \times X \to X$) and numerical (or scalar) multiplication (a map of $F \times X \to X$), which satisfy the commutative, associative, and distributive properties.*

We are primarily interested in spaces of functions and will take F to be either $\mathbb{R}$ or $\mathbb{C}$.

Examples A1.2.

(1) $X = \mathbb{R}^n$ with the operations $(+, \cdot)$, the usual componentwise operations, and with $F = \mathbb{R}$.

(2) $X = C([0, 1])$, all continuous, complex-valued functions on the interval $[0, 1]$. Addition is pointwise, that is, if $f, g \in X$ then $(f+g)(x) = f(x)+g(x)$, and scalar multiplication for $F = \mathbb{C}$ is also pointwise.

(3) $X = l^p$, where l^p consists of all infinite sequences of complex numbers: $x \equiv (x_1, ..., x_n, ...)$, $x_i \in \mathbb{C}$ such that $\sum_{i=1}^{\infty} |x_i|^p < \infty$, for $1 \leq p < \infty$. Again, addition and scalar multiplication are defined componentwise.

Problem A1.1. Show that l^p is an LVS. (*Hint*: Use the Minkowski inequality:

$$\left(\sum_{i=1}^{\infty} |x_i + y_i|^p \right)^{\frac{1}{p}} \leq \left(\sum_{i=1}^{\infty} |x_i|^p \right)^{\frac{1}{p}} + \left(\sum_{i=1}^{\infty} |y_i|^p \right)^{\frac{1}{p}}. \bigg)$$

Definition A1.3. *Let X be an LVS over F, $F = \mathbb{R}$ or $\mathbb{C}$. A norm $\|\cdot\|$ on X is a map $\|\cdot\| : X \to \mathbb{R}_+ \cup \{0\}$ such that*

(i) $\|x\| \geq 0$ and if $\|x\| = 0$ then $x = 0$ (positive definiteness);

(ii) $\|\lambda x\| = |\lambda| \, \|x\|$, $\lambda \in F$ (homogeneity);

(iii) $\|x + y\| \leq \|x\| + \|y\|$, $x, y \in X$ (triangle inequality).

Examples A1.4.

(1) $X = \mathbb{R}^n$ and define $\|x\| = \left(\sum_{i=1}^{n} |x_i|^2\right)^{1/2}$, $x \in \mathbb{R}^n$. This is a norm, called the Euclidean norm, and has the interpretation of the length of the vector x.

(2) $X = C([0, 1])$ can be equipped with many norms, for example, the L^p-norms: $\|f\|_p \equiv \left(\int_0^1 |f(x)|^p dx\right)^{1/p}$, $f \in X, 1 \leq p < \infty$, and the sup-norm $(p = \infty)$: $\|f\|_\infty \equiv \sup_{x \in [0,1]} |f(x)|$, $f \in X$.

(3) $X = l^p$ has a norm given by $\|x\| \equiv \left(\sum_{i=1}^{\infty} |x_i|^p\right)^{1/p}$, $x \in X$.

Problem A1.2. Show that the maps $X \to \mathbb{R}^+$ claimed to be norms in Examples A1.4 are norms.

Definition A1.5. *An LVS X with a norm $\|\cdot\|$ is a normed linear vector space (NLVS), that is, an NLVS X is a pair $(X, \|\cdot\|)$ where X is an LVS and $\|\cdot\|$ is a norm on X.*

A1.2 Elementary Topology in Normed Vector Spaces

Elementary topology studies relations between certain families of subsets of a set X. A *topology* for a set X is built out of a distinguished family of subsets, called the open sets. In an NLVS X, there is a standard construction of these open sets.

Definition A1.6. *Let X be an NLVS, and let $a \in X$. An open ball about a of radius r, denoted $B_r(a)$, is defined by*

$$B_r(a) \equiv \{x \in X | \ \|x - a\| < r\}.$$

In analogy with the Euclidean metric on $\mathbb{R}^n$ (see (1), Examples A1.4), we interpret $\|x - a\|$ as "the distance from x to a," and so $B_r(a)$ consists of all points $x \in X$ lying within distance r from a. The open balls in X are the fundamental building blocks of open sets.

Definition A1.7. *A subset $E \subset X$, X an NLVS, is open if for each $x \in E$ there exists an $r > 0$ (depending on x) such that $B_r(x) \subset E$. A neighborhood of $x \in X$ is an open set containing x.*

Example A1.8.

Let $X = \mathbb{R}^n$ with the Euclidean metric. Then

$$B_\epsilon(0) = \left\{ x \in \mathbb{R}^n | \left(\sum_{i=1}^{n} |x_i|^2 \right)^{\frac{1}{2}} < \epsilon \right\},$$

simply a ball of radius ϵ about the origin. Note that $B_\epsilon(0)$ does not include its boundary, that is, $\left\{ x | \left(\sum_{i=1}^{n} x_i^2 \right)^{1/2} = \epsilon \right\}$. It is easy to check that $B_\epsilon(0)$ is an open set. If we were to include the boundary, the resulting set would not be open.

Definition A1.9. *A subset $B \subset X$, X an NLVS, is closed if $X \setminus B \equiv \{x \in X | x \notin B\}$ (the complement of B in X) is open.*

Sequences provide a convenient tool for studying many properties of subsets of an NLVS X, including open and closed sets.

Definition A1.10. *Let X be an NLVS.*

(1) *A sequence $\{x_n\}$ in X converges to $x \in X$ if $\|x_n - x\| \to 0$ as $n \to \infty$.*

(2) *A sequence $\{x_n\}$ in X is Cauchy if $\|x_n - x_m\| \to 0$ as $n, m \to \infty$.*

Problem A1.3. Prove that

(1) every convergent sequence is a Cauchy sequence,

(2) every convergent sequence is uniformly bounded, that is, if $\{x_n\}$ is convergent, then there exists an M, $0 < M < \infty$, such that $\|x_n\| \leq M$ for all $n = 1, 2, \ldots$.

(Hints:

(1) *Use an $\epsilon/2$-argument and the identity: $x_n - x_m = x_n - x + x - x_m$.*

(2) *Use the triangle inequality to show that*

$$| \|x_n\| - \|x_m\| | \leq \|x_n - x_m\|.)$$

A nice feature of NLVS is that sequences can be used to characterize closed sets.

Proposition A1.11. *A subset $E \subset X$, X an NLVS, is closed if for any convergent sequence $\{x_n\}$ in E, $\lim_{n\to\infty} x_n \in E$.*

Proof.

(1) Suppose E is closed and $\lim_{n\to\infty} x_n = x \notin E$ but $x_n \in E$. Since $X \setminus E$ is open, there is an $\epsilon > 0$ such that $B_\epsilon(x) \cap E = \phi$. But $x_n \to x$, so for this ϵ, there is an N such that $n > N \Rightarrow \|x_n - x\| < \epsilon \Rightarrow x_n \in B_\epsilon(x)$. So we have a contradiction, and $x \in E$.

(2) Conversely, suppose E contains the limit of all its convergent sequences. Suppose $X \setminus E$ is not open; then there is a point $y \in X \setminus E$ such that for each $\epsilon > 0$, $B_\epsilon(y) \cap E \neq \phi$. Let $\epsilon_n \equiv 1/n$ and choose $y_n \in B_{\epsilon_n}(y) \cap E$. Then $\{y_n\}$ is a sequence in E and $\lim_{n\to\infty} y_n = y$. But $y \notin E$, hence we get a contradiction (i.e., $X \setminus E$ is open so E is closed). □

Definition A1.12. *If $M \subset X$, X an NLVS, then $\bar{M}$, the closure of M in X, is $M \cup \{$limits of all nonconstant convergent sequences in $M\}$. The set of points that are limits in X of nonconstant sequences in M are the accumulation points or cluster points of M.*

Remark A1.13. By definition, $\bar{M} \supset M$ and $\bar{M}$ is closed. $\bar{M}$ is, in fact, the smallest closed set containing M.

Remark A1.14. Exercise (1) in Problem A1.3 is significant because the converse is *not* true in every NLVS X. That is, if $\{x_n\}$ is a Cauchy sequence in X, it is not necessarily true that $\{x_n\}$ is convergent. As an example, consider $X = C([0, 1])$ in the $\|\cdot\|_p$, $1 \le p < \infty$, norm (see Example A1.4 (2)). We construct a sequence $f_n \in X$ by

$$f_n(x) = \begin{cases} 0, & 0 \le x \le \frac{1}{2}, \\ n\left(x - \frac{1}{2}\right), & \frac{1}{2} \le x \le \frac{1}{2} + \frac{1}{n}, \\ 1, & \frac{1}{2} + \frac{1}{n} \le x \le 1. \end{cases}$$

Clearly, $f_n \in C([0, 1])$ and, for $n \ge m$ and $p = 1$,

$$\begin{aligned} \|f_n - f_m\|_1 &= \int_0^1 |f_n(x) - f_m(x)| dx \\ &= \int_{\frac{1}{2}}^{\frac{1}{2}+\frac{1}{n}} (n - m)(x - 1/2) dx + \int_{\frac{1}{2}+\frac{1}{n}}^{\frac{1}{2}+\frac{1}{m}} [1 - m(x - 1/2)] dx \\ &= \frac{1}{2}\left(\frac{1}{m} - \frac{1}{n}\right), \end{aligned}$$

and so $\{f_n\}$ is a Cauchy sequence. It is easy to check that for any p, $\{f_n\}$ is Cauchy. Now for any $x \in [0, 1]$, we can compute $\lim_{n\to\infty} f_n(x)$. The limit function is

$$f(x) = \begin{cases} 0, & 0 < x \le \frac{1}{2}, \\ 1, & \frac{1}{2} < x \le 1. \end{cases}$$

This limit function $f \notin C([0, 1])$ (i.e., f is discontinuous), and so $C([0, 1])$ does *not* contain the limit of all its Cauchy sequences! Later we will discuss the manner in which we can extend $C([0, 1])$ to a larger NLVS (called the completion) to include the limit of all the Cauchy sequences in $C([0, 1])$.

A1.3 Banach Spaces

Definition A1.15. *An NLVS X is complete if every Cauchy sequence in X converges to an element of X. A complete NLVS is called a Banach space.*

Remark A1.16. Every finite-dimensional NLVS is complete and hence a Banach space. This follows from the fact that both $\mathbb{R}$ and $\mathbb{C}$ are complete and from the fact that every finite-dimensional LVS is isomorphic to $\mathbb{R}^N$ or $\mathbb{C}^N$ for some N. For this reason, the term "Banach space" is usually used to denote an infinite-dimensional NLVS.

Examples A1.17.

(1) $X = \mathbb{R}^N$ with the Euclidean norm $\|\cdot\|_e$ is complete.

(2) $X = C([0, 1])$ is complete with the sup-norm $\|\cdot\|_\infty$, but it is not complete with the p-norm for $1 \le p < \infty$; see Remark A1.14.

(3) l^p is complete; we prove this in Proposition A1.18.

Problem A1.4. Prove statements (1) and (2) in Examples A1.17. (*Hint:* For (2), use standard results on uniform convergence.)

Proposition A1.18. *The NLVS l^p as defined in (3) of Examples A1.2, with the $\|\cdot\|_p$-norm, $1 \le p < \infty$, is a Banach space.*

Proof. Let $\{x^{(n)}\}$, $x^{(n)} \equiv (x_1^{(n)}, x_2^{(n)}, \ldots) \in l^p$, be a Cauchy sequence, that is,

$$\|x^{(n)} - x^{(m)}\|_p = \left[\sum_{i=1}^{\infty} |x_i^{(n)} - x_i^{(m)}|^p\right]^{\frac{1}{p}} \to 0 \tag{A1.1}$$

as $m, n \to \infty$. We must show (1) that $\{x^{(n)}\}$ is convergent to some x, and (2) that $x \in l^p$.

(1) Condition (A1.1) implies that each sequence of real (or complex) numbers $\{x_i^{(n)}\}_{n=1}^{\infty}$ (fixed i) is Cauchy, and from the completeness of $\mathbb{R}$ (or $\mathbb{C}$) it is convergent. Hence there is an $x_i \in \mathbb{R}$ (or $\mathbb{C}$) such that $x_i = \lim_{n\to\infty} x_i^{(n)}$. Let $x \equiv (x_1, x_2, \ldots)$. We show that $x^{(n)} \to x$ in the l^p-norm. From the definition of $\{x^{(n)}\}$, for each $\epsilon > 0$ there is an $N_\epsilon \in \mathbb{N}$ such that $m, n > N_\epsilon$ implies $\left[\sum_{i=1}^{k} |x_i^{(n)} - x_i^{(m)}|^p\right]^{1/p} < \epsilon$. Now we take the limit of the left side as $n \to \infty$ (fixed k) and get $S_m(k) = \left[\sum_{i=1}^{k} |x_i - x_i^{(m)}|^p\right]^{1/p} < \epsilon$. As a sequence in k, $S_m(k)$ is monotonic (i.e., $S_m(k) \le S_m(k')$ if $k < k'$) and bounded, and hence $\lim_{k\to\infty} S_m(k)$ exists and is less than ϵ. This is the l^p-norm of $(x - x^{(m)})$, that is, $\|x - x^{(m)}\|_p < \epsilon$. Hence, $x^{(m)} \to x$.

(2) To prove that $x \in l^p$, we use Minkowski's inequality (Problem A1.1) and begin by considering a finite sum as above. We write

$$\left(\sum_{i=1}^{k} |x_i|^p\right)^{\frac{1}{p}} \le \left(\sum_{i=i}^{k} |x_i^{(n)} - x_i|^p\right)^{\frac{1}{p}} + \left(\sum_{i=1}^{k} |x_i^{(n)}|^p\right)^{\frac{1}{p}}.$$

Since $x^{(n)} \in l^p$ and $\|x^{(n)} - x\|_p \to 0$, given $\epsilon > 0$, we can choose $n > N_\epsilon$ such that

$$\lim_{k\to\infty} \left(\sum_{i=1}^{k} |x_i|^p \right)^{\frac{1}{p}} \leq \|x^{(n)}\|_p + \epsilon,$$

and hence $\|x\|_p < \infty$ (i.e., $x \in l^p$). □

We now introduce some other topological ideas associated with Banach spaces.

Definition A1.19. *Let X be an NLVS. A subset $A \subset X$ is dense in X if each point $x \in X$ can be approximated arbitrarily closely by points in A, namely, for each $x \in X$ there is a sequence $\{a_n\} \subset A$ such that $\lim_{n\to\infty} a_n = x$. Equivalently, for any $x \in X$ and any $\epsilon > 0$ there exists $a \in A$ such that $\|x - a\| < \epsilon$.*

Examples A1.20.

(1) The rational numbers Q are dense in the real numbers.

(2) The *Weierstrass approximation theorem* states that the set of all polynomials in x on $[0, 1]$ (i.e., functions of the form $\sum_{i=1}^{k} c_i x^i + c_0$) is dense in $C([0, 1])$ with the sup-norm.

(3) Let X be a Banach space, and let $A \subset X$ be a subset that is not closed. Then A is dense in $\bar{A}$, its closure.

Definition A1.21. *A Banach space X is separable if it contains a countable, dense subset. (A set is countable if its elements can be put in one-to-one correspondence with the set of integers $\mathbb{Z}$.)*

Examples A1.22.

(1) The set of polynomials on $[0, 1]$ with rational coefficients is a dense, countable set in $C([0, 1])$ with the sup-norm, so this Banach space is separable.

(2) The set of all elements $x \in l^p$ with $x_i \in \mathbb{Q}$ is a dense, countable set in l^p, so l^p is separable.

(3) Simply for reference, the space $L^\infty(\mathbb{R})$, the LVS of all (essentially) bounded functions on $\mathbb{R}$ with the (ess) sup-norm, is nonseparable.

Remark A1.23. All the spaces used in the chapters are *separable*, unless specifically mentioned.

A1.4 Compactness

We introduce another important topological notion associated with Banach spaces (which is, of course, an important notion in its own right).

Definition A1.24. *Let X be a (separable) Banach space. A subset $Y \subset X$ is called compact (respectively, relatively compact) if any infinite subset of Y contains a convergent sequence and the limit of this sequence belongs to Y (respectively, to X).*

Note that if $Y \subset X$ is relatively compact, then $\bar{Y}$, the closure of Y in X, is compact. Moreover, if Y is compact, then Y is a closed subset of X. This follows since if $\{y_n\}$ is a Cauchy sequence in Y, it is an infinite subset of Y. Since X is complete, $\{y_n\}$ is convergent and the limit $y = \lim_{n\to\infty} y_n \in Y$ by definition of compactness. Hence, we have the next result.

Proposition A1.25. *Let X be a Banach space. If $Y \subset X$ is compact, then it is closed and bounded (i.e., $\exists M > 0$ such that $\|y\| \leq M \;\; \forall y \in Y$).*

Proof. It remains to show that Y is bounded. Suppose not; then for each $n \in \mathbb{Z}$, there exists $y_n \in Y$ such that $\|y_n\| > n$. Then $\{y_n\}$ is an infinite subset of Y. By compactness, we can choose a subsequence $\{y_{\sigma(n)}\}$ of $\{y_n\}$, where $\sigma : \mathbb{Z}_+ \to \mathbb{Z}_+$ is an order-preserving map, which converges. Then, as Y is compact, $\exists y \in Y$ such that $\|y_{\sigma(n)}\|$ converges to $\|y\|$. But, $\|y_{\sigma(n)}\| \geq \sigma(n) \geq n \to \infty$, and so we get a contradiction. Hence Y is bounded. □

It is a well-known result that the converse of Proposition A1.25 is not true: There are many closed and bounded subsets of certain Banach spaces which are not compact.

Examples A1.26.

(1) The problem just described cannot occur in $\mathbb{R}^N$ or $\mathbb{C}^N$: the *Heine–Borel theorem* states that a set is compact if and only if it is closed and bounded.

(2) Let $X = C([0, 1])$, and let $Y \subset X$ be the subset of all functions in X which, together with their derivative, are uniformly bounded, that is, $Y = \{f \in X| \; \|f\|_\infty < M \text{ and } \|f'\|_\infty < M\}$. The *Ascoli–Arzelà theorem* states that Y is compact in X.

We now want to relate the definition in A1.24, which depends on the metric structure of X, to the general topological notion of compactness.

Definition A1.27. *Let $A \subset X$, X any topological space. A collection of open subsets $\{V_i\}$ of X is called an open covering of A if $A \subset \cup_i V_i$.*

Theorem A1.28. *A subset $K \subset X$, X a Banach space, is compact if and only if any (countable) open covering of K contains a finite subcovering of K.*

Proof.

(1) Suppose $K \subset X$ is compact, and let $\{V_i\}$ be an open covering of K. Suppose on the contrary that there exists no finite subcovering. Then $K \setminus \cup_{i=1}^n V_i$ is

nonempty for any n. Let $x_n \in K \setminus \cup_{i=1}^{n} V_i$. By the compactness of K, there exists a convergent subsequence $\{x_{n'}\}$ and $x = \lim_{n' \to \infty} x_{n'} \in K$. Since $\{V_i\}$ covers K, $x \in V_j$, for some j. Since V_j is open, $x_{n'} \in V_j$ for all large n', say $n' > N_0$. Then $\{V_1, \ldots, V_{N_0}, V_j\}$ (where $x_{n'} \in V_n$) is a finite covering of K, giving a contradiction.

(2) Suppose K has the finite covering condition but is not compact. Let $\{x_n\}$ be an infinite (countable) sequence in K with no limit points. The sets $U_n \equiv \{x_n, x_{n+1}, \ldots\}$ are closed (U_n has no limit points) and $\cap_n U_n = \cap_n U_n = \phi$. Then U_n^c is open and $\{U_n^c\}$ forms an open cover of K. Then there exists a finite subcovering $\{U_{n'}^c\}$ of K with $\cap_{n'} U_{n'} = \phi$, which is certainly not true by construction. □

Appendix 2. The Banach Spaces $L^p(\mathbb{R}^n)$, $1 \leq p < \infty$

We present the basic theory of this family of Banach spaces. The reader can find any of the proofs not presented here in any standard text on real or functional analysis; see, for example, Reed and Simon, Volume I [RS1], Royden [Ro], or Rudin [R].

A2.1 The Definition of $L^p(\mathbb{R}^n)$, $1 \leq p < \infty$

The most important Banach spaces for us are the L^p-spaces, $1 \leq p < \infty$. The value $p = 2$ plays an especially important role, and we will describe its properties in greater detail later. There are two basic ways to obtain the L^p-spaces, and both are useful. The first, which we will sketch here, uses the idea of the completion of an NLVS. The second involves Lebesgue integration theory, so we will merely mention the results.

Definition A2.1. *The support of a function $f\colon \Omega \subset \mathbb{R}^n \to \mathbb{C}$, denoted by supp f, is the closed set that is the complement of the union of all open sets on which f vanishes.*

Problem A2.1. Prove that supp $f = \overline{\{x \in \Omega \mid f(x) \neq 0\}}$.

Definition A2.2. *The set of infinitely differentiable functions on $\mathbb{R}^n$ with bounded (and hence compact) supports is denoted by $C_0^\infty(\mathbb{R}^n)$. If $\Omega \subset \mathbb{R}^n$ and is open, then $C_0^\infty(\Omega)$ consists of all infinitely differentiable functions with compact supports in Ω.*

We note that it is easy to see that $C_0^\infty(\mathbb{R}^n)$ is an LVS over $\mathbb{C}$ or $\mathbb{R}$. We will always take LVS over $\mathbb{C}$ in this section. For each p, $1 \le p < \infty$, and any $f \in C_0^\infty(\Omega)$, define the nonnegative quantity

$$\|f\|_p \equiv \left[\int_\Omega |f(x)|^p \, dx\right]^{\frac{1}{p}}, \tag{A2.1}$$

which exists as a Riemann integral because supp f is compact and because f is bounded on its support. We see that $(C_0^\infty(\Omega), \|\cdot\|_p)$ is an NLVS.

Problem A2.2. Prove that for any open $\Omega \subset \mathbb{R}^n$, $(C_0^\infty(\Omega), \|\cdot\|_p)$ is an NLVS for $1 \le p < \infty$. (*Hint: Use the simple identity,* $|f+g|^p \le 2^p(|f|^p + |g|^p)$.)

Our goal is to associate with $X_p \equiv (C_0^\infty(\Omega), \|\cdot\|_p)$ a Banach space that contains X_p as a dense set. Recall that we have already seen in Remark A1.14 that X_p is not complete. There are Cauchy sequences in X_p whose limits do not exist in X_p. Hence, we want to enlarge X_p by adding to it "the limits of all Cauchy sequences in X_p." This will result in a "larger" space, called the *completion* of X_p. It is a Banach space that contains a copy of X_p as a dense subset. We will sketch the idea of the completion of an NLVS. In this manner, we obtain the Banach spaces in which we are interested.

Definition A2.3. *For any open $\Omega \subset \mathbb{R}^n$ and $1 \le p < \infty$, the Banach space $L^p(\Omega)$ is the completion of $(C_0^\infty(\Omega), \|\cdot\|_p)$.*

Let $(X, \|\cdot\|)$ be an NLVS that is not complete. We show how to obtain a Banach space $\bar{X}$ with a dense subset $\tilde{X}$, such that $\tilde{X}$ is isomorphic to X. That is, the sets $\tilde{X}$ and X are in one-to-one correspondence, and if $\tilde{x} \in \tilde{X}$ corresponds to $x \in X$, then $\|\tilde{x}\| = \|x\|$ (the norm on the left is the induced norm on $\bar{X}$).

Step 1. Consider the collection of all Cauchy sequences in X. Let $\alpha \equiv \{x_n\}$ and $\beta \equiv \{y_n\}$ denote any two Cauchy sequences. We say that two Cauchy sequences are *equivalent* if $\lim_{n\to\infty} \|x_n - y_n\| = 0$.

Problem A2.3. Show that (i) this defines an equivalence relation on the set of Cauchy sequences, and (ii) if $\{x_n\}$ is Cauchy, then $\lim_{n\to\infty} \|x_n\|$ exists.

Let $[\alpha]$ denote the equivalence class of α. We define a set $\bar{X}$ to be the set of all equivalence classes of Cauchy sequences in X. The set $\bar{X}$ has a distinguished subset $\tilde{X}$, which consists of all equivalence classes of constant sequences (i.e., if $x \in X$, then let $\alpha_x \equiv \{x_n = x\}$). We see that $[\alpha_x]$ consists of all Cauchy sequences converging to x. Hence, $\tilde{X}$ is isomorphic with X.

Problem A2.4. Show that $\bar{X}$ is an LVS, and verify that $\tilde{X}$ is isomorphic to X.

Step 2. We define a norm on $\bar{X}$ as follows. Let $[\alpha] \in \bar{X}$, and take $\{x_n\} \in [\alpha]$. Then define

$$\|[\alpha]\|_{\bar{X}} \equiv \lim_{n\to\infty} \|x_n\| .$$

Problem A2.5. Verify that $\|\cdot\|_{\bar{X}}$ is a norm on $\bar{X}$. (Be sure to check that it is "well defined," that is, independent of the representative chosen.)

Note that if $[\alpha_x] \in \tilde{X}$, then naturally

$$\|[\alpha_x]\|_{\bar{X}} = \|x\|,$$

so that the isomorphism between $\tilde{X}$ and X is isometric. We conclude that $\left(\bar{X}, \|\cdot\|_{\bar{X}}\right)$ is an NLVS with a subset $\tilde{X}$ isometric with X.

Step 3. We sketch the proof that $\left(\bar{X}, \|\cdot\|_{\bar{X}}\right)$ is complete. Suppose $\{[\alpha^N]\}$ is a Cauchy sequence in $\bar{X}$. Let $\{x_k^N\}$ be a representative. Then for $\epsilon > 0$ there exists $\mathcal{N}_\epsilon$ such that for $N, M > \mathcal{N}_\epsilon$, we have

$$\left\|[\alpha^N] - [\alpha^M]\right\|_{\bar{X}} = \lim_{i\to\infty} \|x_i^N - x_i^M\| < \frac{\epsilon}{2} .$$

Hence there exists $i_0 > 0$ such that $\|x_i^N - x_i^M\| < \epsilon/2$ for $i > i_0$. For $N > M$, choose an index $i_N > i_0$ for which $\|x_{i_N}^N - x_i^M\| < \epsilon/2$. We can choose the map $N \to i_N$ to be order-preserving (i.e., $P > Q \Rightarrow i_P > i_Q$). Since $\{x_i^M\}$ is Cauchy, there exists $i_1 > 0$ such that $j, k > i_1$ implies

$$\|x_j^M - x_k^M\| < \frac{\epsilon}{2} .$$

Choose $i_M > \max(i_0, i_1)$. We claim $\{x_{i_N}^N\} \subset X$ is a Cauchy sequence. This follows since

$$\|x_{i_N}^N - x_{i_M}^M\| \leq \|x_{i_N}^N - x_{i_M}^M\| + \|x_{i_N}^M - x_{i_M}^M\| < \epsilon,$$

by definition of i_N. Finally, construct $[\alpha] \equiv \left[\{x_{i_N}^N\}\right] \in \bar{X}$. Then we have for N sufficiently large,

$$\begin{aligned} \left\|[\alpha^N] - [\alpha]\right\|_{\bar{X}} &= \lim_{k\to\infty} \|x_k^N - x_{i_k}^K\| \\ &\leq \lim_{k\to\infty} \left(\|x_{i_N}^N - x_{i_k}^k\| + \|x_k^N - x_{i_N}^N\|\right) \\ &\leq \frac{3\epsilon}{2} . \end{aligned}$$

Problem A2.6. Check that the calculations in Step 3 are independent of the representative of the equivalence class chosen.

Consequently, we know that there exists a Banach space $\bar{X}_p$ that contains a dense isometric copy of $(C_0^\infty(\Omega), \|\cdot\|_p)$. A natural question that arises is: What is the nature of the objects in $\bar{X}_p$? The best way to answer this is by means of the Lebesgue theory of integration. We wish to avoid this discussion, and so we refer the reader to any standard text, for example, Royden [Ro]. However, for the reader familiar with this theory, an element of $\bar{X}_p$ is an equivalence class of measurable functions on Ω having the property that $|f|^p$ is Lebesgue integrable on Ω with $\int_\Omega |f|^p < \infty$. Two measurable functions f and g on Ω are *equivalent* if they differ on a set of Lebesgue measure zero. Up to this equivalence, then, we can interpret elements of $L^p(\Omega)$ as functions such that $|f|^p$ is finitely integrable over Ω.

A2.2 Important Properties of L^p-Spaces

We establish here several basic and important properties of L^p-spaces, which we use throughout the chapters.

1. Density results

For $\Omega \subset \mathbb{R}^n$ open, we have seen that the linear vector space of functions $C_0^\infty(\Omega)$ is dense in $L^p(\Omega)$. Since $L^p(\Omega)$ is a Banach space, this means that for any $f \in L^p(\Omega)$ (where f is understood in the sense described in Section A2.1) and for any $\epsilon > 0$, we can find $g \in C_0^\infty(\Omega)$ such that $\|f - g\|_p < \epsilon$. Consequently, in most cases, it suffices to prove a property for smooth functions. The same density result applies for other classes of functions. We list some of the common classes here:

$$C_0(\Omega) \equiv \text{all continuous functions on } \Omega \text{ with bounded supports;}$$

$$B_0(\Omega) \equiv \text{all bounded functions on } \Omega \text{ with bounded supports.}$$

Problem A2.7. Prove that these two classes of functions are dense in $L^p(\Omega)$.

When $\Omega = \mathbb{R}^n$, we define the Schwartz class functions $\mathcal{S}(\mathbb{R}^n)$ as follows:

$$\mathcal{S}(\mathbb{R}^n) \equiv \{f \in C^\infty(\mathbb{R}^n) \mid \text{if } g \text{ is any partial derivative of } f, \text{ then for any polynomial } p, \lim_{\|x\|\to\infty} |g(x)p(x)| = 0\}.$$

The Schwartz functions form a particularly nice class of functions. By the same argument as in Problem A2.7, these functions form a dense set in $L^p(\mathbb{R}^n)$.

2. *The Hölder Inequality*

It is easy to see that the product of two L^p-functions need not be in L^p. For example, take the space $L^1(\mathbb{R})$ and consider the function

$$f(x) = \begin{cases} |x|^{-\frac{1}{2}}, & |x| < 1, \\ 0, & |x| \geq 1 . \end{cases}$$

Then, $f \in L^1(\mathbb{R})$, but $f^2 \notin L^1(\mathbb{R})$. We do, however, have the possibility that fg belongs to some $L^{p'}$-space if f and g belong to some other L^s- and L^t-spaces.

Theorem A2.4 (The Hölder inequality). *Let $p, q, r \in \mathbb{R}^+$ be such that $1/p+1/q = 1/r$. If $f \in L^p$ and $g \in L^q$, then $fg \in L^r$ and*

$$\|fg\|_r \leq \|f\|_p \, \|g\|_q \, .$$

Proof.

(1) We first prove the following inequality for any $a, b \in \mathbb{R}_+$ and any $p, q \in \mathbb{R}_+$ such that $1/p + 1/q = 1$:

$$ab \leq p^{-1}a^p + q^{-1}b^q \, . \tag{A2.2}$$

Consider the function

$$a = F(b) = b^{q-1}$$

or, equivalently,

$$b = F^{-1}(a) = a^{p-1},$$

where we use the fact that $(p-1)(q-1) = 1$. We consider the rectangle $[0, b_0] \times [0, a_0]$ in the ba-plane, where $a_0, b_0 > 0$ are arbitrary. Then the curve $a = F(b)$ divides the rectangle into two regions, I and II. The area of region I, which lies below the curve $a = F(b)$, is

$$A_{\mathrm{I}} = \int_0^{b_0} F(b)db = q^{-1}b_0^q \, ,$$

whereas for the complementary region,

$$A_{\mathrm{II}} = \int_0^{a_0} F^{-1}(a)da = p^{-1}a_0^p \, .$$

It is easy to see that we always have

$$a_0 b_0 \leq q^{-1}b_0^q + p^{-1}a_0^p \, ,$$

which is just (A2.2).

(2) We next consider $1/p + 1/q = 1/r$. Setting a^r and b^r into (A2.2), we obtain

$$(ab)^r \le p^{-1}a^{rp} + q^{-1}b^{rq},$$

for $1/p + 1/q = 1$. Now set $p' \equiv rp$ and $q' = rq$ to obtain

$$(ab)^r \le r(p'^{-1}a^{p'} + q'^{-1}b^{q'}), \tag{A2.3}$$

for $1/p' + 1/q' = 1/r$.

(3) If either f or g vanishes almost everywhere, the inequality is trivial. So suppose $f \in L^p$ and $g \in L^q$ are nonzero. Let $\hat{f} \equiv f\|f\|_p^{-1}$ and $\hat{g} \equiv g\|g\|_q^{-1}$. Substituting these into (A2.3) and integrating, we obtain

$$\begin{aligned} \|\hat{f}\hat{g}\|_r^r &\le rp^{-1}\|\hat{f}\|_p^p + rq^{-1}\|\hat{g}\|_q^q \\ &\le r(p^{-1} + q^{-1}) \le 1. \end{aligned}$$

By the definition of $\hat{f}$ and $\hat{g}$, we see that this is simply the desired inequality. □

Proposition A2.5. *Let $f \in L^p$, $1 \le p < \infty$. Then*

$$\|f\|_p = \sup_{g \in L^q} \left\{ \|fg\|_1 \, \|g\|_q^{-1} \mid g \neq 0 \text{ and } \frac{1}{p} + \frac{1}{q} = 1 \right\}. \tag{A2.4}$$

Proof. The Hölder inequality with $r = 1$ implies that the right side of (A2.4) is bounded above by $\|f\|_p$. Assume $f \neq 0$. Now let $g \equiv c|f|^{p-1}$. By a simple calculation,

$$\|fg\|_1 = c\|f\|_p^p,$$

so if we take $c \equiv \|f\|_p^{1-p}$, we get equality in (A2.4). Finally, we check that for such $g \in L^q$,

$$\begin{aligned} \|g\|_q &= \|f\|_p^{1-p} \left[\int |f(x)|^{q(p-1)} \, dx \right]^{\frac{1}{q}} \\ &= \|f\|_p^{1-p} \, \|f\|_p^{p/q} = 1 \end{aligned}$$

since $(p-1) = p/q$. □

3. The Minkowski Inequality

We used a basic inequality in Problem A2.2 to show that L^p is an LVS. This car be refined so that the constant does not depend on p.

Theorem A2.6 (The Minkowski inequality). *Let* $1 \leq p < \infty$ *and* $f, g \in L^p$. *Then*

$$\|f + g\|_p \leq \|f\|_p + \|g\|_p .$$

Proof. The case $p = 1$ is obvious. We use Proposition A2.5 to obtain

$$\begin{aligned} \|f+g\|_p &= \sup_{h \in L^q} \left\{ \|(f+g)h\|_1 \ \|h\|_q^{-1} \mid h \neq 0 \text{ and } \frac{1}{p} + \frac{1}{q} = 1 \right\} \\ &\leq \sup_{h \in L^q} \left\{ (\|fh\|_1 + \|gh\|_1) \|h\|_q^{-1} \right\} \\ &\leq \|f\|_p + \|g\|_p . \end{aligned}$$

□

Problem A2.8. Formulate and prove the Minkowski and Hölder inequalities for l^p, $1 \leq p < \infty$ (see Examples A1.2 and Problem A1.1).

4. Lebesgue Dominated Convergence

This is one of the most useful results of Lebesgue integration theory. It is, of course, applicable to Riemann integrable functions. Suppose we have a sequence of functions f_n in $L^p(\mathbb{R}^n)$ such that f_n converges pointwise to f almost everywhere. We want to know if $f_n \to f$ in $L^p(\mathbb{R}^n)$.

Theorem A2.7 (Lebesgue dominated convergence). *Suppose* $f_n \in L^p(\mathbb{R}^n)$ *and* $f_n \to f$ *almost everywhere. If there exists* $g \in L^p(\mathbb{R}^n)$ *such that* $|f_n|^p \leq |g|^p$ *almost everywhere, then* $f \in L^p$ *and* $f_n \to f$ *in* L^p.

We need one other result about sequences of functions that converge in the L^p-norm.

Theorem A2.8. *Suppose that* $\{f_n\}$ *is a sequence of functions in* $L^1(\mathbb{R}^n)$ *that converges to* f *in the* L^1*-norm. Then, there is a subsequence* $\{f_{n(k)}\}$ *that converges pointwise almost everywhere to* f.

We refer the reader to [RS1] or [Ro] for the proof.

Appendix 3. Linear Operators on Banach Spaces

A3.1 Linear Operators

Definition A3.1. *Let X and Y be LVS. A linear operator A from X to Y is a linear map defined on a linear subspace $D(A) \subset X$, called the domain of A, into Y.*

We write $A : D(A) \to Y$. A linear map has the property that for $\lambda \in \mathbb{C}$ (or $\mathbb{R}$) and $f, g \in D(A)$,

$$A(\lambda f + g) = \lambda Af + Ag.$$

Note that $D(A)$ is a linear space by definition, so $f, g \in D(A)$ implies that $f + \lambda g \in D(A)$.

Examples A3.2.

(1) Let $X = Y = \mathbb{R}^n$. Any linear transformation $A : \mathbb{R}^n \to \mathbb{R}^n$ (i.e., $D(A) = \mathbb{R}^n$) can be written in the form $(Ax)_i = \sum_{j=1}^{n} A_{ij} x_j$, for some n^2 quantities $A_{ij} \in \mathbb{R}$. This matrix representation of A depends on the choice of basis for $\mathbb{R}^n$. The numbers x_i are the coefficients of x relative to this basis.

(2) Let $X = Y = C([0, 1])$, and let $a \in X$. Define $A : X \to X$ by $(Af)(x) = a(x)f(x)$, $f \in X$. Then A is a linear operator with $D(A) = X$. The operator A is "multiplication by the function a" and is called a *multiplication operator*.

(3) Let $X = Y = C([0, 1])$ and let $D(A) = C^1([0, 1])$, the LVS of continuous functions that are differentiable and whose first derivatives are continuous. Note that $D(A) \neq X$. Define an operator A, for $f \in D(A)$, by $(Af)(x) = (df/dx)(x)$. Then A: $C^1([0, 1]) \to C([0, 1])$ is a linear operator.

(4) Let $X = Y = C([0, 1])$ and let $K(s, t)$ be continuous in s and t on $[0, 1] \times [0, 1]$. Let $f \in X$, and define $(Af)(t) = \int_0^1 K(t, s) f(s) ds$. It is easy to check that this integral defines an element of X for each $f \in X$. Hence, we may take $D(A) = X$, and A is a linear operator.

(5) Let $X = C([0, 1])$ and $Y = \mathbb{R}$. For each $f \in X$, define $Af = \int_0^1 f(s) ds$. Then $A: X \to Y$ with $D(A) = X$ is a linear operator.

(6) Let $X = Y = C([0, 1])$. The map $(Af)(x) = f^2(x)$, defined on all of X, *is not linear.* This is an example of a nonlinear operator.

Problem A3.1. Verify all the statements in Examples A3.2.

Definition A3.3. *A linear operator* $A : D(A) \subset X \to Y$ *is densely defined if* $D(A)$ *is dense in* X.

Remark A3.4. Note that the definition of a linear operator includes both the domain and the action of the operator on elements in the domain. If the domain is changed, the operator is also changed. All the operators in Examples A3.2 are densely defined. A linear operator that is not densely defined has many undesirable properties (for example, we shall see that it does not have a uniquely defined adjoint). For this reason, we will work only with densely defined operators. Moreover, we will assume that X is an NLVS. If $D(A) \subset X$ is not dense, we will replace X by $\overline{D(A)} \equiv X_1$, the closure of $D(A)$ in X, which is a Banach space. We then consider the operator A with dense domain $D(A)$ in X_1.

It is possible to define algebraic operations on the family of linear operators from X to Y. Assume X and Y have the same field of scalars F. In particular, we have

(1) *Numerical multiplication*: If $\lambda \in F$, and $x \in D(A)$, we define λA by $(\lambda A)(x) = \lambda(Ax)$. Thus, if $\lambda \neq 0$, $D(\lambda A) = D(A)$, and if $\lambda = 0$, λA extends to the zero operator on X.

(2) *Addition*: Let A, B be two linear operators on X into Y, and suppose $f \in D(A) \cap D(B)$. We define $A + B$ on f by $(A + B)f = Af + Bf$. In some cases it may happen that $D(A) \cap D(B) = \{0\}$, so we cannot define $A + B$. In general, $D(A) \cap D(B)$ is a nonempty subset of $D(A)$ and $D(B)$ (although it may not be dense). For example, consider the operator A defined in part (3) of Examples A3.2. Let $B \equiv d^2/dx^2$, with $D(B) = C^2([0, 1])$. Then, $D(A + B) = D(B) \subset D(A)$ and is dense.

(3) *Multiplication*: Let A, B be linear operators from X into Y, and let $x \in D(B)$ be such that $Bx \in D(A)$. Then we can define AB on the set of all such x by $(AB)x = A(Bx)$. Note that if A and B are everywhere defined operators from X into X, then AB and BA are everywhere defined on X.

A3.2 Continuity and Boundedness of Linear Operators

We will now explicitly assume X and Y are NLVS, and we will write $A : X \to Y$ even when A is defined only on $D(A) \subset X$. Moreover, by operator we will always mean *linear operator.*

Definition A3.5. *An operator $A : X \to Y$ is continuous if for each convergent sequence $\{x_n\} \subset D(A)$ with $\lim_{n\to\infty} x_n = x \in D(A)$, we have $\lim_{n\to\infty} Ax_n = A(\lim_{n\to\infty} x_n) = Ax$.*

Problem A3.2. Show that A is continuous if for each sequence $\{x_n\} \subset D(A)$ with $\lim_{n\to\infty} x_n = 0$, we have $\lim_{n\to\infty} Ax_n = 0$.

Example A3.6. With reference to part (2) of Example A3.2, the operator A, multiplication by $a \in C([0, 1])$, is continuous on $X = C([0, 1])$ with *any* norm on X as given in Example A1.4 (2). This follows from the fact that a is bounded: Let $\sup_{s\in[0,1]} |a(s)| = \|a\|_\infty$. Then, it follows that

$$\|Af\|_\infty \leq \|a\|_\infty \|f\|_\infty$$

and

$$\|Af\|_p \leq \|a\|_\infty \|f\|_p, \quad 1 \leq p < \infty.$$

So if $f_n \to 0$ in any of these norms, $Af_n \to 0$.

Definition A3.7. *An operator $A : X \to Y$ is bounded if there exists a $K > 0$ such that for any $f \in D(A)$,*

$$\|Af\|_Y \leq K\|f\|_X.$$

Remarks A3.8.

(1) If should be noted that both continuity and boundedness of an operator depend on the norms of X and Y.

(2) If both X and Y are finite-dimensional, any operator $A : X \to Y$ is bounded and continuous.

Proposition A3.9. *Let X be an NLVS.*

(1) *Any bounded operator on X is continuous.*

(2) *Any bounded operator $A : X \to Y$, Y a Banach space, can be extended to all of X, that is, there is a unique bounded operator $\tilde{A}$ with $D(\tilde{A}) = X$ such that $\tilde{A}f = Af$, $f \in D(A)$.*

Proof.

(1) Let $\{x_n\} \subset D(A)$ be a sequence such that $x_n \to 0$. Since A is bounded, $\|Ax_n\| \leq K\|x_n\| \to 0$, so $Ax_n \to 0$.

(2) Let $x \in X \setminus D(A)$. Since $D(A)$ is dense, there is a sequence $\{x_n\} \subset D(A)$ such that $x_n \to x$. Because A is bounded, it follows that $\{Ax_n\}$ is Cauchy in Y. Since Y is a Banach space, the sequence $\{Ax_n\}$ converges to some $y \in Y$. Now define $\tilde{A}$ on X by $\tilde{A}x \equiv Ax$, $x \in D(A)$, and $\tilde{A}x \equiv \lim_{n\to\infty} Ax_n$, for $x = \lim_{n\to\infty} x_n \notin D(A)$ and $\{x_n\} \subset D(A)$. □

Problem A3.3. Complete the proof of Proposition A3.9 by showing that $\tilde{A}$ is well defined (i.e., independent of the sequence), that $\tilde{A}$ is linear, that $\tilde{A}$ is unique, and that $\tilde{A}$ is bounded.

Theorem A3.10. *Let $A : X \to Y$ be a linear operator. Then the following two statements are equivalent:*

(1) *A is continuous;*

(2) *A is bounded.*

Proof.

(2) ⇒ (1) If A is bounded, we observed that A is continuous; in fact, we can extend A to a bounded operator on X and the extension (which is bounded) is continuous.

(1) ⇒ (2) Suppose A is continuous but unbounded. This means the exists a sequence $\{x_n\}$ in $D(A)$, with $\|x_n\| < M$, such that $\|Ax_n\| \geq n\|x_n\|$. Let $y_n \equiv x_n\|Ax_n\|^{-1}$. Then $\{y_n\} \subset D(A)$, $\|y_n\| \leq n^{-1}$, so $y_n \to 0$, but $\|Ay_n\| = 1$; that is, $\lim_{n\to\infty} Ay_n \neq 0$ but $y_n \to 0$. This contradicts the continuity of A, so A must be bounded. □

Definition A3.11. *Let $A : X \to Y$ be bounded. Then $\|A\|$, the norm of A, is defined by*

$$\|A\| \equiv \inf\{K \geq 0 \,|\, \|Ax\|_Y \leq K\|x\|_X, \text{ for all } x \in X\},$$

which is equivalent to

$$\|A\| \equiv \sup_{\substack{x\in X \\ \|x\|\neq 0}} \|x\|_X^{-1}\|Ax\|_Y.$$

Problem A3.4. Prove the equivalence of these two definitions.

Note that we have assumed that if A is bounded, then $D(A) = X$, as follows from Proposition A3.9. Moreover, we have

$$\|Ax\| \leq \|A\| \, \|x\|, \quad x \in X.$$

Examples A3.12 (with reference to Examples A3.2).

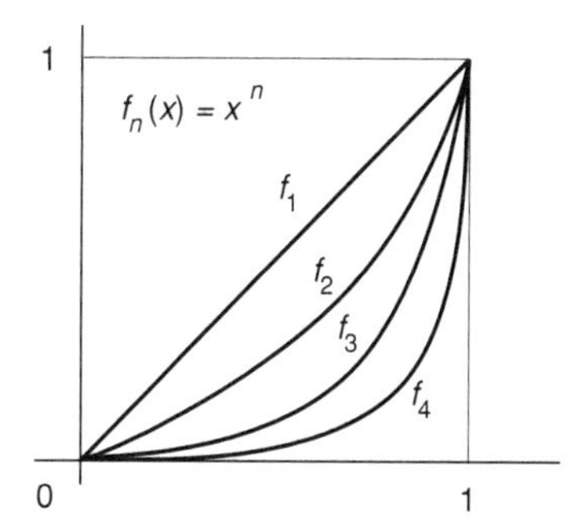

FIGURE A3.1 The functions f_n.

(1) All linear operators $A : \mathbb{R}^n \to \mathbb{R}^n$ are bounded. Note that there are actually many norms on the set of $n \times n$ matrices. The most common are $\|A\|_\infty = \sup_{i,j} |a_{ij}|$, and $\|A\|_{\text{tr}} = \text{Tr}(A^*A)^{1/2}$. The norm defined in Definition A3.11 is the operator norm. This norm is distinguished by the fact that it satisfies $\|A^*A\| = \|A\|^2$, where $A^* = \bar{A}^T$ and A^T is the transpose of A.

(2) Let $Af = af$, with $a, f \in X \equiv C([0, 1])$. The multiplication operator A is bounded in all norms on $C([0, 1])$ (although this space is a Banach space only in the sup-norm). In the Banach space $(C([0, 1]), \|\cdot\|_\infty)$, we have $\|A\| = \|a\|_\infty$. To see this, note that for any $f \in X$, $\|Af\|_\infty = \|af\|_\infty \leq \|a\|_\infty \|f\|_\infty$, and so $\|A\| \leq \|a\|_\infty$. Upon choosing $f = 1 \in X$, we have $\|Af\|_\infty = \|a\|_\infty$, so $\|a\|_\infty \leq \|A\|$.

(3) Let $Af = f'$, $f \in C^1([0, 1])$, and $X = C([0, 1])$. Then A is unbounded. To see this, we can, for example, note by Theorem A3.9 that if A was bounded we could extend it to all X, but there are plenty of functions in X for which f' does not exist! Alternately, consider a sequence of functions f_n whose graphs are given in Figure A3.1. Then each $f_n \in C^1([0, 1])$, $\|f_n\| = 1$, but $\|f_n'\|_\infty \to \infty$ as $n \to \infty$. Consequently, there exists no finite $K > 0$ such that $\|Af_n\| \leq K \|f_n\|_\infty = K$, for all n.

(4) Let $Af(t) = \int_0^1 K(t, s)f(s)ds$, where $X = C([0, 1])$ and K is continuous on $[0, 1] \times [0, 1]$. In the sup-norm on X, $\|A\| = \sup_{s \in [0,1]} \int_0^1 |K(s, t)|dt$.

Problem A3.5. Compute the bound on A given in Example A3.12 (4).

We now return to the algebraic structure on operators introduced in Section A3.1. There, this structure was complicated by the fact that the domains depended on the operators. In light of Proposition A3.9, we do not encounter this problem if we restrict ourselves to bounded operators since we can then take the domain to be all of X.

Definition A3.13. *Let X and Y be two Banach spaces. We denote by $\mathcal{L}(X, Y)$ the family of all bounded linear operators from X to Y.*

Let us make a few remarks. First, in this definition, the space X need not be a Banach space; an NLVS will do. Second, we will always assume that a bounded operator on X has been extended so that its domain is all of X. Finally, if $Y = X$, we write $\mathcal{L}(X)$ for $\mathcal{L}(X, X)$).

Proposition A3.14. *If $A, B \in \mathcal{L}(X, Y)$ and $c \in \mathbb{C}$ (or $\mathbb{R}$), then cA and $A + B \in \mathcal{L}(X, Y)$ and*

$$\begin{aligned} \|cA\| &= |c|\|A\|, \\ \|A + B\| &\leq \|A\| + \|B\|. \end{aligned}$$

Moreover, if $A \in \mathcal{L}(X, Y)$ and $B \in \mathcal{L}(Y, Z)$, then $BA \in \mathcal{L}(X, Y)$ and

$$\|AB\| \leq \|A\| \, \|B\|.$$

Problem A3.6. Prove Proposition A3.14.

Corollary A3.15. *$\mathcal{L}(X, Y)$ is an LVS over $\mathbb{C}$ (or $\mathbb{R}$).*

It follows from the last two results that $\| \cdot \|$, as defined in Definition A3.11, is a norm on $\mathcal{L}(X, Y)$. This follows from Proposition A3.14 and the facts that $\|A\| \geq 0$ and $\|A\| = 0$ if and only if $A = 0$. This shows that $\mathcal{L}(X, Y)$ is an NLVS. Hence we can ask if $\mathcal{L}(X, Y)$ is complete in this norm.

Theorem A3.16. *Let X be an NLVS and Y be a Banach space. Then $\mathcal{L}(X, Y)$ is a Banach space with the norm $\| \cdot \|$ given in Definition A3.11.*

Proof. Let $\{A_n\}$ be a Cauchy sequence in $\mathcal{L}(X, Y)$. We must show that there exist $A \in \mathcal{L}(X, Y)$ such that $A = \lim_{n \to \infty} A_n$. Take any $x \in X$ and consider the sequence $\{A_n x\}$. Then we have

$$\|A_n x - A_m x\| \leq \|A_n - A_m\| \, \|x\|,$$

by linearity and boundedness. Hence, $\{A_n x\}$ is a Cauchy sequence in Y, and as Y is a Banach space, it converges. Let $y = \lim_{n \to \infty} A_n x$, and define $A : X \to Y$ by $Ax = y$.

Problem A3.7. Continue the proof of Theorem A3.16. Show that A is well defined on X, linear, and bounded. (*Hint*: Use the fact that $|\|A_n\| - \|A_m\|| \leq \|A_n - A_m\|$, so $\{ \|A_n\| \}$ is Cauchy and hence converges. For boundedness of A, use

$$\|Ax\| \leq \lim_{n \to \infty} \|A_n x\| \leq \left(\lim_{n \to \infty} \|A_n\| \right) \|x\|.)$$

Completion of the proof. We show that $\lim_{n \to \infty} A_n = A$:

$$\|A - A_n\| = \sup_{x \neq 0} \|(A_n - A)x\| \, \|x\|^{-1},$$

and

$$\begin{aligned} \|(A_n - A)x\| = \|A_n x - Ax\| &\leq \lim_{m\to\infty} \|A_n x - A_m x\| \\ &\leq \left(\lim_{m\to\infty} \|A_n - A_m\|\right) \|x\|. \end{aligned}$$

So, by choosing n suitably large and using the Cauchy property of $\{A_n\}$, we have $\lim_{n\to\infty} \|A_n - A\| = 0$. □

A3.3 The Graph of an Operator and Closure

We introduce a tool useful in the study of operators.

Definition A3.17. *Let $A : X \to Y$ with domain $D(A) \subset X$. The graph of A, denoted $\Gamma(A)$, is the subset of $X \times Y \equiv \{(x, y) | x \in X, y \in Y\}$ given by*

$$\Gamma(A) \equiv \{ (x, Ax) \mid x \in D(A) \}.$$

Problem A3.8. Show that $\Gamma \subset X \times Y$ is a linear subspace (we assume that X and Y have the same field of scalars). Note that $X \times Y$ is an LVS under componentwise addition and scalar multiplication.

We denote by $\{x\} \times Y \equiv \{(x, y) | y \in Y\} \subset X \times Y$. One may ask when a subset $\Gamma \subset X \times Y$ is the graph of some linear operator $A : X \to Y$. Of course, Γ must be a subspace of $X \times Y$. A convenient condition on Γ is the following.

Proposition A3.18. *A linear subspace $\Gamma \subset X \times Y$ is the graph of some linear operator A if and only if*

$$(\{0\} \times Y) \cap \Gamma = \{(0, 0)\}. \tag{A3.1}$$

Proof.

(1) If there exists a linear operator A such that $\Gamma = \Gamma(A)$ and $(0, y) \in \Gamma$, then $y = A(0) = 0$, so condition (A3.1) holds.

(2) Assume (A3.1) holds. Let $D(A) = \{x \mid (x, y) \in \Gamma, \text{for some } y \in Y\}$. For each $x \in D(A)$, define $Ax = y$, where $(x, y) \in \Gamma$. We want to show that A is well defined and linear. Suppose there exist $y, y' \in Y$ such that (x, y) and (x, y') are in Γ. Since Γ is linear, $(x, y) - (x, y') = (0, y - y') \in \Gamma$ and by (A3.1), $y = y'$. Hence, A is well defined. Second, if $x_1, x_2 \in D(A)$ and $\lambda \in \mathbb{C}$, then $(x_1 + x_2, Ax_1 + Ax_2) \in \Gamma$. This implies that $(x_1 + x_2, A(x_1 + x_2)) \in \Gamma$, so $A(x_1 + x_2) = Ax_2 + Ax_2$. Since $\lambda(x_1, Ax_1) = (\lambda x_1, \lambda Ax_1)$, $A(\lambda x_1) = \lambda Ax_1$. Hence, A is linear. □

Definition A3.19. *Let X, Y be NLVS. For each $(x, y) \in X \times Y$, define*

$$\|(x, y)\| = \|x\|_X + \|y\|_Y.$$

Lemma A3.20. $(X \times Y, \|\cdot\|)$ *is an NLVS.*

Proof. We have seen that $X \times Y$ is an LVS. For the norm, we must check the conditions of Definition A1.3:

(i) $\|(x, y)\| \geq 0$ since $\|x\|_X \geq 0$, $\|y\|_Y \geq 0$ and it follows that $\|(x, y)\| = 0$ if and only if $x = 0 = y$.

(ii) $\|\lambda(x, y)\| = \|(\lambda x, \lambda y)\| = |\lambda|(\|x\|_X + \|y\|_Y) = |\lambda|\|(x, y)\|$.

(iii)
$$\begin{aligned} \|(x, y) + (x', y')\| &= \|(x + x', y + y')\| \\ &= \|x + x'\|_X + \|y + y'\|_Y \\ &\leq (\|x\|_X + \|y\|_Y) + (\|x'\|_X + \|y'\|_Y) \\ &= \|(x, y)\| + \|(x', y')\|. \end{aligned}$$

□

Problem A3.9. Prove that the map $X \times Y \to \mathbb{R}$ defined by

$$\|(x, y)\|_E \equiv [\, \|x\|_X^2 + \|y\|_Y^2 \,]^{\frac{1}{2}}$$

is a norm on $X \times Y$. Show that the topology on $X \times Y$ determined by $\|\cdot\|_E$ is *equivalent* to the topology induced by the norm in Definition A3.19. This means that an open set in one topology contains an open set in the other.

Problem A3.10. Show that if $A \in \mathcal{L}(X, Y)$, then $\Gamma(A) \subset X \times Y$ is closed. (We will see later that the converse is not true). (*Hint*: $\Gamma(A)$ is closed if for any convergent sequence $\{(x_n, Ax_n)\}$ in $\Gamma(A)$, $\lim_{n\to\infty}(x_n, Ax_n) = (x, y) \in \Gamma(A)$, that is, $x \in D(A)$ and $Ax = y$.)

Definition A3.21. *Let A be an operator on X with $D(A)$. An operator $\bar{A}$ with domain $D(\bar{A})$ is called the closure of A if $\Gamma(\bar{A}) = \overline{\Gamma(A)}$. An operator A is said to be closable if it has a closure.*

Remark A3.22. Suppose A is a linear operator with graph $\Gamma(A)$. It may happen (and does!) that $\overline{\Gamma(A)}$ is *not* the graph of any operator (precisely how this may happen is given in Proposition A3.18). In this case A is not closable. Suppose, however, that A is closable. Then the closure $\bar{A}$ is unique and satisfies $\bar{A}x = Ax$, $x \in D(A)$. This follows directly from Definition A3.21.

An important connection between everywhere defined operators (i.e., $D(A) = X$) and boundedness is given by the next result.

Theorem A3.23. (Closed graph theorem). *Let X, Y be Banach spaces. Let A be defined on X (i.e., $D(A) = X$). If A is closed (i.e., $A = \bar{A}$), then A is bounded.*

The proof of this is given in Reed and Simon, Volume I [RS1]. As the proof requires more machinery, the Baire category theorem, we will not give it here

Note that there are everywhere defined, unbounded operators. By Theorem A3.23, these will not be closed!

Examples A3.24.

(1) Let $X = (C([0, 1]), \|\cdot\|_\infty)$ and $A = d/dx$ with $D(A) = C^1([0, 1])$, as in Example A3.2 (3). We claim that $(A, D(A))$ is closed. To see this, let (f_n, Af_n) be a convergent sequence in $\Gamma(A)$. Then $f_n \to f$ in X and $Af_n = f_n' \to g$ in X. By a standard result on uniform convergence, $f \in C^1([0, 1])$ and $f' = g$. Consequently, $f \in D(A)$ and $(f, Af) \in \Gamma(A)$. We can modify this example slightly to obtain a closable operator whose closure is $(A, D(A))$. Simply take $A_1 = d/dx$ and $D(A_1) = C^2([0, 1])$.

Problem A3.11. Prove that $(A_1, D(A_1))$ is closable and that its closure is $(A, D(A))$.

(2) Here is an example of an operator that is *not closable*. Let $X = L^2(\mathbb{R})$, and choose some $f_0 \in X$ with $\|f_0\| = 1$. Define B, on $D(B) = C_0^\infty(\mathbb{R})$, by $Bf = f(0)f_0$. Then $(B, D(B))$ is a densely defined operator and is not closable. Consider, for example, a function $h \in C_0^\infty([-1, 1])$, $h \geq 0$, $h(0) = 1$, and $\int h = 1$. Let $h_n(x) \equiv h(nx)$. Then $\|h_n\| = 1/n \to 0$ as $n \to \infty$. However, $Bh_n = h_n(0)f_0 = f_0$, so $\|Bh_n\| = 1$. This shows that $(0, f_0) \in \overline{\Gamma(B)}$, and so it cannot be the graph of an operator, by Proposition A3.18.

A3.4 Inverses of Linear Operators

Let X and Y be Banach spaces (over $\mathbb{C}$ for simplicity), and let A be a linear operator from X to Y. We associate with A three linear subspaces:

$D(A) \subset X$, *domain of* A;

$\text{Ran}(A) \subset Y$, *range of* A, $\text{Ran}(A) \equiv \{y \in Y \mid y = Ax, \text{ some } x \in D(A)\}$;

$\ker A \subset X$, *kernel of* $\ker A \equiv \{x \in D(A) \mid Ax = 0\}$.

Definition A3.25. *An operator $A^{-1} : Ran(A) \to D(A)$ is called the inverse of A if $A^{-1}A = id_{D(A)}$ (the identity map on $D(A)$) and $AA^{-1} = id_{\text{Ran}(A)}$.*

Theorem A3.26. *An operator A has an inverse if and only if*

$$\ker A = \{0\}.$$

Proof.

(1) $\Leftarrow$ If $\ker A = \{0\}$, then for any $y \in \text{Ran}(A)$, there exists a unique $x \in D(A)$ such that $Ax = y$, for suppose $x_1, x_2 \in D(A)$ and $Ax_1 = y = Ax_2$. Then,

by linearity, $Ax_1 - Ax_2 = A(x_1 - x_2) = 0$, so $x_1 - x_2 \in \ker A$. But $\ker A = \{0\}$, whence $x_1 = x_2$. Because of this we can define an operator $A^{-1} : \mathrm{Ran}(A) \to D(A)$ by $A^{-1}y \equiv x$, where $Ax = y$. The operator A^{-1} is well defined and linear. Moreover, $A^{-1}y = A^{-1}(Ax) = x$, so $A^{-1}A = id_{D(A)}$, and for $y \in \mathrm{Ran}(A)$, $AA^{-1}y = Ax = y$, so $AA^{-1} = id_{\mathrm{Ran}(A)}$.

(2) $\Rightarrow$ Suppose $A^{-1} : \mathrm{Ran}(A) \to D(A)$ exists. If $x \in \ker A$, then $x = A^{-1}Ax = 0$, so $\ker A = \{0\}$. □

As this theorem shows, the condition for the existence of A^{-1} is simply that $\ker A = \{0\}$. This guarantees that the inverse map is a function. However, we have little information about A^{-1}. In particular, we would like A^{-1} to be bounded on all of Y.

Definition A3.27. *A linear operator A is invertible if A has a bounded inverse defined on all of Y.*

Problem A3.12. Prove that if A is invertible, then A^{-1} is unique.

Example A3.28. Let $X = C([0, 1])$, and define A by

$$(Af)(t) = \int_0^t f(s)ds, \quad f \in X.$$

This operator A has an inverse by the fundamental theorem of calculus: $A^{-1} \equiv d/dt$. But A is *not invertible*, since A^{-1} is not bounded!
Warning: According to our definition, A may have an inverse but *not* be invertible.

Theorem A3.29. *Let A and B be bounded, invertible operators. Then AB is invertible and $(AB)^{-1} = B^{-1}A^{-1}$.*

Problem A3.13. Prove Theorem A3.29.

Theorem A3.30. *Let T be a bounded operator with $\|T\| < 1$. Then $1 - T$ is invertible, and the inverse is given by an absolutely convergent Neumann series:*

$$(1 - T)^{-1} = \sum_{n=0}^{\infty} T^n,$$

that is, $\lim_{N\to\infty} \sum_{k=0}^{N} T^k$ converges in norm to $(1 - T)^{-1}$.

Proof. Since $\sum_{k=0}^{N} \|T^k\| < \sum_{k=0}^{N} \|T\|^k$, and the series $\sum_k \|T\|^k$ converges (as $\|T\| < 1$, it is simply a geometric series), the sequence of bounded operators $\sum_{k=0}^{N} T^k$ is norm Cauchy. Since Y is a Banach space, this Cauchy sequence converges to a bounded operator. Now we compute

$$(1 - T)\sum_{k=0}^{\infty} T^k = \sum_{k=0}^{\infty}(T^k - T^{k+1}) = 1,$$

where the manipulations are justified by the norm convergence of the power series. Similarly,

$$\left(\sum_{k=0}^{\infty} T^k\right)(1 - T) = 1,$$

so the series defines $(1 - T)^{-1}$. □

Theorem A3.31. *Let A be invertible, and let B be bounded with $\|B\| < \|A^{-1}\|^{-1}$. Then $A + B$ is invertible and*

$$(A + B)^{-1} = (1 + A^{-1}B)^{-1}A^{-1} = A^{-1}(1 + BA^{-1})^{-1}.$$

Proof. Write

$$A + B = A(1 + A^{-1}B),$$

and note that $\|A^{-1}B\| \leq \|A^{-1}\| \, \|B\| < 1$, so by Theorem A3.30, $1 + A^{-1}B$ is invertible (take $T = -A^{-1}B$). By Theorem A3.29, the product $A(1 + A^{-1}B)$ is invertible and the first formula is obtained. By writing

$$A + B = (1 + BA^{-1})A,$$

we obtain the second formula. □

Example A3.32. Let $X = Y = C([0, 1])$, and let $K(t, s)$ be a continuous function on $[0, 1] \times [0, 1]$ (see Example A3.2 (4)). We define a bounded operator K by

$$(Kf)(t) = \int_0^t K(t, s) f(s) ds.$$

Consider the equation on $C([0, 1])$:

$$f - \lambda K f = f_0,$$

where $f_0 \in C([0, 1])$ is fixed. We claim that the equation has a unique solution for any $f_0 \in C([0, 1])$ if $|\lambda| < \|K\|^{-1} = \left(\sup_{0 \leq t \leq 1} \int_0^t |K(t, s)| ds\right)^{-1}$. Indeed, in that case $\|\lambda K\| < 1$, so $1 - \lambda K$ is invertible by Theorem A3.30 and we have

$$f = (1 - \lambda K)^{-1} f_0,$$

which is easily seen to be a solution. It is unique since $(1 - \lambda K)$ is invertible. Moreover, the Neumann expansion for $(1 - \lambda K)^{-1}$ gives

$$f = \sum_{i=0}^{\infty} \lambda^i K^i f_0,$$

which is useful since it provides a norm-approximate solution

$$f_N = \sum_{i=0}^{N} \lambda^i K^i f_0.$$

This problem can also be solved by applying the Fredholm alternative, Theorem 9.12.

A3.5 Different Topologies on $\mathcal{L}(X)$

Let X be a Banach space. We recall that $\mathcal{L}(X)$, the LVS of all bounded operators from X into itself, is a Banach space with the norm defined in Definition A3.11. According to section 2 of Appendix 1, the norm induces a topology on $\mathcal{L}(X)$, called the *norm* or *uniform* topology. There are, however, other important topologies on $\mathcal{L}(X)$, and we discuss the weak and strong topologies here. As these topologies are not induced by a metric, they are not completely described by the convergences of sequences. However, it is this property that will be important for us, so we simply give the conditions for a sequence in $\mathcal{L}(X)$ to converge weakly or strongly.

Definition A3.33. *Let $\{B_n\}$ be a sequence of bounded operators (i.e., $B_n \in \mathcal{L}(X)$).*

(1) *B_n converges uniformly or in norm to $B \in \mathcal{L}(X)$ if $\|B_n - B\| \to 0$, as $n \to \infty$.*

(2) *B_n converges strongly to $B \in \mathcal{L}(X)$ if for all $x \in X$, $\|B_n x - Bx\| \to 0$, as $n \to \infty$.*

We note that since $\mathcal{L}(X)$ is a Banach space, if $\{B_n\}$ converges uniformly, then the limit always exists and belongs to $\mathcal{L}(X)$. To describe the third notion of convergence, weak convergence, we must introduce the LVS of linear functionals on a Banach space X.

Definition A3.34. *A map $l : X \to \mathbb{R}$ is called a bounded linear functional if l is linear on X, that is, for any $x, y \in X$ and $\lambda \in \mathbb{R}$ (or $\mathbb{C}$)*

$$l(\lambda x + y) = \lambda l(x) + l(y),$$

and $\exists K_l > 0$ such that for all $x \in X$, $|l(x)| \leq K_l \|x\|$. The set of all bounded linear functionals on X is denoted by X^*, the dual of X.

It is an elementary fact that X^* is a Banach space. Here, we are interested in the following definition.

Definition A3.35. *A sequence $\{B_n\}$ in $\mathcal{L}(X)$ converges weakly to $B \in \mathcal{L}(X)$ if for all $l \in X^*$ and $x \in X$, $[\, l(B_n x) - l(Bx)\,] = l(B_n x - Bx) \to 0$, as $n \to \infty$.*

Problem A3.14. Prove that uniform convergence $\Rightarrow$ strong convergence $\Rightarrow$ weak convergence.

Appendix 4. The Fourier Transform, Sobolev Spaces, and Convolutions

A4.1 Fourier Transform

A very important transformation on $L^2(\mathbb{R}^n)$ is the Fourier transform. We develop the main points of the theory here.

Definition A4.1. *For any $f \in \mathcal{S}(\mathbb{R}^n)$, define the Fourier transform $\hat{f}(k)$ by*

$$\hat{f}(k) \equiv (2\pi)^{\frac{-n}{2}} \int e^{-ik\cdot x} f(x)dx\,, \tag{A4.1}$$

where $k \cdot x = \sum_{i=1}^n k_i x_i$.

Problem A4.1. Show that $\hat{f} \in \mathcal{S}(\mathbb{R}^n)$ and that $f \to \hat{f}$ is a linear transformation on $\mathcal{S}(\mathbb{R}^n)$. Verify that (A4.1) can be extended to $L^1(\mathbb{R}^n)$.

Since $\mathcal{S}$ is dense in $L^2(\mathbb{R})$, we would like to extend the Fourier transform to all of $L^2(\mathbb{R}^n)$. By Appendix 3, we can do this provided the transformation is bounded on a dense set in $L^2(\mathbb{R}^n)$. Consider the set $\mathcal{E}$ of all finite linear combinations of functions of the form $p(x)e^{-\alpha(x-\beta)^2/2}$, for $\alpha > 0$, $\beta \in \mathbb{R}$, and p any polynomial in x. Such functions are in $\mathcal{S}(\mathbb{R}^n)$. We compute the Fourier transform of a typical one. Without loss of generality, we can take $\alpha = 1$ and $\beta = 0$. Then, setting $g(x) = p(x)e^{-x^2/2}$, we have

$$\hat{g}(k) = (2\pi)^{-\frac{n}{2}} \int e^{-ik\cdot x} p(x) e^{-\frac{x^2}{2}} dx\,.$$

Due to the simple fact that

$$(\partial/\partial k_i)e^{-ik\cdot x} = -ix_i e^{-ik\cdot x}, \tag{A4.2}$$

we have

$$\hat{g}(k) = (2\pi)^{-\frac{n}{2}} \int \left(p(iD)e^{-ik\cdot x}\right) e^{-\frac{x^2}{2}} dx,$$

where D stands for any partial derivative in k_i. By the fast convergence of the integral, we can write

$$\hat{g}(k) = (2\pi)^{-\frac{n}{2}} p(iD) \int e^{-ik\cdot x} e^{-\frac{x^2}{2}} dx.$$

The resulting integral can be computed explicitly by contour deformation.

Problem A4.2. Prove the identity

$$\int e^{-ik\cdot x} e^{-\frac{x^2}{2}} d^n x = (2\pi)^{\frac{n}{2}} e^{-\frac{k^2}{2}},$$

where $\int_{-\infty}^{\infty} e^{-x^2} dx = \pi^{1/2}$.

Consequently, we obtain

$$\hat{g}(k) = p(iD)e^{-\frac{k^2}{2}}.$$

We want to compute the L^2-norm of $\hat{g}$:

$$\begin{aligned} \|\hat{g}\|^2 &= \int |\hat{g}(k)|^2 dk \\ &= \int \left|p(iD)e^{-\frac{k^2}{2}}\right|^2 dk \\ &= (\psi, p(iD)p(iD)\psi), \end{aligned}$$

where we have integrated by parts and write $\psi(k) \equiv e^{-k^2/2}$. Retracing the above steps, the inner product can be written as

$$\begin{aligned} &\int dk \left((2\pi)^{-\frac{n}{2}} \int e^{-ik\cdot x} e^{-\frac{x^2}{2}} dx\right)^* \left(p(iD)p(iD)e^{-\frac{k^2}{2}}\right) \\ &= \int dk \left((2\pi)^{-\frac{n}{2}} p(iD) \int e^{-ik\cdot x} e^{-\frac{x^2}{2}} dx\right)^* \left(p(iD)e^{-\frac{k^2}{2}}\right) \\ &= \int dk \left((2\pi)^{-\frac{n}{2}} \int e^{-ik\cdot x} g(x) dx\right)^* \left(p(iD)e^{-\frac{k^2}{2}}\right) \\ &= \int dx\, g(x)^* \cdot (2\pi)^{-\frac{n}{2}} \int dk\, e^{ik\cdot x} p(iD)e^{-\frac{k^2}{2}} \\ &= \|g\|_2^2. \end{aligned}$$

The reader can verify that the integration by parts and the interchange of the order of integration are all valid due to the absolute convergence of the integral and the superexponential vanishing of the integrand as $\|x\|$ or $\|k\| \to \infty$. Because of the linearity of the map $f \in \mathcal{E} \to \hat{f} \in \mathcal{E}$, we have shown for any $f \in \mathcal{E}$,

$$\|f\|_2 = \|\hat{f}\|_2 .$$

Now $\mathcal{E}$ is dense in $L^2(\mathbb{R}^n)$, since the Hermite functions form a basis for $L^2(\mathbb{R}^n)$. Hence, we have proven the next result.

Theorem A4.2. *The map $f \in \mathcal{S}(\mathbb{R}^n) \to \hat{f} \in \mathcal{S}(\mathbb{R}^n)$, defined by the absolutely convergent integral (A4.1), extends to a bounded, linear transformation F on $L^2(\mathbb{R}^n)$ with*

$$\|Ff\|_2 = \|f\|_2 , \tag{A4.3}$$

for any $f \in L^2(\mathbb{R}^n)$. Furthermore, F is invertible with inverse F^{-1} defined on $\mathcal{E}$ by

$$(F^{-1}g)(x) = (2\pi)^{-\frac{n}{2}} \int e^{ik\cdot x} g(k)dk,$$

so that

$$g(x) = (2\pi)^{-\frac{n}{2}} \int e^{ik\cdot x} \hat{g}(k)dk . \tag{A4.4}$$

Remarks A4.3.

(1) The identity (A4.3) is referred to as the Plancherel theorem.

(2) Formula (A4.4) is called the Fourier inversion formula. Formulas (A4.1) and (A4.4) are not defined for arbitrary $f \in L^2(\mathbb{R}^n)$. It can be shown, however, that the functions defined for $R > 0$ and $f \in L^2(\mathbb{R}^n)$ by

$$\hat{f}_R(k) \equiv (2\pi)^{-\frac{n}{2}} \int_{\|x\|<R} e^{-ik\cdot x} f(x)dx$$

actually converge strongly to Ff as $R \to \infty$.

(3) Because F is an invertible isometry, it is a unitary transformation on $L^2(\mathbb{R}^n)$. As a consequence, we have

$$(Ff, Fg) = (f, g),$$

for any $f, g \in L^2(\mathbb{R}^n)$.

Finally, we note that the simple identity (A4.2) and the related identity

$$\left(\frac{\partial}{\partial x_i}\right) e^{-ik\cdot x} = -ik_i e^{-ik\cdot x} \tag{A4.5}$$

show the utility of the Fourier transform for solving constant coefficient partial differential equations on $\mathbb{R}^n$. In particular, if $g \in \mathcal{S}(\mathbb{R}^n)$, then denoting the Laplacian by $-\Delta$,

$$-\Delta \equiv \sum_{i=1}^{n} \left(-\frac{\partial^2}{\partial x_i^2} \right),$$

we obtain

$$(-\Delta g)^\wedge(k) = \|k\|^2 \hat{g}(k). \tag{A4.6}$$

Using the inversion formula, we can write the left side of (A4.6) as

$$(-F\Delta F^{-1})\hat{g}(k) = \|k\|^2 \hat{g}(k)$$

or

$$-F\Delta F^{-1} = \|k\|^2, \tag{A4.7}$$

valid on the domain of the Laplacian in $L^2(\mathbb{R}^n)$. This identity will be useful in our study of Sobolev spaces in the next section.

A4.2 Sobolev Spaces

We collect here basic results on the Sobolev spaces $H^{s,p}(\mathbb{R}^n)$ associated with $\mathbb{R}^n$. There are two approaches to this material: distributional derivatives and the Fourier transform. To introduce the notion of a weak or distributional derivative, let us first consider $f \in C_0^\infty(\mathbb{R}^n)$ and suppose g is simply in $C^1(\mathbb{R}^n)$. Then the integration-by-parts formula is

$$\int f(x) \left(\frac{\partial g}{\partial x_i} \right)(x) = -\int \left(\frac{\partial f}{\partial x_i} \right)(x) g(x). \tag{A4.8}$$

Note that the right side of (A4.8) makes sense even if g is not differentiable.

Definition A4.4. *Let $g \in L^p(\mathbb{R}^n)$. We say that g has a weak or distributional derivative with respect to x_i in L^p if there exists $\phi \in L^p(\mathbb{R}^n)$ such that for any $f \in C_0^\infty(\mathbb{R}^n)$,*

$$\int f(x)\phi(x)dx = -\int \left(\frac{\partial f}{\partial x_i} \right)(x) g(x) dx.$$

By Hölder's inequality, Theorem A2.4, both sides converge. We call ϕ the L^p-derivative of g with respect to x_i. In a similar manner, we can define higher-order, weak L^p-derivatives.

We can now define Sobolev spaces as the completion of $C_0^\infty(\mathbb{R}^n)$ with respect to the following norm. Let $\sigma = (\sigma_1, \ldots, \sigma_n)$ denote an n-tuple of positive integers, and let $|\sigma| \equiv \sum_{i=1}^{n} \sigma_i \geq 0$. For $1 \leq p < \infty$ and $s \in \mathbb{N}$, we define for any $f \in C_0^\infty(\mathbb{R}^n)$,

$$\|f\|_{s,p} = \left[\sum_{|\sigma| \leq s} \sum_{i=1}^{n} \left\| \frac{\partial^{\sigma_i} f}{\partial x_i^{\sigma_i}} \right\|_p^p \right]^{\frac{1}{p}} . \tag{A4.9}$$

Problem A4.3. Show that $\left(C_0^\infty(\mathbb{R}^n), \|\cdot\|_{s,p}\right)$ is an NLVS.

Definition A4.5. *The pth Sobolev space of order s, $H^{s,p}(\mathbb{R}^n)$, is the Banach space obtained as the completion of $\left(C_0^\infty(\mathbb{R}^n), \|\cdot\|_{s,p}\right)$.*

As in the case of $L^p(\mathbb{R}^n)$, it can be shown that $H^{s,p}(\mathbb{R}^n)$ is the set of all functions $f \in L^p$ that have distributional L^p-derivatives up to and including order s. Additionally, when $p = 2$, we write $H^s(\mathbb{R}^n)$ for the Sobolev space of order s.

Problem A4.4. Show that $H^s(\mathbb{R}^n)$ is a Hilbert space.

The Fourier transform and formulas (A4.2) and (A4.5) allow us to characterize $H^s(\mathbb{R}^n)$ as follows. For $f \in \mathcal{S}(\mathbb{R}^n)$, we define

$$\|f\|_s^2 \equiv \int \left(1 + \|k\|^2\right)^s \left|\hat{f}(k)\right|^2 dk \,. \tag{A4.10}$$

These two norms (A4.9) and (A4.10) are *equivalent* in that they have the same Cauchy sequences (see Problem A3.9). Consequently, the completion yields the same Banach space $H^s(\mathbb{R}^n)$.

Problem A4.5. Verify that the norms (A4.9) and (A4.10) on $\mathcal{S}(\mathbb{R}^n)$ are equivalent.

Functions in the Sobolev spaces $\mathcal{S}(\mathbb{R}^n)$ become more regular with increasing index s. We will state one version of a theorem of this type. This is an example of a body of results called the *Sobolev embedding theorems*. These theorems are used to prove the regularity of solutions to elliptic differential equations.

Theorem A4.6. *Any function $f \in H^{s+k}(\mathbb{R}^n)$, for $s > n/2$ and $k \geq 0$, can be represented by a function in $C^k(\mathbb{R}^n)$.*

A4.3 Convolutions

We recall that for two functions $f, g \in C_0^\infty(\mathbb{R}^n)$, the *convolution* of f and g, denoted by $f * g$, is defined by

$$(f * g)(x) = \int f(x-y)g(y)dy\,.$$

To extend this definition to L^p-spaces, we can use Young's inequality.

Theorem A4.7. (Young's inequality). *Let p, q, and r be nonnegative real numbers, and suppose $1/p + 1/q = 1 + 1/r$. Then if $f \in L^p(\mathbb{R}^n)$ and $g \in L^q(\mathbb{R}^n)$, the convolution $f * g \in L^r(\mathbb{R}^n)$ and satisfies*

$$\|f * g\|_r \leq \|f\|_p \, \|g\|_q\,. \tag{A4.11}$$

Convolutions of functions in $\mathcal{S}(\mathbb{R}^n)$ behave nicely under the Fourier transform. For $f, g \in \mathcal{S}(\mathbb{R}^n)$, we have

$$\widehat{fg}(k) = \hat{f} * \hat{g}(k)\,. \tag{A4.12}$$

Problem A4.6. Verify the identity (A4.12).

References

[AF] Agler, J., and Froese, R.: Existence of Stark Ladder Resonances, *Commun. Math. Phys.* **100**, 161–171 (1985).

[AC] Aguilar, J., and Combes, J. M.: A Class of Analytic Perturbations for One-Body Schrödinger Hamiltonians, *Commun. Math. Phys.* **22**, 269–279 (1971).

[Ag1] Agmon, S.: *Lectures on Exponential Decay of Solutions of Second-Order Elliptic Equations*, Princeton University Press, Princeton, NJ, 1982.

[Ag2] Agmon, S.: *Lectures on Elliptic Boundary Value Problems*, Van Nostrand, Princeton, NJ, 1965.

[AgHeSk] Agmon, S., Herbst, I. W., and Skibsted, E.: Perturbation of Embedded Eigenvalues in the Generalized N-Body Problem, *Commun. Math. Phys.* **122**, 411–438 (1989).

[Ah] Ahia, F.: Lower Bounds on Width of Stark Resonances in One Dimension, *Letts. Math. Phys.* **24**, 21–29 (1992).

[AG] Akhiezer, N. I., and Glazman, I. M.: *The Theory of Linear Operators in Hilbert Space*, Vols. 1 & 2, Frederick Ungar, New York, 1963.

[AJS] Amrein, W. O., Jauch, J. M., and Sinha, K. B.: *Scattering Theory in Quantum Mechanics*, Lecture Notes and Supplements in Physics, Vol. 16, W. A. Benjamin, Inc., London, 1977.

[Ar] Arsen'ev, A. A.: On the Singularities of Analytic Continuations and Resonances Properties of a Solution of the Scattering Problem for the Helmholtz Equation, *Zh. Vyohisl. Mat. i Mat. Fiz.* **12**, 112–138 (1972).

[AH] Ashbaugh, M., and Harrell, E. M.: Perturbation Theory for Shape Resonances and Large Barrier Potentials, *Commun. Math. Phys.* **83**, 151–170 (1982).

[Av] Avron, J. E.: Bender-Wu Formulas for the Zeeman Effect in Hydrogen, *Ann. Phys.* **131**, 73–94 (1981).

[AvHe] Avron, J. E., and Herbst, I. W.: Spectral and Scattering Theory of Schrödinger Operators Related to the Stark Effect, *Commun. Math. Phys.* **52**, 239–254 (1977).

[AHS] Avron, J. E., Herbst, I. W., and Simon, B.: Schrödinger Operators with Magnetic Fields, I. General Interactions, *Duke Math. J.* **45**, 847 (1978).

[BFS] Bach, V., Fröhlich, J., and Sigal, I.M.: Mathematical Theory of Nonrelativistic Matter and Radiation, *Letters in Math. Phys.***34**, 183–201 (1995).

[B1] Balslev, E.: Analytic Scattering Theory for 2-body Schödinger operators, *J. Funct. Anal.* **29**, 375–396 (1978).

[B2] Balslev, E.: Local Spectral Deformation Technique for Schrödinger Operators, *J. Funct. Anal.* **58**, 79–105 (1984).

[B3] Balslev, E.: Analyticity Properties of Eigenfunctions and the Scattering Matrix, *Commun. Math. Phys.* **114**, 599 (1988).

[BS1] Balslev, E., and Skibsted, E.: Resonance Theory of Two-Body Schrödinger Operators, *Ann. Inst. Henri Poincaré* **51**, 2 (1989).

[BS2] Balslev, E., and Skibsted, E.: Asymptotic and Analyticity Properties of Resonance Functions, *Ann. Inst. Henri Poincaré* **53**, 123–137 (1990).

[BB] Babbitt, D., and Balslev, E.: Local Distortion Techniques and Unitarity of the S-Matrix for the 2-Body Problem, *J. Math. Anal. Appl.* **54**, 316–347 (1976).

[BC] Balslev, E., and Combes, J. M.: Spectral Properties of Many Body Schödinger Operators with Dilation Analytic Interactions, *Commun. Math. Phys.* **22**, 280–294 (1971).

[Be] Beale, J. T.: Scattering Frequencies of Resonators, *Commun. Pure Appl. Math.* **26**, 549–563 (1973).

[Ben] Bentosela, F.: Bloch Electrons in Constant Electric Field, *Commun. Math. Phys.* **68**, 173–182 (1979).

[BCDSSW] Bentosela, F., Carmona, R., Duclos, P., Simon, B., Soulliard, B., and Weder, R.: Schrödinger Operators with an Electric Field and Random or Deterministic Potentials, *Commun. Math. Phys.* **88**, 387–397 (1983).

[BG] Bentosela, F., and Grecchi, V.: Stark Wannier Ladders, *Commun. Math. Phys.* **142**, 169–192 (1991).

[Br] Briet, P.: Bender Wu Formula for the Zeeman Effect, *J. Math. Phys.* **36**, 3871 (1995).

[BCD1] Briet, P., Combes, J. M., and Duclos, P.: On the Location of Resonances for Schrödinger Operators in the Semiclassical Limit, I. Resonance Free Domains, *J. Math. Anal. Appl.* **126**, 90–99 (1987).

[BCD2] Briet, P., Combes, J. M., and Duclos, P.: On the Location of Resonances for Schrödinger Operators, II. Barrier Top Resonances, *Commun. Partial Diff. Equations* **12**, 201–222 (1987).

[BCD3] Briet, P., Combes, J. M., and Duclos, P.: Spectral Stability under Tunneling, *Commun. Math. Phys.* **126**, 133–156 (1989).

[BHM1] Brown, R. M., Hislop, P. D., and Martinez, A.: Lower Bounds on the Interaction between Cavities Connected by a Thin Tube, *Duke Math. J.* **73**, 163–176, (1994).

[BD] Buslaev, V. S., and Dmitrieva, L. A.: Spectral Properties of the Bloch Electrons in External Fields, Berkeley preprint PAM-477, University of California, 1989.

[CS] Carmona, R., and Simon, B.: Pointwise Bounds on Eigenfunctions and Wave Packets in N-Body Quantum Systems, *Commun. Math. Phys.* **80**, 59–98 (1981).

[CDS] Combes, J. M., Duclos, P., and Seiler, R.: Krein's Formula and One-Dimensional Multiple Wells, *J. Funct. Anal.* **52**, 257–301 (1983).

[CDKS] Combes, J. M., Duclos, P., Klein, M., and Seiler, R.: The Shape Resonance, *Commun. Math. Phys.* **110**, 215–236 (1987).

[CH] Combes, J. M., and Hislop, P. D.: Stark Ladder Resonances for Small Electric Fields, *Commun. Math. Phys.* **140**, 291–320 (1991).

[CH1] Combes, J. M., and Hislop, P. D.: Localization for Some Continuous, Random Hamiltonians in d-Dimensions, *J. Funct. Anal.* **124**, 149–180 (1994).

[CT] Combes, J. M., and Thomas, L.: Asymptotic Behavior of Eigenfunctions for Multiparticle Schrödinger Operators, *Commun. Math. Phys.* **34**, 251–276 (1973).

[Cy] Cycon, H. L.: Resonances Defined by Modified Dilations, *Helv. Phys. Acta* **58**, 969–981 (1985).

[CFKS] Cycon, H. L., Froese, R. G., Kirsch, W., and Simon, B.: *Schrödinger Operators, with Application to Quantum Mechanics and Global Geometry*, Springer-Verlag, New York, 1986.

[Da] Davies, E. B.: Metastable States of Molecules, *Commun. Math. Phys.* **75**, 263–283 (1980).

[DeBH] DeBièvre, S., and Hislop, P. D.: Spectral Resonances for the Laplace-Beltrami Operator, *Ann. Inst. Henri Poincaré* **48**, 105–146 (1988).

[DeBHS] DeBièvre, S., Hislop, P. D., and Sigal, I. M.: Scattering Theory for the Wave Equation on Non-Compact Manifolds, *Reviews in Mathematical Physics* **4**, 575–618 (1992).

[DHSV] Deift, P., Hunziker, W., Simon, B., and Vock, E.: Pointwise Bounds on Eigenfunctions and Wave Packets in N-Body Quantum Systems IV, *Commun. Math. Phys.* **64**, 1–34 (1978).

[EG] Evans, L. C., and Gariepy, R. F.: *Measure Theory and Fine Properties of Functions*, CRC Press: Boca Raton, FL, 1991.

[Fef] Fefferman, C. L.: The N-Body Problem in Quantum Mechanics, *Commun. Pure Appl. Math.* **39 S**, S67–S109 (1986).

[Fe] Fernandez, C. A.: Resonances in Scattering by a Resonator, *Indiana Univ. Math. J.* **34**, 115–125 (1985).

[FL] Fernandez, C. A., and Lavine, R.: Lower Bounds for Resonance Widths in Potential and Obstacle Scattering, *Commun. Math. Phys.* **128**, 263–284 (1990).

[FrHe] Froese, R. G., and Herbst, I. W.: Exponential Bounds and Absence of Positive Eigenvalues for N-Body Schrödinger Operators, *Commun. Math. Phys.* **87**, 429–447 (1982).

[FHHO1] Froese, R., Herbst, I., Hoffmann-Ostenhof, M., Hoffmann-Ostenhof, T.: On the Absence of Positive Eigenvalues for One-Body Schrödinger Operators. *J. d'Anal. Math.* **41**, 272–284 (1982).

[FHHO2] Froese, R., Herbst, I., Hoffmann-Ostenhof, M., and Hoffmann-Ostenhof, T.: L^2-Exponential Lower Bounds to Solutions of the Schrödinger Equation, *Commun. Math. Phys.* **87**, 265–286 (1982).

[FrHi] Froese, R. G., and Hislop, P. D.: Spectral Analysis of Second Order Elliptic Operators on Noncompact Manifolds, *Duke Math. J.* **58**, 103–129 (1989).

[FrSp] Fröhlich, J., and Spencer, T.: Absence of Diffusion in the Anderson Tight Binding Model for Large Disorder or Low Energy, *Commun. Math. Phys.* **88**, 151–184 (1983).

[Gad1] Gadyl'shin, R. R.: On Acoustic Helmholtz Resonator and on Its Electromagnetic Analogy, *J. Math. Phys.* **35**, 3464–3481 (1994).

[Gad2] Gadyl'shin, R. R.: Asymptotics of Scattering Frequencies with Small Imaginary Parts for an Acoustic Resonator, *Math. Model. Numer. Anal.* **28**, 761–780 (1994).

[Ga] Gamow, G.: Zur Quantentheorie des Atomkernes, *Z. Phys.* **51**, 204–212 (1928).

[G1] Gérard, C.: Resonances in Atom-Surface Scattering, *Commun. Math. Phys.* **126**, 263–290 (1989).

[G2] Gérard, C.: Resonance Theory for Periodic Schrödinger Operators, *Bull. Soc. Math. Fr.* **118**, 27–54 (1990).

[G3] Gérard, C.: Semiclassical Resolvent Estimates for Two- and Three-Body Schrödinger Operators, *Commun. Partial Diff. Eq.* **15**, 1161–1178 (1990).

[GG] Gérard, C., and Grigis, A.: Precise Estimates of Tunneling and Eigenvalues Near a Potential Barrier, *J. Diff. Eq.* **72**, 149–177 (1988).

[GM1] Gérard, C., and Martinez, A.: Principe d'Absorption Limite pour les Opérateurs de Schrödinger à Longue Portée, *C. R. Acad. Sci. Paris* **306**, 121–123 (1988).

[GM2] Gérard, C., and Martinez, A.: Prolongement Méromorphic de la Matrice de Scattering pour les Problèmes à Deux Corps à Longue Portée, *Ann. Inst. Henri Poincaré* **31**, 81–110 (1989).

[GMR] Gérard, C., Martinez, A., and Robert, D.: Breit–Wigner Formulas for the Scattering Phase and the Total Cross-Section in the Semiclassical Limit, *Commun. Math. Phys.* **121**, 323–336 (1989).

[GS] Gérard, C., and Sigal, I. M.: Space–Time Picture of Semiclassical Resonances, *Commun. Math. Phys.* **145**, 281–328 (1992).

[GSj] Gérard, C., and Sjöstrand, J.: Semiclassical Resonances Generated by a Closed Trajectory of Hyperbolic Type, *Commun. Math. Phys.* **108**, 391–421 (1987).

[Gr] Graf, G. M.: The Mourre Estimate in the Semiclassical Limit, *Lett. Math. Phys.* **20**, 47–54 (1990).

[GrGr] Graffi, S., and Grecchi, V.: Resonances in the Stark Effect and Perturbation Theory, *Commun. Math. Phys.* **62**, 83–96 (1978)

[GC] Gurney, R. W., and Condon, E. U.: *Nature* **122** 439 (1928).

[Ha] Hagedorn, G.: A Link between Scattering Resonances and Dilation-Analytic Resonances in Few-Body Quantum Mechanics, *Commun. Math. Phys.* **65**, 181–201 (1979).

[Hal] Halmos, P.: *Introduction to Hilbert Space and the Theory of Spectral Multiplicity*, Chelsea Publishing Company, New York, 1957.

[H1] Harrell, E. M.: Double Wells, *Commun. Math. Phys.* **75**, 239–261 (1980).

[H2] Harrell, E. M.: General Lower Bounds for Resonances in One Dimension, *Commun. Math. Phys.* **86**, 221–225 (1982).

[HaSi] Harrell, E. M., and Simon, B.: Mathematical Theory of Resonances Whose Widths Are Exponentially Small, *Duke Math. J.* **47**, 845 (1980).

[Hel] Helffer, B.: *Semiclassical Analysis of Schrödinger Operators and Applications*, Springer Lecture Notes in Mathematics No. 1336, Springer-Verlag, Berlin, 1988.

[HeM] Helffer, B., and Martinez, A.: Comparaison entre les Diverses Notions de Résonances, *Helv. Phys. Acta* **60**, 992–1003 (1987).

[HR] Helffer, B., and Robert, D.: Puits de Potentiel Généralisés et Asymptotique Semi-classique, *Ann. Inst. Henri Poincaré* **41**, 291–331 (1984).

[HSj1] Helffer, B., and Sjöstrand, J.: Multiple Wells in the Semi-classical Limit I. *Commun. in P.D.E.* **9**, 337–408 (1984).

[HSj2] Helffer, B., and Sjöstrand, J.: Puits Multiples en Limite Semi-classique, II. Interaction de Moléculaire, Symétries, Perturbation. *Ann. Inst. Henri Poincaré* **42**, 127–212 (1985).

[HSj3] Helffer, B., and Sjöstrand, J.: Multiple Wells in the Semi-classical Limit, III. Interaction through Non-resonant Wells, *Math. Nachrichte* **124**, 263–313 (1985).

[HSj4] Helffer, B., and Sjöstrand, J.: Resonances en Limite Semi-classique. *Memoire de la SMF*, Nos. 24–25, **114** (1986).

[He1] Herbst, I. W.: Schrödinger Operators with External Homogeneous Electric and Magnetic Fields, in *Rigorous Atomic and Molecular Physics*, Proc. fourth Int. School of Math. Phys., Enrice, G. Velo and A. S. Wightman, eds., Plenum, New York, 1981.

[He2] Herbst, I. W.: Perturbation Theory for the Decay Rate of Eigenfunctions in the Generalized N-Body Problem, *Commun. Math. Phys.* **158**, 517–536 (1993).

[He3] Herbst, I. W.: Dilation Analyticity in Constant Electric Field, I. The Two-Body Problem, *Commun. Math. Phys.* **64**, 279–298 (1979).

[He4] Herbst, I. W.: Exponential Decay in the Stark Effect, *Commun. Math. Phys.* **75**, 197–205 (1980).

[HH] Herbst, I. W., and Howland, J.: The Stark Ladder and Other One-Dimensional External Field Problems, *Commun. Math. Phys.* **80**, 23–42 (1981).

[HiSm] Hirsch, M., and Smale, S.: *Differential Equations, Dynamical Systems, and Linear Algebra*, Academic Press, Boston, 1974.

[H] Hislop, P. D.: Singular Perturbations of Dirichlet and Neumann Domains and Resonances for Obstacle Scattering, *Astérisque* **210**, 197–216 (1992).

[His] Hislop, P. D.: The Geometry and Spectra of Hyperbolic Manifolds, *Proceedings of the Indian Academy of Sciences; Math. Sci.* **104**, 715–776 (1994).

[HM] Hislop, P. D., and Martinez, A.: Scattering Resonances of a Helmholtz Resonator, *Indiana Univ. Math. J.* **40**, 767–788 (1991).

[HN] Hislop, P. D., and Nakamura, S.: Semiclassical Resolvent Estimates, *Ann. Inst. Henri Poincaré* **51**, 187–198 (1989).

[HS1] Hislop, P. D., and Sigal, I. M.: Shape Resonances in Quantum Mechanics, in *Differential equations and Mathematical Physics*, I. Knowles, ed., Lect. Notes in Math. Vol. 1285, Springer-Verlag, New York, 1986.

[HS2] Hislop. P. D., and Sigal, I. M.: Semiclassical Theory of Shape Resonances in Quantum Mechanics, *Memoirs of the AMS* **399**, 1–123 (1989).

[HOAM] Hoffmann–Ostenhof, M., Hoffmann–Ostenhof, T., Ahlichs, R., and Morgan, J.: On Exponential Fall-Off of Wave Functions and Electron Densities, in *Mathematical Problems in Theoretical Physics*, Lecture Notes in Physics **116**, K. Osterwalder, Ed., Springer-Verlag, Berlin, 1980.

[Ho1] Howland, J.: Puiseux Series for Resonances at an Embedded Eigenvalue, *Pacific J. Math.* **55**, 157–176 (1974).

[Ho2] Howland, J. S.: Imaginary Part of a Resonance in Barrier Penetration, *J. Math. Anal. Appl.* **86**, 507–517 (1982).

[Hu1] Hunziker, W.: Schrödinger Operators with Electric or Magnetic Fields, in *Mathematical Problems in Theoretical Physics*, Lecture Notes in Physics, Vol. 116, K. Osterwalder, ed., Berlin, Springer-Verlag, 1979.

[Hu2] Hunziker, W.: Distortion Analyticity and Molecular Resonance Curves, *Ann. Inst. Henri Poincaré* **45**, 339–358 (1986).

[Hu3] Hunziker, W.: Resonances, Metastable States, and Exponential Decay Laws in Perturbation Theory, *Commun. Math. Phys.* **132**, 177–188 (1990).

[Hu4] Hunziker, W.: Notes on Asymptotic Perturbation Theory for Schrödinger Eigenvalue Problems, *Helv. Phys. Acta* **61**, 257–304 (1988).

[Hu5] Hunziker, W., and Pillet, C. A.: Degernerate Asymptotic Perturbation Theory, *Commun. Math. Phys.* **90**, 219–233 (1983).

[JP] Jakšić, V., and Pillet, C. A.: On a Model for Quantum Friction I, Fermi's Golden Rule and Dynamics at Zero Temperature, *Ann. Inst. Henri Poincaré* **62**, 47–68 (1995).

[J1] Jensen, A.: Local Distortion Technique, Resonances and Poles of the S-Matrix, *J. Math. Anal. Appl.* **59**, 505–513 (1977).

[J2] Jensen, A.: Resonances in an Abstract Analytic Scattering Theory, *Ann. Inst. Henri Poincaré* **33**, 209–223 (1980).

[J3] Jensen, A.: High Energy Resolvent Estimates for Generalized N-Body Schrödinger Operators, *Publ. RIMS, Kyoto Univ.* **25**, 155–167 (1989).

[J4] Jensen, A.: Bounds on Resonances for Stark–Wannier and Related Hamiltonians, *J. Oper. Th.* **19**, 69–80 (1988).

[J5] Jensen, A.: Perturbation results for Stark Effect Resonances, *J. Reine Angew. Math.* **394**, 168–179 (1989).

[JLMS1] Jona-Lasinio, G., Martinelli, F., and Scoppola, E.: New Approach in the Semi-classical Limit of Quantum Mechanics, I. Multiple Tunnelings in One-Dimension, *Commun. Math. Phys.* **80**, 223 (1981).

[JLMS2] Jona-Lasinio, G., Martinelli, F., and Scoppola, E.: Multiple Tunnelings in d-Dimensions: A Quantum Particle in a Hierarchical Model, *Ann. Inst. Henri Poincaré* **43**, 2 (1985).

[KR] Kaidi, N., and Rouleux, M.: Multiple Resonances in the Semi-Classical Limit, *Commun. Math. Phys.* **133**, 617–634 (1990).

[K] Kato, T.: *Perturbation Theory for Linear Operators*, second edition, Springer–Verlag, Berlin, 1980.

[KS1] Kirsch, W., and Simon, B.: Universal Lower Bounds on Eigenvalue Splitting for One-Dimensional Schrödinger Operators, *Commun. Math. Phys.* **97**, 453–460 (1985).

[KS2] Kirsch, W., and Simon, B.: Comparison Theorems for the Gap of Schrödinger Operators, *J. Funct. Anal.* **75**, 396–410 (1987).

[Kl1] Klein, M.: On the Absence of Resonances for Schrödinger Operators with Nontrapping Potentials in the Classical Limit, *Commun. Math. Phys.* **106**, 485–494 (1986).

[Kl2] Klein, M.: On the Mathematical Theory of Predissociation, *Ann. of Physics* **178**, 48–73 (1987).

[Klp] Klopp, F.: Resonances for Perturbations of a Semi-classical Periodic Schrödinger Operator, *Arkiv för Math.* **32**, 323–371 (1994).

[Kn] Knopp, A.: *Theory of Analytic Functions*, Dover Publications, Mineola, NY, 1966.

[LL] Landau, L. D., and Lifshitz, E. M.: *Quantum Mechanics*, Pergamon Press, Oxford, 1977.

[La] Lavine, R.: Spectral Density and Sojourn Times, in *Atomic Scattering Theory*, J. Nuttall, ed., University of Western Ontario Press, London, Ontario, 1978.

[LP] Lax, P. D., and Phillips, R. S.: *Scattering Theory*, revised edition, Academic Press, Inc., Boston, 1989.

[Li] Lieb, E. H.: Stability of Matter, *Review of Modern Physics* **48**, 553–569 (1976).

[Ma] Marsden, J. E.: *Basic Complex Analysis*, W. H. Freeman and Company, San Francisco, 1973.

[M1] Martinez, A.: Estimations de l'Effet Tunnel pour le Double Puits, I. *J. Math. Pures Appl.* **66**, 195–215 (1987).

[M2] Martinez, A.: Estimations de l'Effet Tunnel pour le Double Puits, II. Etats Hautement Exités, *Bull. Soc. Math. Fr.* **116**, 199–229 (1988).

[M3] Martinez, A.: Résonances dans l'Approximation de Born–Oppenheimer I, *J. Diff. Eq.* **91**, 204–234 (1991).

[M4] Martinez, A.: Résonances dans l'Approximation de Born–Oppenheimer II. Largeur des Résonances, *Commun. Math. Phys.* **135**, 517–530 (1991).

[MR] Martinez, A., and Rouleux, M.: Effet Tunnel entre Puits Dégenères, *Comm. Partial Differential Equations* **13**, 1157–1187 (1988).

[Mas] Maslov, V. P., and Nasaikinskii, V. E.: *Asymptotics of Operator and Pseudo-Differential Equations*, Contemporary Soviet Mathematics, Consultants Bureau, New York, 1988.

[Mo1] Morawetz, C.: Exponential Decay for Solutions of the Wave Equation, *Commun. Pure Appl. Math.* **19**, 439–444 (1966).

[Mo2] Morawetz, C.: *Notes on Time Decay and Scattering for Some Hyperbolic Problems*, SIAM, Philadelphia, PA, 1975.

[N1] Nakamura, S.: A Remark on Eigenvalue Splittings for One-Dimensional Double-Well Hamiltonians, *Letters in Math. Phys.* **11**, 337–340 (1986).

[N2] Nakamura, S.: Scattering Theory for the Shape Resonance Model I, *Ann. Inst. Henri Poincaré* **50**, 115–131 (1989).

[N3] Nakamura, S.: Scattering Theory for the Shape Resonance Model II, *Ann. Inst. Henri Poincaré* **50**, 132–142 (1989).

[N4] Nakamura, S.: Shape Resonances for Distortion Analytic Schrödinger Operators, *Commun. Part. Diff. Eqns.* **14**, 1385–1419 (1989).

[N5] Nakamura, S.: Distortion Analyticity for Two-Body Schrödinger Operators, *Ann. Inst. Henri Poincaré* **53**, 149–157 (1990).

[N6] Nakamura, S.: A Note on the Absence of Resonances for Schrödinger Operators, *Letters in Math. Phys.* **16**, 217–233 (1988).

[NN1] Nenciu, A., and Nenciu, G.: Dynamics of Bloch Electrons in External Electric Fields, I. Bounds for Interband Transitions and Effective Wannier Hamiltonians, *J. Phys.* **A14**, 2817–2827 (1981).

[NN2] Nenciu, A., and Nenciu, G.: Dynamics of Bloch Electrons in External Electric Fields, II. The Existence of Stark Wannier Ladder Resonances, *J. Phys.* **A15**, 3313–3328 (1982).

[Ne] Newton, R. G.: *Scattering Theory of Waves and Particles*, second edition, Springer-Verlag, Berlin, 1982.

[OC] O'Connor, T.: Exponential Decay of Bound State Wave Functions, *Commun. Math. Phys.* **32**, 319–340 (1973).

[OY] Okamoto, T., and Yajima, K.: Complex Scaling Technique in Nonrelativistic QED, *Ann. Inst. Henri Poincaré* **42**, 311–327 (1985).

[Op] Oppenheimer, R. J.: Three Notes on the Quantum Theory of Aperiodic Effects, *Phys. Rev.* **31**, 66–81 (1928).

[Or] Orth, A.: Quantum Mechanical Resonance and Limiting Absorption: The Many Body Problem, *Commun. Math. Phys.* **126**, 559–573 (1990).

[Ou] Outassourt, A.: Analyse Semi-classique pour l'Opérateur de Schrödinger à Potential Périodique, *J. Funct. Anal.* **72**, 1 (1987).

[Per] Persson, A.: Bounds for the Discrete Part of the Spectrum of a Semi-bounded Schrödinger Operator, *Math. Scand.* **8**, 143–153 (1960).

[PS] Petkov, V., and Stoyanov, L.: *Geometry of Reflecting Rays and Inverse Spectral Problems*, John Wiley & Sons, Chichester, 1992.

[RS1] Reed, M., and Simon, B.: *Methods of Modern Mathematical Physics, I. Functional Analysis*, second edition, Academic Press, New York, 1980.

[RS2] Reed, M., and Simon, B.: *Methods of Modern Mathematical Physics, II. Fourier Analysis, Self-Adjointness*, Academic Press, New York, 1975.

[RS3] Reed, M., and Simon, B.: *Methods of Modern Mathematical Physics, III. Scattering Theory*, Academic Press, New York, 1979.

[RS4] Reed, M., and Simon, B.: *Methods of Modern Mathematical Physics, IV. Analysis of Operators*, Academic Press, New York, 1978.

[R1] Robert, D.: *Autour de l'Approximation Semiclassique*, Birkhauser, Basel, 1983.

[RT1] Robert, D., and Tamura, H.: Semiclassical Bounds for Resolvents of Schrödinger Operators and Asymptotics for Scattering Phases, *Commun. Partial Diff. Eq.* **9**, 1017–1058 (1984).

[RT2] Robert, D., and Tamura, H.: Semiclassical Estimates for Resolvents and Asymptotics for Scattering Phases, *Ann. Inst. Henri Poincaré* **46**, 415–442 (1987).

[Ro] Royden, H.: *Real Analysis*, second edition, MacMillan Publishing Co., New York, 1968.

[R] Rudin, W.: *Real and Complex Analysis*, McGraw-Hill Book Co., New York, 1966.

[Sch] Schrödinger, E.: *Collected Papers on Wave Mechanics*, London and Glasgow, Blackie and Son Limited, 1928.

[ST] Segur, H., and Tanveer, S., eds.: *Asymptotics Beyond All Orders*, Birkhauser, Boston, 1991.

[Sig] Sigal, I. M.: *Scattering Theory for Many-Body Quantum Mechanical Systems–Rigorous Results*, Lecture Notes in Mathematics Vol. 1011, Springer-Verlag, Berlin, 1983.

[Si1] Sigal, I. M.: Geometric Methods in the Quantum Many-Body Problem. Non-existence of Very Negative Ions. *Commun. Math. Phys.* **85**, 309–324 (1982).

[Si2] Sigal, I. M.: Complex Transformation Method and Resonances in One-Body Quantum Systems. *Ann. Inst. Henri Poincaré* **41**, 103–114 (1984); and Addendum **41**, 333 (1984).

[Si3] Sigal, I. M.: A Generalized Weyl Theorem and the L^p-Spectra of Schödinger Operators, *J. Operator Theory* **13**, 119–129 (1985).

[Si4] Sigal, I. M.: General Characteristics of Non-Linear Dynamics, in *Spectral and Scattering Theory*, M. Ikawa, ed., Marcel Dekker, Inc., New York, 1994.

[Si5] Sigal, I. M.: Sharp Exponential Bounds on Resonance States and Width of Resonances, *Advances in Math.* **9**, 127–166 (1988).

[Sim1] Simon, B.: Pointwise Bounds on Eigenfunctions and Wave Packets in N-Body Quantum Systems I, *Proc. Amer. Math. Soc.* **42**, 395–401 (1974).

[Sim2] Simon, B.: Pointwise Bounds on Eigenfunctions and Wave Packets in N-Body Quantum Systems II, *Proc. Amer. Math. Soc.* **45**, 454–456 (1974).

[Sim3] Simon, B.: Pointwise Bounds on Eigenfunctions and Wave Packets in N-Body Quantum Systems III, *Trans. Amer. Math. Soc.* **208**, 317–329 (1975).

[Sim4] Simon, B.: *Trace Ideals and Their Applications*, Cambridge University Press, Cambridge, England, 1979.

[Sim5] Simon, B.: Semiclassical Analysis of Low Lying Eigenvalues, I. Non-degenerate Minima: Asymptotic Expansions, *Ann. Inst. Henri Poincaré* **38**, 296–307 (1983).

[Sim6] Simon, B.: Semiclassical Analysis of Low Lying Eigenvalues, II. Tunneling, *Ann. Math.* **120**, 89–118 (1984).

[Sim7] Simon, B.: Semiclassical Analysis of Low Lying Eigenvalues, III. Width of the Ground State Band in Strongly Coupled Solids, *Ann. of Phys.* **158**, 415–442 (1984).

[Sim8] Simon, B.: Semiclassical Analysis of Low Lying Eigenvalues, IV. The Flea on the Elephant, *J. Funct. Anal.* **63**, 123 (1984).

[Sim9] Simon, B.: The Theory of Resonances for Dilation Analytic Potentials and the Foundations of Time Dependent Perturbation Theory, *Ann. Math.* **97**, 247–274 (1973).

[Sim10] Simon, B.: Resonances and Complex Scaling: A Rigorous Overview, *Int. J. Quant. Chem.* **14**, 529–542 (1978).

[Sim11] Simon, B.: The Definition of Molecular Resonance Curves by the Method of Exterior Complex Scaling. *Phys. Letts.* **71A**, 211–214 (1979).

[Sim12] Simon, B.: Large Orders and Summability of Eigenvalue Perturbation Theory: A Mathematical Overview, *Int. J. Quant. Chem.* **21**, 3–25 (1982).

[Sim13] Simon, B.: Schrödinger Semigroups, *Bull. Am. Math. Soc.* **7**, 447–526 (1982).

[SWYY] Singer, I. M., Wong, B., Yau, S. T., and Yau, S. S. T.: An Estimate of the Gap of the First Two Eigenvalues for a Schrödinger Operator, *Ann. Scuola Norm. Sup. Pisa Cl. Sci.* **12**, 319–333 (1985).

[Sk1] Skibsted, E.: Truncated Gamow Functions, α-Decay, and the Exponential Law, *Commun. Math. Phys.* **104**, 591–604 (1986).

[Sk2] Skibsted, E.: Truncated Gamow Function and the Exponential Law, *Ann. Inst. Henri Poincaré* **46**, 131–153 (1987).

[Sk3] Skibsted, E.: On the Evolution of Resonance States, *J. Math. Anal. Appl.* **141**, 27–48 (1989).

[St] Stone, M. H.: *Linear Transformations in Hilbert Space and Their Applications to Analysis*, Amer. Math. Soc., Providence, RI, 1963.

[Th3] Thirring, W.: *A Course in Mathematical Physics 3, Quantum Mechanics of Atoms and Molecules*, E. M. Harrell, trans., Springer-Verlag, New York, 1979.

[Th4] Thirring, W.: *A Course in Mathematical Physics 4, Quantum Mechanics of Large Systems*, E. M. Harrell, trans., Springer-Verlag, New York, 1983.

[T1] Titchmarsh, E. C.: *The Theory of Functions*, second edition, Cambridge University Press, Cambridge, England, 1939.

[T2] Titchmarsh, E. C.: *Eigenfunction Expansions Associated with Second-Order Differential Equations*, Part I, second edition, Part II, Clarendon Press, Oxford, England, 1962 and 1958.

[VH] Vock, E., and Hunziker, W.: Stability of Schrödinger Eigenvalue Problems, *Commun. Math. Phys.* **38**, 281–302 (1982).

[vN] von Neumann, J.: *The Mathematical Foundations of Quantum Mechanics*, Princeton University Press, Princeton, NJ, 1955.

[W1] Wang, X. P.: Time-Decay of Scattering Solutions and Resolvent Estimates for Semiclassical Schrödinger Operators, *J. Diff. Eq.* **71**, 348–395 (1988).

[W2] Wang, X. P.: Barrier Resonances in Strong Magnetic Fields, *Commun. Partial Diff. Eq.* **17**, 1539–1566 (1992).

[Wx] Waxler, R.: The Time Evolution of a Class of Metastable States, to appear in *Commun. Math. Phys.* 1995.

[Y] Yafaev, D. R.: The Eikonal Approximation and the Asymptotics of the Total Cross-Section for the Schrödinger Equation, *Ann. Inst. Henri Poincaré* **44**, 397–425 (1986).

[Z] Zhislin, G.: Discussion of the Spectrum of the Schrödinger Operator for Systems of Several Particles, *Tr. Mosk. Mat. Obs.* **9**, 81–128 (1960).

Index

Applied Mathematical Sciences

(continued from page ii)